KEYS TO Lichens OF NORTH AMERICA

REVISED AND EXPANDED

IRWIN M. BRODO

Photographs by Sylvia Duran Sharnoff and Stephen Sharnoff

Drawings by Susan Laurie-Bourque

Keys to Lichens of North America: Revised and Expanded

Keys to Lichens of North America: Revised and Expanded

Irwin M. Brodo

Photographs by Sylvia Duran Sharnoff and Stephen Sharnoff
Drawings by Susan Laurie-Bourque

Published with the collaboration of the Canadian Museum of Nature, Ottawa, Ontario, Canada.

Yale University Press
New Haven and London

Yale University Press books may be purchased in quantity for educational, business, or promotional use. For information, please e-mail sales.press@yale.edu (U.S. office) or sales@yaleup.co.uk (U.K. office).

Printed in the United States of America.

Library of Congress Control Number: 2015939463
ISBN 978-0-300-19573-6 (paperback : alk. paper)
A catalogue record for this book is available from the British Library.

10 9 8 7 6 5 4 3 2

Dedicated to my teachers

JOSEPH J. COPELAND, who set me on the botanical path and gave me a love of field work,

THOMAS EISNER, who showed me how science works by being such an inspired scientist,

HENRY A. IMSHAUG, who made me into a lichenologist.

Contents

Preface

Many users of *Lichens of North America* (LNA) (Brodo et al. 2001) have asked that the keys to groups, genera, and species be reproduced as a separate publication bound in such a way as to make them convenient to use in university courses and workshops. This would free the larger volume to be used as more of a reference, somewhat removed from the ravages of splashed reagents and dusty packets. The present compilation was initially prepared to answer those requests.

As time passed, however, with the inevitable changes of classification and names, description of new species, redefinition of old species, and the discovery of "problems" in the keys published in LNA, it became clear that a full revision of the keys would be appropriate. With Yale University Press's support, I began an expansion of the keys to include a much larger percentage of the North American lichen flora.

The keys are reproduced more or less as they appeared in LNA, except that they are grouped together in alphabetical order by genus. The keys were then thoroughly revised based on literature that has appeared since the original book manuscript was submitted in 1998, as well as additional studies of my own on many genera using specimens in Ottawa (CANL) and other herbaria, especially NY (New York Botanical Garden). Many corrections and nomenclatural updates were made, and hundreds of additional species were incorporated into the text. Brand new keys to dozens of genera were constructed including, among others, *Acarospora, Biatora, Chrysothrix, Fuscidea, Lecidea, Micarea, Phlyctis, Placynthiella, Pyrrhospora, Sarcogyne, Sphaerophorus,* and *Xylographa*. This book covers 2028 species and 22 additional infraspecific taxa in 382 genera, which is 42% of the presently known lichen flora (4881 species, according to Esslinger [2014]). In LNA, 1086 species were in the keys, with ca. 800 species covered as main entries and another 700 mentioned in the notes.

In this version, species illustrated in LNA are in **boldface**. Author citations have been put in the keys rather than in the index. In most cases, I have followed Esslinger (2014). Since the lichen keys were written to be used in conjunction with the full descriptions, photographs, maps, and discussions found in LNA, they will seem short on detail to those attempting to use them alone. This is especially true of species illustrated in LNA, and I recommend using these keys as a complement to the book. To make reference to LNA easier despite name changes to many lichens as well as reidentification or reclassification of many vouchers used for the color plates, almost all the names that were used in LNA also appear in this volume, either as key entries, synonyms, or in "Notes" or comments, and every name in this book is listed in the Index so it can be found.

These keys do not include every lichen in North America. I have tried to include all the common or conspicuous species as well as all the species mentioned, but not illustrated, in LNA, except if the species is rare and unlikely to be encountered by most collectors. An exception is made for some small genera where including all species regardless of rarity made more sense. As users of these keys gain more experience in collecting, they will undoubtedly want to move to more comprehensive (and technical) monographs and regional treatments, and in many cases these references are given.

Lichenicolous fungi are not treated in this book, but some parasitic lichens (e.g., in *Caloplaca, Diploschistes,* and *Opegrapha*) are included. Some non-lichenized "stubbles" are covered in the *Calicium* key.

As in LNA, I have tried to avoid unnecessary jargon and concentrate on clarity. All keys, however, need a glossary to define technical terms and to explain how the author uses them. I have therefore included in this book the original glossary found in LNA together with the figures and photographs that illustrate the terms. The figure numbers also refer to the same figures found in LNA, with the exception of Plate 925, spore types in *Rinodina*, which is new. With the inclusion of so many additional species, I found it necessary to include more chemical characters including the use of *p*-phenylenediamine (PD), something I avoided in LNA. Although there is no key to lichen photobionts in the book, the most important algae are illustrated in Figure 1 and listed with their defining characteristics in the Glossary. Good keys to photobionts are found in Nash et al. (2002) and Smith et al. (2009).

The simplified format lends itself well to periodic updates and corrections, so I encourage users of the keys to contact me about problems or errors they encounter (ibrodo@mus-nature.ca).

Acknowledgments

I would like to thank Stephen Sharnoff for permission to use the photographs from *Lichens of North America* that illustrate the glossary. I also thank Jean Thomson Black at Yale University Press for giving me permission to use this format with students at Eagle Hill, Maine, and in several workshops, to test its usefulness as a separate publication. Michele LeBlanc assembled the figures and plates for these editions, and I much appreciate her diligence and skill. John Sheard generously sent some of his original photos of *Rinodina* spores which were assembled into Plate 925. Early drafts of some or all of the keys were sent to certain individuals for comments and trials, and I am extremely grateful for their corrections and suggestions: Ted Ahti (*Cladonia*), Kerry Knudsen (*Acarospora, Polysporina, Sarcogyne,* California lichens), Colin Freebury (Saskatchewan), John Sheard (*Rinodina*), Jason Hollinger (*Lecanora*), James Lendemer (*Lepraria*), Chris Lewis (Ontario), Katie Glew (Northwest), Caleb Morse (the Midwest), Diane Haugland (Alberta), Stephen Clayden (*Stereocaulon* and Maritime Canada). Thanks also go to Nicolas Magain at Duke for his unpublished observations on the *Peltigera polydactylon* group. The manuscript benefited from the expert editing skills of Kate Davis at the Canadian Museum of Nature, as well as Susan Goods and Roger Bull, also at the Museum. Susan checked citations and helped with the index, and Roger assembled the figures and plates into a printable file. I thank Fred Brown, bassist in our klezmer band and a professional indexer, for showing me how to prepare an index in MS-Word and giving me the confidence to do it. I am grateful for the help and hospitality I received at the NYBG from Richard Harris, Bill Buck, and James Lendemer while working there on the eastern lichens not found in CANL. The book also benefited from the many excellent suggestions made by reviewers Kerry Knudsen and Richard Harris.

Identification Keys to Genera and Major Groups

Key to Keys

1. Fruiting body a mushroom or fleshy, ephemeral, club-shaped stalk with a mushroom-like consistency........2
1. Fruiting body a perennial ascoma (apothecium, perithecium, or lirella); or thallus without any fruiting bodies 3

2. Fruiting body a pale to dark yellow, gilled mushroom, growing out of a lichenized thallus consisting of either dark green spherical granules or green lobed squamules ....*Lichenomphalia*
2. Fruiting body club-shaped, usually unbranched, white to orange, growing out of a gelatinous, dark green, barely lichenized algal film .... *Multiclavula*

3. Thallus erect, shrubby, tufted, or pendent, at least in part (sometimes with a crust-like or scaly basal thallus as well); branches or stalks round, angular, or flat in cross section, but if flattened, then both surfaces similar in color and texture..4
3. Thallus composed entirely of flat lobes, branches, or scales, with clearly different upper and lower surfaces, **OR** crust-like and firmly attached to the substrate over the entire lower surface, or growing within the substrate..5

4. Thallus relatively large, more than 5 mm long or high, with branches or stalks mostly more than 0.2 mm in diameter (figs. 2h-j)...Fruticose lichens (Key A)
4. Thallus forming small tufts or clumps rarely exceeding 5 mm long or high, consisting of fine, hair-like elements rarely more than 0.2 mm in diameter (e.g., *Polychidium, Ephebe*, or *Coenogonium*), **OR** forming tiny cushions or clumps of erect branches not more than 5 mm high, with cyanobacteria ....Dwarf fruticose lichens (Key B)

5. Thallus consisting of a crust that grows with its entire lower surface in intimate contact with the surface and cannot be removed from the substrate intact (i.e., without leaving part of the lichen's lower surface behind); thallus can be smooth and continuous, powdery, cracked into angular patches (areoles), extremely thin (even growing within the substrate), or quite thick and warty, sometimes lobed at the margins (fig. 2a-d) ....6 (Crustose lichens)
5. Thallus consisting of more or less flat dorsiventral lobes or scales, ascending or closely appressed to the substrate, more or less easily removed from the substrate with the lower surface intact (fig. 2e-g) 9

6. Fruiting bodies absent .... Sterile crustose lichens (Key C)
6. Fruiting bodies (ascomata) present...7

7. Ascomata flask-shaped, perithecia or resembling perithecia (opening by a deeply concave or pit-like ostiole) (figs. 17, 12h, 13d), superficial or buried in the thallus or in a fertile wart (pseudostroma) with only the ostiole visible .... Crustose perithecial lichens (Key D)

[Note: Pycnidia can sometimes resemble perithecia, but they contain hundreds of conidia rather than asci; see fig. 18.]

7. Ascomata apothecia: (1) disk- or cup-shaped (figs. 12a-c; 13a-c,f), sometimes raised on a stubby or slender stalk (fig. 12f); (2) buried in thalline warts (as in *Pertusaria subpertusa* or *P. xanthodes*) (fig. 12g); or (3) elongated and sometimes branched (lirellae)(figs. 12e; 13e) 8

8. Ascomata lirellae, short and ellipsoid to quite long, script-like, and sometimes branched .... Crustose script lichens (Key E)

8. Ascomata apothecia (disk- or cup-shaped or buried within thalline warts)........Crustose disk lichens (Key F)

9.(5) Thallus composed of separate or overlapping scales (squamules) that are rarely more than 5 mm long or broad (fig. 2e) .... Squamulose lichens (Key G)

9. Thallus composed of narrow or broad, flat lobes; thallus generally exceeding 5 mm long or broad (fig. 2f-g) .... 10 (Foliose lichens)

10. Thallus attached to the substrate by a holdfast at a single point, either central, giving it the shape of an umbrella (fig. 2g), or at one edge, giving it the shape of a fan. (Lobes sometimes crowded and overlapping, obscuring the attachment.) .... Umbilicate or fan-shaped lichens (Key H)

10. Thallus attached to the substrate at numerous points, either with hair-like attachment organs (rhizines), a fuzzy tomentum, or directly by the lower surface (fig. 2f) 11

11. Thallus, when wet, translucent and jelly-like, and typically black to very dark gray, brown, or olive (e.g., *Collema*); photobiont blue-green, usually distributed more or less uniformly within the thallus, i.e., not confined to a definite layer (fig. 7) .... Jelly lichens (Key I)

11. Thallus remaining opaque when wet, color various; photobiont green or blue-green, confined to a definite layer within the thallus (fig. 4) 12

12. Dry thallus orange, yellow, greenish yellow, or yellowish green (e.g., *Xanthoria, Candelaria,* and *Flavoparmelia*) .... Orange to yellow or yellowish green foliose lichens (Key J)

12. Dry thallus shades of white, gray, pale green, olive, brown, or black, without a yellowish tint .... Non-umbilicate, non-jelly, non-yellow foliose lichens (Key K)

KEY A: FRUTICOSE LICHENS

1. Thallus bright yellow, bright chartreuse, or orange, at least in part 2
1. Thallus pale greenish yellow, yellowish green, green, white, gray, brown, olive, or black 6

2. Medulla yellow; cortex and medulla K–; branches distinctly flattened throughout, vertically foliose; on the ground in arctic or alpine sites .... ***Vulpicida tilesii* (Ach.) J.-E. Mattsson & M. J. Lai**
2. Medulla white; cortex K– or K+ red-purple; branches round to angular, at least at the tips; on trees or rocks, or, if on the ground, not arctic-alpine 3

3. Thallus bright yellow to chartreuse, K–; on trees and wood, mostly in inland, montane localities, western .... *Letharia*
3. Thallus orange, at least on exposed parts, K+ dark red-purple 4

4. On trees, shrubs, or soil .... *Teloschistes*
4. Attached directly to rock 5

5. Thallus entirely coralloid-branched, with cylindrical branches; California .... *Caloplaca*
5. Thallus with flattened, foliose base, becoming divided and fruticose at lobe tips, often with granular soredia or blastidia at tips; branches sometimes hairy at tips; apothecia rare but large; orange pycnidia are frequent and conspicuous on the branches; on calcareous rock, arctic-alpine .... *Seirophora contortuplicata* (Ach.) Fröden (syn. *Teloschistes contortuplicatus*)

6.(1) Thallus pendent or almost pendent (i.e., growing downward or outward); mostly on trees, shrubs, or vertical surfaces of rocks (includes hair and beard lichens) 7
6. Thallus erect or prostrate (i.e., basically growing upward at least initially); mostly on the ground or on horizontal surfaces of rocks (includes lichens with upright stalks or podetia) 25

7. Thallus greenish yellow or yellowish green (containing usnic acid in the cortex) 8
7. Thallus shades of white, gray, brown, olive, or black (lacking usnic acid) 13

8. Branches with a tough, single, central cord (figs. 8b, 36) .... *Usnea*
8. Branches with a more or less uniform medulla, without a single, central cord 9

9. Pseudocyphellae present 10
9. Pseudocyphellae absent 11

10. Branches flattened in cross section, at least at the base .... *Ramalina*
10. Branches round in cross section, except sometimes at the axils (fig. 8a) .... *Alectoria*

11. Thallus soft and pliable, with a thin cortex (fig. 8d) .... *Evernia*
11. Thallus stiff because of a thick, tough cortex 12

12. Pycnidia black, abundant, appearing as black dots on the margins and sometimes the surface of the thallus; restricted to coastal California.. *Niebla*
12. Pycnidia, when present, pale and inconspicuous; widespread.. *Ramalina*

13.(7) Branches distinctly flattened, at least at the base or tips 14
13. Branches round or angular in cross section throughout (sometimes grooved or twisted, or flattened at the axils) 23

14. Thallus brown to greenish black .... *Kaernefeltia*
14. Thallus pale gray to smoky brownish gray 15

15. Cilia present on the branches or apothecia 16
15. Cilia absent 18

16. Branches 10-25 mm long; soralia C+ faint pink (gyrophoric acid) .... ***Hypotrachyna catawbiense* (Degel.) Divakar et al.** (syn. *Everniastrum catawbiense*)
16. Branches 40-150 mm long; medulla and soralia (if present) C– 17

17. Cortex K+ yellow (atranorin) .... *Heterodermia* (see *Physcia* key)
17. Cortex and medulla K–; lobes without soredia .... ***Anaptychia crinalis* (Schaerer) Vězda** (syn. *A. setifera*)

18. Fruiting bodies almost spherical, black, powdery masses formed close to the tips of the flattened branches; on mossy trees or rocks in Pacific Northwest coastal forests .... *Bunodophoron melanocarpum* (Sw.) Wedin (see couplet 31)

18. Fruiting bodies round or elongate apothecia, never powdery; not found in Pacific Northwest coastal forests 19

19. On trees or wood in the southwest or Appalachian-Great Lakes regions; cortex K+ yellow (atranorin); lower surface gray to black in part .... *Pseudevernia*

19. On rocks and shrubs in coastal California; cortex K–; lower and upper surfaces the same color 20

20. Thallus cortex C–, PD– 21

20. Thallus cortex C+ red or PD+ red 22

21. Branches distinctly flattened, 0.5-2.5 mm wide, with longitudinal ridges at the base; branch tips with clumps of short, colorless hairs, and developing granular soredia; apothecia infrequent, orange, spores colorless, polarilocular .... *Seirophora californica* (Sipman) Fröden (syn. *Teloschistes californicus*)

21. Branches cylindrical to flattened at the axils and base, mostly under 0.5 mm wide, ± smooth or with some hairs at the tips, without soredia; apothecia often present, black, spores brown, *Physconia*-type .... *Tornabea scutellifera* (With.) J. R. Laundon

22. Cortex C+ red, PD– (lecanoric acid) .... *Roccella*

22. Cortex C–, PD+ red (protocetraric acid) .... *Dendrographa*

23.(13) Thallus almost white, sometimes darkening to dark yellowish gray; coastal rocks and shrubs in California; cortex PD+ red .... *Dendrographa*

23. Thallus shades of gray or brown, or almost black; cortex PD– or PD+ yellow (except rarely in some dark species of *Bryoria*); on bark or wood, rarely on rock 24

[Note: See also *Tornabea scutellifera* above, in which the branches can sometimes be mostly cylindrical.]

24. Thallus with thick main branches and much finer secondary and tertiary branches, gray-green, often browned at the tips; fruiting body spherical, at the branch tips, with a mazaedium .... *Sphaerophorus*

24. Thallus gradually tapering, mostly uniform in color; fruiting bodies disk or cup-shaped apothecia with a persistent hymenium .... *Bryoria* key

25.(6) Stalks or branches loosely or densely filled with hyphae or cartilaginous supporting tissues 26

25. Stalks or branches hollow, at least in part 64

26. Stalks or branches distinctly flattened at least at the base or tips 27
26. Stalks or branches mostly round or angular in cross section (sometimes flattened at the axils) 37

27. Thallus yellow-green (containing usnic acid) 28
27. Thallus gray, brown, or black (lacking usnic acid) 30

28. Pseudocyphellae present, slightly depressed, on the lower surface of the branches; lobes thin, entirely flattened and erect .... *Flavocetraria*
28. Pseudocyphellae absent, although the branches sometimes have raised, white warts; lobes relatively thick 29

29. Lobes irregularly branched, strongly curled and twisted, usually with white warts on the margins or tips of the branches; growing unattached over bare soils in the interior high plateaus .... ***Rhizoplaca haydenii* (Tuck.) W. A. Weber**
29. Lobes regularly dichotomous, not curled, generally round at least toward the base, without white warts; on soil in arctic or alpine sites .... ***Allocetraria madreporiformis* (Ach.) Kärnefelt & Thell**

30.(27) Thallus gray or gray-green 31
30. Thallus brown to black 32

31. On limestone, in arid southcentral U.S; branches more or less uniform in width and very regularly dichotomous .... ***Speerschneidera euploca* (Tuck.) Trevisan**
31. On mossy trees or rocks, in humid coastal forests of the Pacific Northwest; main branches broad, giving rise to clusters of much narrower branches, irregularly dichotomous .... *Bunodophoron melanocarpum* (Sw.) Wedin (see couplet 18)

32. Directly on rock 33
32. On soil, heath, or mosses (sometimes over rock) 35

33. Thallus abundantly branched, forming wool-like cushions .... ***Pseudephebe minuscula* (Nyl. *ex* Arnold) Brodo & D. Hawksw.**
33. Thallus with relatively few strap-shaped branches; forming small clumps 34

34. On limestone; photobiont blue-green; thallus jelly-like when wet .... *Lichinella* key

34. On siliceous rock; photobiont green; thallus unchanged when wet; montane west coast and northern Rockies .... ***Cornicularia normoerica* (Gunn.) du Rietz**

35.(32) Branches very flat, regularly dichotomous, with broad, pruinose patches (pseudocyphellae) on lower surface; growing unattached (vagrant), rolling freely over the soil and alpine and subalpine heath; Alaska to N.W.T. and northern B.C. .... ***Masonhalea richardsonii* (Hooker) Kärnefelt**

35. Branches usually grooved or inrolled at the margins; pseudocyphellae linear or irregular in shape, not broad and pruinose; growing in more or less fixed clumps, not rolling freely over the soil or heath 36

36. Medulla C+ pink (gyrophoric acid); lobe margins more or less even, without projections .... *Cetrariella*

36. Medulla C–; lobe margins with stalked pycnidia or cilia .... *Cetraria*

37.(26) Thallus yellowish green (containing usnic acid) 38

37. Thallus white, gray, brown, olive, or black (lacking usnic acid) 45

38. Thallus prostrate; arctic-alpine 39

38. Thallus at least partly erect 40

39. Branches soft and pliable, with a thin cortex .... soil-dwelling form of ***Evernia divaricata* (L.) Ach**

39. Branches stiff, with a thick cortex .... *Alectoria*

40. Branches tall, typically 50-80(-120) mm high; thallus conspicuously blackened at the branch tips; raised white pseudocyphellae abundant .... ***Alectoria ochroleuca* (Hoffm.) A. Massal.**

40. Branches short, usually under 30 mm high; thallus color relatively uniform from base to tips; white dots and bumps (pseudocyphellae) absent 41

41. Branches very slender, less than 0.3 mm in diameter, covered with cottony granules .... *Leprocaulon* (see *Lepraria* key)

41. Branches thicker than 0.3 mm; surface relatively smooth at least at the base 42

42. In arctic or alpine sites in western mountains; not cushion-forming .... ***Allocetraria madreporiformis* (Ach.) Kärnefelt & Thell**

42. Not arctic-alpine; forming rounded cushions or clumps 43

43. Rolling unattached over bare soils in the interior high plateaus; cortex C–, KC+ yellow (usnic acid alone) .... *Rhizoplaca*

43. On seashore rocks in California 44

44. Branches soft, with a loose medulla and algae confined to a thin layer; branch tips becoming granular or sorediate; apothecia along branches, not at tips; cortex C+ orange, KC+ red-orange (xanthones as well as usnic acid) .... ***Lecanora phryganitis* Tuck.**

44. Branches stiff, with a tough cartilaginous medulla more or less filled with algae; surface not disintegrating into granules or soredia; apothecia at branch tips; cortex C–, KC+ gold (usnic acid alone) .... *Cladidium bolanderi* (Tuck.) B. D. Ryan

45.(37) Branching frequently 46

45. Unbranched, or branching at most once or twice 59

46. Thallus black or very dark brown to almost black 47

46. Thallus reddish brown, yellowish brown, olive, gray, or white 49

47. Attached directly to rock .... *Pseudephebe*

47. On soil or heath 48

48. Pseudocyphellae pale, raised, usually conspicuous; thallus pale at the base; cortex and usually medulla C+ pink, PD+ yellow .... ***Gowardia nigricans* (Ach.) P. Halonen et al.** (syn. *Alectoria nigricans*)

48. Pseudocyphellae dark, level with surface or depressed, inconspicuous; thallus dark throughout; cortex C–, PD–; medulla PD+ red .... ***Bryoria nitidula* (Th. Fr.) Brodo & D. Hawskw.**

49.(46) Thallus shades of brown or olive 50

49. Thallus white to dark greenish gray 53

50. Photobiont blue-green .... ***Dendriscocaulon intricatulum* (Nyl.) Henssen**

50. Photobiont green 51

51. Pseudocyphellae absent .... *Sphaerophorus*

51. Pseudocyphellae white, conspicuous 52

52. Pseudocyphellae usually raised like small bumps, C+ pink; medulla usually C+ pink as well (olivetoric acid) .... **_Bryocaulon divergens_ (Ach.) Kärnefelt**

52. Pseudocyphellae depressed, C–; medulla C– .... *Cetraria* (couplet 4)

53.(49) Photobiont blue-green .... **_Dendriscocaulon intricatulum_ (Nyl.) Henssen**

53. Photobiont green 54

54. Branches very slender, less than 0.3 mm in diameter, covered with cottony granules .... *Leprocaulon* (see *Lepraria* key)

54. Branches thicker than 0.3 mm; surface relatively smooth at least at the base 55

55. Thallus growing as prostrate cushions up to 30 mm in diameter on soil in the arid interior of North America .... *Circinaria* (see *Aspicilia* key)

55. Thallus colonies usually larger than 30 mm, not on soil in the arid interior of North America; pseudocyphellae, if present, not round and depressed 56

56. Cortex C+ red; on coastal rocks in southern California; fruiting bodies large, lecanorine, with heavily pruinose disks .... **_Schizopelte californica_ Th. Fr.**

56. Cortex C–; fruiting bodies not pruinose 57

57. Stalks with a tough, cartilaginous central core, more or less covered with granular, cylindrical, or scaly outgrowths (phyllocladia) (fig. 8g) .... *Stereocaulon*

57. Stalks filled with white or orange medulla, without a cartilaginous central core; without phyllocladia (fig. 8e) 58

58. Branch tips blackened; pseudocyphellae usually conspicuous .... **_Gowardia nigricans_ (Ach.) P. Halonen et al.** (syn. *Alectoria nigricans*)

58. Branch tips sometimes browned but not blackened; pseudocyphellae absent .... *Sphaerophorus* key

59.(45) On conifer branches in subalpine sites; fruiting body a mazaedium at the tip of a stubby, barrel-shaped stalk .... **_Tholurna dissimilis_ (Norman) Norman**

59. On soil, rock, or moss; fruiting body not a mazaedium 60

60. Stalks prostrate, white, growing from the edge of a crustose thallus; cortex C+ red

.... ***Ochrolechia frigida* (Sw.) Lynge**

60. Stalks erect, white to gray or olive-gray; cortex C– or C+ purple 61

61. Stalks white, smooth or with longitudinal furrows, not growing from a primary thallus, without fruiting bodies of any kind; cortex C+ violet (fading quickly) .... ***Siphula ceratites* (Wahlenb.) Fr.**

61. Stalks almost white to gray or olive-gray, without furrows, developing from a crustose or granular primary thallus; cortex C– 62

62. Cephalodia absent .... *Baeomyces* key (*Baeomyces*, *Dibaeis*)

62. Lumpy pink to dark gray cephalodia present on primary thallus, sometimes on stalks 63

63. Apothecia usually present at the summit of the stalks, black, cylindrical or almost spherical; stalks with or without soredia .... *Pilophorus*

63. Apothecia extremely rare, brown; granular soredia at the tips of the stalks ....***Stereocaulon pileatum* Ach.**

64.(25) Photobiont blue-green; thallus olive-brown; stalks more or less cylindrical, mostly less than 2 mm high; on rock .... *Peltula*

64. Photobiont green; thallus not olive-brown; stalks taller than 2 mm 65

65. Lacking a basal primary thallus 66

65. With a scaly or crustose basal primary thallus 71

66. Stalks abundantly branched 67

66. Stalks unbranched or branched only once or twice 68

67. Thallus less than 20 mm tall; branches relatively stocky, usually violet or pinkish due to pruina; interior cavity containing webby hyphae .... ***Dactylina ramulosa* (Hooker) Tuck.**

67. Thallus over 20 mm tall, without pruina; interior cavity empty, lined with a cartilaginous stereome (fig. 8f) .... *Cladonia*

68. Thallus C+ red (erythrin); stalks inflated, brittle; on coastal rocks in southern California ....*Schizopelte parishii* (Hasse) Ertz & Tehler (syn. *Hubbsia parishii*)

68. Thallus C–, or, if C+ pink, then arctic-alpine 69

69. Stalks white, slender, 1-2.5 mm in diameter, usually partly or entirely prostrate .... ***Thamnolia vermicularis*** **(Sw.) Ach. *ex* Schaerer** (syn. *Thamnolia subuliformis*)

69. Stalks yellowish, brown, or greenish gray, erect 70

70. Stalks inflated, thin-walled and brittle, with no stereome; entirely without squamules .... ***Dactylina arctica*** **(Richardson) Nyl.**

70. Stalks more or less slender, not inflated, stiff, supported by a cartilaginous stereome in all but a few species (see p. 137), sometimes bearing squamules .... *Cladonia*

71.(65) Primary thallus crustose, granular, not at all squamulose .... ***Pycnothelia papillaria*** **Dufour**

71. Primary thallus squamulose .... *Cladonia*

[Occasionally, some species of reindeer lichens can produce an ephemeral, whitish, crustose thallus when growing on wood.]

KEY B: DWARF FRUTICOSE LICHENS

1. Thallus pale: gray, greenish, yellowish green, olive, or orange 2

1. Thallus dark: black, olive, reddish, or brown 9

2. Thallus orange, pigment in upper cortex, with a green algal layer, K+ dark purple 3

2. Thallus gray, greenish, yellowish green, olive or, if orange, K–, hair-like algal filaments in a fungal envelope 4

3. Branches cylindrical, not sorediate... *Caloplaca*

3. Branches flattened to somewhat cylindrical, sorediate.. *Xanthoria* key, couplet 14

4. Thallus forming cottony tufts of fine, branching filaments or hairs; photobiont green or blue-green.............5

4. Thallus with more or less erect, fine stalks; photobiont green 6

5. On subtropical and tropical trees and shrubs; tufts cottony, sometimes flattened, pale green to orange, almost white when old; disk-shaped apothecia common, yellow to orange; photobiont green (*Trentepohlia*) .... *Coenogonium*

5. On twigs and branches on the very humid west coast; thallus forming small, foam-like clumps, pale olive or olive-gray; photobiont blue-green (*Hyphomorpha*)................................ *Leptogidium* (see *Polychidium* key)

6. Stalks covered with cottony granules or soredia giving the clumps the appearance of a pale gray to whitish or yellowish, leprose, crustose lichen .... *Leprocaulon* (see *Lepraria* key)

6. Stalks without granules or soredia...7

7. Thallus continuous, thin and membranous to verruculose; on bark or moss 8

7. Thallus cylindrical, dark olive or olive-gray; on rock .... *Peltula*

8. Stalks thallus-colored, pale greenish-gray, like long, slender isidia, often branched, containing green algae; apothecia sometimes produced on basal crust; on bark in the Pacific Northwest .... ***Loxosporopsis corallifera*** **Brodo, Henssen & Imshaug**

8. Stalks colorless, brown or black, without algae, usually unbranched, sometimes fan-shaped, producing conidia at the tips.. *Gyalideopsis* key

9.(1) Thallus dark red-brown, with abundant, large, red-brown apothecia; growing in patches of moss over rock; photobiont blue-green (*Nostoc*)... ***Polychidium muscicola*** **(Sw.) Gray**

9. Thallus black to shades of brown, olive, or olive-gray; apothecia or perithecia inconspicuous, rare, or absent; photobiont blue-green or green 10

10. On bark or moss; thallus olive to olive-gray or brownish gray; photobiont blue-green 11

10. On rock; thallus mostly black or dark brown to very dark olive, at least when dry 12

11. On moss or mossy trunks; branches thick, 0.1-0.6 mm in diameter, spiny, dark olive-gray to brownish; photobiont *Nostoc* .... ***Dendriscocaulon intricatulum*** **(Nyl.) Henssen**

11. On twigs and branches in humid localities on the west coast; branches hair-like, less than 0.1 mm in diameter, not spiny; thallus forming small, foam-like clumps, pale olive or olive-gray; photobiont *Scytonema* .... *Leptogidium* (see *Polychidium* key)

12. Thallus like black wool, the hair-like filaments 10-20 µm thick; contain green algae or cyanobacteria; on shaded, sometimes wet rock faces 13

12. Thallus forming tufts or clumps of much thicker filaments (40-140 µm) or forming tiny cushions of erect branches, containing cyanobacteria; on dry or wet rocks, usually exposed 15

13. Containing cyanobacteria (*Scytonema*); on calcareous rock; widespread but infrequently collected .... *Thermutis velutina* (Ach.) Flotow

13. Containing the green alga *Trentepohlia*; on siliceous rock 14

14. Cells enveloping algal filaments long and straight (fig. 34b) .... ***Racodium rupestre* Pers.**
14. Cells enveloping algal filaments irregular and knobby (fig. 34a) ...*Cystocoleus ebeneus* (Dillwyn) Thwaites

15.(12) Thallus forming irregular patches or felt-like mats (when abundant), not rounded cushions; usually on wet rocks near a stream or lake shore; lacking any blue-green, rhizoid-like hyphae at the base of the clumps; photobiont *Stigonema* (fig. 31).. ***Ephebe lanata* (L.) Vainio**
15. Thallus forming rounded clumps or cushions, 5-15 mm across and 6 mm high 16

16. Branches short filaments, under 0.2 mm across; on dry or wet rock; blue-green, rhizine-like hyphae developing at the base of the cushions; photobiont filamentous, *Stigonema* .... ***Spilonema revertens* Nyl.**
16. Branches erect, becoming hollow when older, 0.15-0.5 mm thick, dark brown, without rhizine-like hyphae; photobiont in packets of single cells, with gelatinous sheaths that are K– or golden, Chlorococcaceae; southwestern, on siliceous rocks .... *Pterygiopsis cava* M. Schultz

KEY C: STERILE CRUSTOSE LICHENS For a more complete key, see Lendemer (2010).

1. Soredia and isidia absent 2
1. Soredia or isidia present 10

2. On bark; older parts of thallus covered with hollow pustules that break up into granular or soredia-like schizidia; thallus K+ yellow, PD+ orange (thamnolic acid) .... ***Variolaria pustulata* (Brodo & W. L. Culb.) Lendemer et al.** (syn. *Loxospora pustulata*)
2. On wood, soil, mosses, dead vegetation, or rock 3

3. On rock; thallus cortex C+ red (lecanoric acid) .... *Dirina*
3. On wood, soil, mosses, or dead vegetation 4

4. Thallus yellow to orange, or greenish yellow to yellow-green 5
4. Thallus without yellow or orange tint 6

5. Thallus K– .... *Lecanora*
5. Thallus K+ purple; on soil .... *Fulgensia*

6. Thallus pale 7

6. Thallus dark 9

7. Thallus cortex and medulla PD+ red (fumarprotocetraric acid); thallus almost white, with tall verrucae; arctic-alpine .... ***Pertusaria dactylina* (Ach.) Nyl.**

7. Thallus cortex and medulla PD– or PD+ yellow (fumarprotocetraric acid absent); arctic-alpine or temperate 8

8. Thallus C+ pink or red, PD– (gyrophoric acid; baeomycesic acid absent) .... ***Trapeliopsis granulosa* (Hoffm.) Lumbsch** (form lacking soredia)

8. Thallus C–, PD+ yellow or orange (baeomycesic, stictic, or thamnolic acid; gyrophoric acid absent) .... *Baeomyces* key

9.(6) Thallus dark steel gray, to dark greenish gray; verrucose, C+ pink .... ***Trapeliopsis flexuosa* (Fr.) Coppins & P. James**

9. Thallus dark brown, granular, C+ pink or C– .... *Placynthiella*

10.(1) Isidia present 11

10. Isidia absent 18

11. Thallus orange-yellow, K+ purple; southeastern U.S.; on bark .... *Caloplaca epiphora* (Tayl.) Dodge

11. Thallus olive, white, gray or brownish, K– or K+ yellow 12

12. Thallus olive, distinctly lobed; photobiont blue-green; directly on rock in and near southeastern Alaska .... ***Vestergrenopsis isidiata* (Degel.) E. Dahl**

12. Thallus white to gray-green, or reddish brown, not lobed at the edge; photobiont green; on bark, wood, mosses, or dead vegetation, temperate or tropical 13

13. On exposed wood, soil, or peat; thallus brown .... *Placynthiella*

13. On bark, rock, mosses, or dead vegetation; thallus pale, white to gray-green 14

14. Thick red or white prothallus present; non-corticate "isidia" subspherical (granular) to cylindrical; coastal plain of southeastern U.S. .... *Cryptothecia* key

14. Prothallus present or absent, never red; true isidia cylindrical; arctic-alpine, or temperate 15

15. On mosses or dead vegetation in arctic-alpine habitats; thallus white, verrucose; isidia thick and crowded; medulla PD+ red .... ***Pertusaria dactylina* (Ach.) Nyl.**

15. On bark, moss, or rock in temperate forests; thallus yellowish gray to greenish gray, medulla PD– 16

16. On bark in the coastal Pacific Northwest, thallus more or less smooth; isidia long and slender, often branched; thallus C–... ***Loxosporopsis corallifera* Brodo, Henssen & Imshaug**

16. On bark, moss, or rock, eastern; thallus C– or C+ red 17

17. Thallus pale gray, smooth to verrucose; isidia short and stout; thallus C+ pink (gyrophoric acid); on rocks and tree bark; Appalachians north to New England .... ***Ochrolechia yasudae* Vainio**

17. Thallus dark greenish gray; isidia short, slender; thallus C– (2'-*O*-methylperlatolic acid); on moss, bark, or mossy rock; eastern .... *Pertusaria globularis* (Ach.) Tuck.

18.(10) Soredia comprising entire thallus 19

18. Soredia in delimited soralia, or diffuse on older parts of the thallus 23

19. Thallus white, pale green, yellowish green, gray, or brownish, rarely very pale yellowish 20

19. Thallus greenish yellow, yolk-yellow, lemon yellow, yellow-orange, or orange-yellow, bright yellow, or orange 21

20. Thallus yellowish green to pale green; prothallus white, forming a fibrous fringe; thallus C–, KC+ gold, K, PD– (usnic acid and zeorin present); on trunks of hardwoods, especially maples, rarely forest boulders; mainly Appalachian-Great Lakes region .... ***Lecanora thysanophora* R. C. Harris**

[Note: *Lecanora expallens* Ach. is a similar species on bark and wood mainly in coastal areas. It is C+ orange, KC+ dark orange (thiophanic and usnic acids and zeorin).]

20. Thallus greenish to gray, rarely very pale yellowish; fibrous white prothallus absent (although a white or brown cottony hypothallus may be present); thallus KC– (usnic acid rare); shaded tree bases, rocks, wood, and moss, typically in humid situations .... *Lepraria*

21. Soredia K+ purple (anthraquinones present, calycin absent) .... *Caloplaca*

21. Soredia K– (anthraquinones absent, calycin present) 22

[Note: If soredia are C+ orange (xanthones), consider *Lecanora expallens* or *Pyrrhospora quernea*.)

22. Thallus uniformly leprose, the soredia not clustered or associated with granules; in shaded, humid habitats .... *Chrysothrix* key

22. Thallus usually producing soredia in small clumps, originating from the breakup of tiny areoles, but the soredia can become confluent in well-developed specimens; very common on exposed deciduous trees .... *Candelariella*

23.(18) Thallus containing yellow pigments, distinctly bright yellow or orange or pale yellow to yellow-green, K+ deep purple or K–, sometimes C+ orange 24

23. Thallus not containing yellow pigments, not distinctly yellow or orange, K– or K+ yellow to red, not C+ orange 29

24. On soil 25

24. On bark, wood, or rock 27

25. Thallus K– .... *Arthrorhaphis*

25. Thallus K+ deep purple 26

26. Areoles producing schizidia from pustules, not truly sorediate .... *Fulgensia*

26. Areoles clearly sorediate .... *Caloplaca tominii* (Savicz) Ahlner

27. Thallus not lobed, entirely crustose, lacking pustules or schizidia .... couplet 58 of *Caloplaca* key
[Note: If K–, C+ orange, see *Lecanora expallens* and *Pyrrhospora quernea*.]

27. Thallus margin distinctly lobed 28

28. Thallus with a lower cortex, producing pustules, granules, or granular schizidia on the upper surface .... ***Rusavskia sorediata* (Vainio) S. Y. Kondr. & Kärnefelt** (syn. *Xanthoria sorediata*)

28. Thallus entirely crustose, lacking a lower cortex .... couplet 22 of *Caloplaca* key

29.(23) Cephalodia present 30

29. Cephalodia absent 32

30. Thallus continuous, flat, with a distinctly lobed margin; cephalodia flat, disk-shaped, on the thallus surface .... *Placopsis*

30. Thallus verrucose or dispersed areolate, not lobed at the margin; cephalodia convex, cushion-shaped, between the areoles 31

31. Growing on rock .... **_Amygdalaria panaeola_ (Ach.) Hertel & Brodo**
31. Growing on bark or wood .... *Coccotrema pocillarium* (Cummings) Brodo

32.(29) On rock 33
32. On bark, wood, or soil 37

33. Soralia KC+ purple (tastes very bitter; picrolichenic acid); common and widespread .... **_Variolaria amara_ Ach.** (syn. *Pertusaria amara*) [see also couplet 49]
33. Soralia KC– or KC+ red 34

34. Thallus C+ red or pink 35
34. Thallus C– 36

35. Thallus C+ pink (gyrophoric acid); soredia white, in irregular patches; common in east, infrequent on west coast .... *Trapelia placodioides* Coppins & P. James
35. Thallus cortex C+ red (erythrin and lecanoric acid); southern U.S. .... *Dirina*

36. Thallus dark blue-gray, thick and areolate with coarse, almost isidioid, gray soredia on areole margins; cortex and medulla K– .... *Caloplaca chlorina* (Flotow) H. Olivier
36. Thallus pale gray, continuous, not areolate; soredia fine, mostly white, in round or irregular patches on thallus surface; cortex or medulla K+ yellow or red, or K– .... *Porpidia*

37.(32) On trees in tropical and subtropical regions; prothallus thick and conspicuous 38
37. In temperate to boreal regions; prothallus absent or present 39

38. Prothallus red or white; thallus not lobed, more or less smooth, although not corticate, and usually cottony or webby only at the thallus edges .... *Cryptothecia* key
38. Prothallus black; thallus clearly lobed, cottony or webby throughout .... **_Crocynia pyxinoides_ Nyl.**

39. Soredia diffuse, not confined to discrete soralia even in young parts of the thallus, or arising as schizidia from the breakup of small hollow pustules 40

39. Soredia at first produced in discrete soralia, or from the disintegration of solid granules or verrucae, sometimes becoming confluent in older parts of the thallus; pustules and schizidia absent 42

40. Thallus PD–, K+ yellow (atranorin and caperatic acid); northeastern .... *Cliostomum leprosum* (Räsänen) Holien & Tønsberg

40. Thallus PD+ yellow or orange, K+ red or yellow 41

41. Thallus PD+ yellow, K+ yellow becoming blood red (norstictic acid) or PD+ yellow, K– (psoromic acid) .... *Phlyctis*

41. Thallus PD+ orange, K+ bright yellow (thamnolic acid) .... *Loxospora* key

42. Thallus cortex or soredia C+ pink or red, KC+ pink (gyrophoric acid) 43

[Note: See also the key to *Biatora*, couplets 3-9 for more sterile sorediate crusts of this type.]

42. Thallus cortex C– or C+ orange (gyrophoric acid absent); soredia KC– or KC+ purple 44

43. Soredia white or yellowish; thallus continuous, verruculose, or more or less smooth .... *Ochrolechia*

43. Soredia pale greenish to dark green; thallus verrucose, areolate, or granular .... *Trapeliopsis*

44. Soredia green or yellow-green; thallus greenish or olive to brownish gray to dark greenish gray 45

44. Soredia white, pale to dark gray, or brownish; thallus pale to dark slate gray or yellowish white 49

45. Thallus distinctly lobed, appearing crustose but lower cortex present .... *Hyperphyscia*

45. Thallus not at all lobed; truly crustose 46

46. Soralia PD+ red (fumarprotocetraric acid); often with brown prothallus; common, boreal to North Temperate .... *Fuscidea arboricola* Coppins & Tønsberg

46. Soralia PD– 47

47. Soredia fine; soralia remaining tiny, round, and discrete; lacking substances; lacking a prothallus .... *Lecania croatica* (Zahlbr.) Kotlov

47. Soredia coarse, soralia soon coalescing into patches of green soredia 48

48. Contains perlatolic and hyperlatolic acids; with or without a brown or black prothallus; infrequent, northeast and northwest .... *Ropalospora viridis* (Tønsberg) Tønsberg

48. Contains divaricatic acid; brown prothallus usually conspicuous, especially when growing on birch, a preferred substrate; Appalachian-Great Lakes .... *Fuscidea pusilla* Tønsberg

49.(44) Soralia KC+ purple; thallus cortex K–, PD– or rarely PD+ red-orange (protocetraric acid) .... ***Variolaria amara* Ach.** (syn. *Pertusaria amara*) [see also couplet 33]

49. Soralia KC–; thallus cortex K+ yellow or K–, PD+ yellow, orange, or red, or PD–. 50

50. Thallus K–, PD–; with a dark prothallus 51

50. Thallus K+ yellow or PD+ yellow or red; prothallus absent or pale 52

51. Thallus with conspicuous blue-black or dark gray prothallus; soredia usually grayish blue at least in part; all grayish blue tissue turning reddish purple in nitric acid; common in Pacific Northwest on alders .... *Mycoblastus caesius* (Coppins & P. James) Tønsberg

51. Thallus with brown prothallus; soredia white to brownish white, nitric acid negative; common in boreal and North Temperate forests on various trees, especially *Betula* .... *Fuscidea pusilla* Tønsberg

52. Thallus and soredia K+ yellow, PD– (or faint yellow) (atranorin) 53

52. Thallus and soredia K+ yellow or K– (or brownish), PD+ yellow, orange, or red 55

53. Thallus lacking zeorin, but frequently containing one or two other unknown triterpenes; soralia typically round, excavate; thallus dark to very pale greenish gray; mainly Canada in east, as well as western montane, New Mexico to Yukon.. *Lecanora impudens* Degel.
[See also *Buellia griseovirens*, couplet 55.]

53. Thallus containing zeorin; East Temperate, rare in Canada 54

54. Thallus well developed, areolate, white to bluish gray, with discrete soralia erupting from areoles containing coarse soredia mostly 35-60 µm in diameter; prothallus absent or white fibrous ....*Lecanora appalachensis* Lendemer & R. C. Harris

54. Thallus thin, continuous to areolate, soralia becoming irregular in shape and finally confluent, with fine soredia mostly 23-37 µm in diameter; prothallus always present, membranous and translucent ....*Lecanora nothocaesiella* R. C. Harris & Lendemer

55.(52) Soredia olive-black, brown, or dark greenish gray, at least in part, PD+ yellow, K+ red (norstictic acid), rarely K+ yellow (atranorin alone or with stictic instead of norstictic acid); on bark and wood, frequent in the west and scattered in the northeast .... *Buellia griseovirens* (Turner & Borrer *ex* Sm.) Almb.

55. Soredia white, PD+ red or orange, K– or K+ yellow 56

56. Thallus within the bark, only forming a white stain, containing *Trentepohlia*; soralia punctiform, scattered, erupting from thallus with very coarse granular soredia, PD+ orange, K+ yellow (stictic acid); common, mainly on hardwoods; East Temperate .... *Nadvornikia sorediata* R. C. Harris

56. Thallus thin or thick, superficial, containing Trebouxioid algae 57

57. Thallus PD+ red or orange, K– or brownish (not thamnolic acid), very thin 58

57. Thallus PD+ orange, K+ deep yellow (thamnolic acid), thin or more frequently thick and verruculose; soralia (actually, sorediate fruiting warts) 59

58. Thallus PD+ red (fumarprotocetraric acid), whitish; soralia powdery; boreal forest, mostly on conifers and birch.. ***Ramboldia cinnabarina* (Sommerf.) Kalb et al.** (syn. *Pyrrhospora cinnabarina*)

58. Thallus PD+ orange (pannarin and zeorin), dark to pale gray; Appalachian-Great-Lakes region, on white cedar (*Thuja*) .... *Megalospora porphyrites* (Tuck.) R. C. Harris

59. Soralia remaining round and discrete; temperate eastern regions, mainly on deciduous trees .... ***Variolaria trachythallina* (Erichson) Lendemer et al.** (syn. *Pertusaria trachythallina*)

59. Soralia coalescing in older parts of thallus; mostly Appalachian-Great Lakes, less frequent in west coast mountains, usually on conifers .... *Loxospora elatina* (Ach.) A. Massal.

KEY D: CRUSTOSE PERITHECIAL LICHENS AND LICHENS WITH ASCOMATA RESEMBLING PERITHECIA

1. Spores 1-celled, ellipsoid and large (mostly 45-150 x 25-60 µm), most commonly with thick walls (2-30 µm); ascomata buried in thalline warts 2

1. Spores 1- celled, ellipsoid and small (less than 45 µm long), with thin walls; **or** 2- to many-celled with either thin or thick walls; ascomata superficial or more or less buried in warts, pseudostromata, or the thallus 4

2. Spore walls conspicuously thick-walled, often layered (fig. 15-l); excipulum poorly defined and always colorless; widespread .... *Pertusaria*

2. Spore walls relatively thin, never layered; excipulum colorless but distinct and well developed; Pacific Northwest 3

3. Ostiole deep and hole-like; hymenium and asci IKI– or orange; thallus cortex and medulla K+ yellow, PD+ orange, UV– (stictic acid); on coastal maritime rocks; locally common .... ***Coccotrema maritimum* Brodo**

3. Ostiole slightly depressed, not forming a hole; hymenium and asci IKI+ blue; thallus area around ostiole C+ pink (gyrophoric acid), medulla UV+ white (alectoronic acid); on maritime or alpine rocks near coast; rare .... *Ochrolechia subplicans* (Nyl.) Brodo

4.(1) Spores numerous in each ascus; inconspicuous, temperate to boreal lichens 5

4. Spores 8 or fewer per ascus 6

5. Perithecium-like ascomata red-brown to black, without pruina; spores subglobose, ellipsoid, or fusiform, 1- to 6-celled; on mosses and plant remains, bark, or wood; asci cylindrical or club-shaped .... *Thelopsis*

5. Yellow pruina present on the thalline covering of the perithecium-like ascomata; spores subglobose to ellipsoid, 1-celled; on rock, wood, soil or decaying lichens; asci pear-shaped (fig. 14o) .... *Thelocarpon*

6. Growing on rock, soil, mosses, dead vegetation, barnacle shells, lichens, or leaves 7

6. Growing on bark or wood 21

7. Spores 1-celled 8

7. Spores 2- or more celled 12

8. Photobiont blue-green; thallus reddish, dark brown, or black and gelatinous when wet; on rock 9

8. Photobiont green; thallus not reddish and gelatinous when wet 11

9. Paraphyses absent, replaced with well-developed periphyses (fig. 17); rare .... *Cryptothele* (see *Pyrenopsis* key)

9. Paraphyses persistent, branched or unbranched; possibly widespread 10

10. Photobiont with sheaths (gelatinous covering of cells) K– or golden (Chroococcaceae) .... *Pterygiopsis* (see Key B)

10. Photobiont with sheaths K+ purple (*Gloeocapsa*) .... *Pyrenopsis*

11.(8) Paraphyses persistent; on soil; thallus thin, continuous, greenish or greenish-brown; widespread but overlooked .... *Thrombium epigaeum* (Pers.) Wallr.

11. Paraphyses absent, disappearing or not developing; on rock .... *Verrucaria*

12.(7) Spores muriform, many-celled 13

12. Spores transversely septate; rarely submuriform 17

13. Ostioles deeply depressed (hole-like), sometimes encircled with radiating ridges; paraphyses persistent; spores dark brown; thallus thick, gray; medulla C+ red .... *Diploschistes*

13. Ostioles level with perithecial surface, slightly depressed, or prominent, without radiating ridges; paraphyses persistent or disintegrating or not developing; spores colorless or brown; thallus thin or thick, pale gray to dark brown; medulla C– 14

14. Paraphyses persistent, branched and anastomosing 15

14. Paraphyses absent or disappearing 16

15. Ascus tip entirely IKI– .... *Thelenella*

15. Ascus tip with a central IKI+ blue column .... *Protothelenella*

16. Algae present in perithecial cavity .... *Staurothele*

16. Algae absent from perithecial cavity .... *Polyblastia* key

17.(12) On mosses, dead vegetation, or leaves; spores 2- to 8-celled, rarely submuriform .... *Porina* key

17. On rock or barnacle shells 18

18. Spores 4- to many-celled; paraphyses persistent .... *Porina*

18. Spores 2-celled (rarely 4-celled); paraphyses persistent or disappearing 19

19. On intertidal calcareous rocks or barnacle shells; thallus light brown to pale orange, often only a membranous stain; photobiont blue-green; spores constricted at the septa; on Atlantic and Pacific coasts .... *Collemopsidium halodytes* (Nyl.) Grube & B. D. Ryan (syn. *Pyrenocollema halodytes*)

19. On non-maritime, mostly calcareous, rocks; thallus pale gray or yellowish white; photobiont green; spores not constricted at the septa; North Temperate to arctic-alpine 20

20. Paraphyses disintegrating or not developing; excipulum carbonized or darkly pigmented; periphyses conspicuous; spores narrow, length to width ratio mostly 2-3:1; mostly arctic-alpine .... *Thelidium*

20. Paraphyses persistent; excipulum pale or colorless; periphyses absent; spores broadly ellipsoid, length to width ratio mostly 2:1 or less; northeastern U.S. and Canada .... *Acrocordia conoidea* (Fr.) Körber

21.(6) Spores very long and narrow, many-celled, not tapering, constricted at the septa giving the spores a segmented, worm-like appearance .... ***Stictis urceolatum* (Ach.) Gilenstam** (*syn. Conotrema urceolatum*)

21. Spores broadly to narrowly ellipsoid, fusiform, or needle-shaped, often tapering, never worm-like............22

22. Ostiole deep, forming a pit or deep hole; exciple pale or carbonized, well developed, surrounded by a thalline envelope or covering giving the ascoma a double-walled appearance, especially when the two walls are not fused 23

22. Ostiole not deep; exciple or excipulum dark or pale, surrounded by thalline tissue or not 25

23 Short, hair-like hyphae resembling periphyses lining the inner face of exciple around the ostiole; ascomata lacking a columella; northeastern and western coasts as well as southeast .... *Thelotrema*

23. Hair-like hyphae resembling periphyses absent; many species with a column of sterile tissue (columella) developing in the ascomatal cavity; southeastern coastal plain...24

24. Exciple (inner wall of ascoma), and columella, when present, black......................................*Ocellularia* key

24. Exciple, and columella, when present, pale to reddish brown..*Myriotrema* key

25.(22) Spores pale to dark brown when mature 26

25. Spores colorless..29

26. Ostioles of the ascomata (perithecia) lateral, not at the summits, often at the end of a long or short neck .... *Lithothelium*

26. Ostioles at the summits of the ascomata..27

27. Walls of perithecium-like ascomata pale; spores muriform ....*Myriotrema* key

27. Walls of perithecia black, carbonaceous 28

28. Spores transversely septate or muriform, with unevenly thickened walls, central spore locules about equal in size, lens-shaped; spores not constricted at the septa .... *Pyrenula*

28. Spores transversely septate, with uniformly thickened walls, spore locules not equal in size; spores constricted at the septa; thallus pale gray, endophloeodal; perithecia partly immersed, wall dark above and pale below; spores fusiform, with pointed ends, 18-24 x 5-9 µm; central U.S. to northeastern Canada and New England. .... *Eopyrenula intermedia* Coppins

29.(25) Spores 1-celled; asci with 8 spores; thallus not at all yellow; perithecia black, not in thalline warts; on bark; spores conspicuously warty on the surface; paraphyses branched; southeastern coastal plain .... *Monoblastia* [See key in Harris (1995)]

29. Spores 2- or more celled 30

30. Perithecia buried in a wartlike pseudostroma, often several perithecia per wart .... *Trypethelium* key

30. Perithecia discrete, not in a pseudostroma 31

31. Spores with unevenly thickened walls .... *Trypethelium* key

31. Spores with uniformly thickened walls 32

32. Spores muriform; thallus thin or endophloeodal 33

32. Spores transversely septate 35

33. Perithecia black, wall black; spores 17-25 x 7-10 µm; thallus endophloeodal; East Temperate .... *Julella fallaciosa* (Stizenb. *ex* Arnold) R. C. Harris

33. Perithecium-like ascomata pale; southeastern 34

34. Ascomata superficial, wall orange to dark brown; spores 12.5-17 x 8.5-11 µm; thallus membranous, brownish, PD– .... *Topelia aperiens* P. M. Jørg. & Vĕzda

34. Ascomata sunken in the thallus; spores 9-20 x 4-6 µm; thallus light gray, PD+ yellow .... *Myriotrema*

35.(32) Spores 6- or more celled 36

35. Spores 2- to 4-celled 37

36. Spores 18-27 x 4-7 µm, (5-)6- to 8(-9)-celled, constricted at the septa; perithecia black or almost black; thallus whitish, within substrate, indistinct; southern coastal plain .... *Polymeridium quinqueseptatum* (Nyl.) R. C. Harris

36. Spores 24-125 µm long, 8- to 14-celled, usually not, or only slightly constricted at the septa; perithecia pale to dark brown or black; thallus yellowish gray to greenish gray, or pale greenish, superficial or within bark .... *Porina* key

37. Spores primarily 4-celled; paraphyses mostly unbranched 38
37. Spores primarily 2-celled; paraphyses slightly to abundantly branched 40

38. Thallus UV+ yellow (lichexanthone); spores 22-32 x 7-10 µm; common in Florida and adjacent Georgia and Alabama .... *Polymeridium catapastum* (Nyl.) R. C. Harris
38. Thallus UV– 39

39. Ostioles prominent, usually at the tip of an off-center to lateral neck; hymenium IKI+ greenish-blue; spores with central cells larger than tip cells, 18-27(-30) x 7-10(-12) µm; northeastern .... *Lithothelium hyalosporum* (Nyl.) Aptroot (syns. *Arthopyrenia hyalospora, Plagiocarpa hyalospora, Pleurotrema solivagum*)
39. Ostiole not prominent, at the summit of the perithecium; hymenium IKI–; spores with all cells approximately equal in length .... *Porina* key

40.(37) Spores narrow, 4.5-6(-7) µm wide, with smooth walls; paraphyses branched and anastomosing; perithecia 0.1-0.5 mm in diameter 41
40. Spores broad, 6-23 µm wide, with ornamented (rough) walls; paraphyses sparsely branched; perithecia 0.25-0.6 mm in diameter 42

41. Spores 2(-4)-celled, cells clearly unequal in size in most spores, narrowly ellipsoid to fusiform, length to width ratio about 2.5- 3:1, 14-20 x 4.5-6(-7) µm; perithecia 0.1-0.25 mm in diameter; East Temperate .... *Anisomeridium polypori* (Ellis & Everh.) M. E. Barr (syn. *A. nyssaegenum*)
[Note: Species of *Arthopyrenia* will also key here.]
41. Spores 2-celled, almost equal in size, ellipsoid, length to width ratio 2-2.5:1, 10-15(-18) x 4.5-7 µm; perithecia 0.2-0.5 mm in diameter; throughout the east, less common along the west coast .... *Anisomeridium biforme* (Borrer) R. C. Harris

42. Spores 33-48(-60) x 15-23 µm; ostioles usually off-center, often on a short neck; central to eastern U.S. .... *Acrocordia megalospora* (Fink) R. C. Harris

42. Spores 11-16.5 x 6-9.5 µm; ostioles ± central, level with perithecial surface; North Temperate, Quebec to North Dakota .... *Acrocordia cavata* (Ach.) R. C. Harris

KEY E: CRUSTOSE SCRIPT LICHENS

1. On rock 2
1. On bark, wood, leaves, or other lichens 4

2. Spores 1-celled; thallus white, K+ red (norstictic acid); western coastal mountains and Gaspé Peninsula (Quebec); very rare .... *Lithographa tesserata* (DC.) Nyl.
2. Spores septate 3

3. Hymenial surface ("disks" of ascomata) pruinose; exciple containing crystals .... *Lecanographa* (*Dendrographa* key)
3. Hymenial surface rarely pruinose; exciple lacking crystals .... *Opegrapha*

4. Lirellae immersed in wart-like pseudostroma 5
4. Lirellae developing directly on the thallus 6

5. Lirellae black, opening with only a narrow fissure, pseudostroma heavily pruinose; spores narrowly ellipsoid, 4- to 6-celled when mature .... *Sarcographa*
5. Lirellae brown, flat, with a completely exposed hymenium; pseudostroma without pruina or lightly pruinose; spores fusiform, 6- to 8-celled .... ***Glyphis cicatricosa* Ach.**

6. Spores 1-celled 7
6. Spores 2- or more celled 8

7. Lirella wall very pale to very dark brown, not carbonized; common and widespread .... *Xylographa*
7. Lirella wall black, carbonized; rare, in Pacific Northwest .... *Ptychographa xylographoides* Nyl.

8. Spores muriform 9
8. Spores only transversely septate 11

9. Spores brown to olive .... *Platygramme* key

9. Spores colorless 10

10. Lirellae with well-developed black wall (although sometimes immersed); asci cylindrical to club-shaped .... *Graphis key*
10. Lirellae lacking any wall; asci balloon-shaped .... *Arthothelium* (*Arthonia* key)

11.(8) Spores 2-celled 12
11. Spores 4- or more celled 13

12. Asci broad, balloon-shaped; exciple not developed .... *Arthonia*
12. Asci long and club-shaped; exciple well developed .... *Melaspilea*

13. Spore walls evenly thickened, cells cylindrical (appearing square) 14
13. Spore walls unevenly thickened, cells lens-shaped 18

14. Ascomata appearing lecanorine (surrounded with thalline tissue); often pruinose 15
14. Ascomata not at all lecanorine in appearance; pruinose or not 17

15. Ascomata black, heavily pruinose; spores fusiform, more than 4.0 µm wide, mostly 4-celled .... *Schismatomma* and *Dendrographa*
15. Ascomata pale brown, not or lightly pruinose; spores fusiform to needle-shaped, under 4.0 µm wide; 6-to10-celled 16

16. On bark, southeastern coastal plain; spores narrow, fusiform to needle-shaped, 7- to 10-celled, (25-)35-50 x 2.5-4.0 µm; ascomata pale red-brown, 0.2-0.7 x 0.2-0.3 mm .... *Enterographa anguinella* (Nyl.) Redinger
16. On evergreen leaves, Pacific Northwest; spores fusiform, 6- to 8-celled, 23-30 x 2.5-4 µm; ascomata tan, only slightly elongated .... *Enterographa oregonensis* Sparrius & Björk

17.(14) Lirellae with persistent carbonized walls (exciple) .... *Opegrapha*
17. Lirellae without distinct walls (exciple absent) .... *Arthonia*

18.(13) Spores pale to dark brown when mature .... *Phaeographis*
18. Spores colorless .... *Graphis* key

KEY F: CRUSTOSE DISK LICHENS

1. Ascomata or pycnidia at the summit of a slender or stout stalk 2
1. Ascomata immersed to superficial, broadly attached or sometimes constricted at the base, but not raised on a conspicuous stalk 5

2. Stalks stout; ascomata more than 0.5 mm in diameter, with flat to convex or hemispherical disks; spores remain within asci at maturity 3
2. Stalks slender, hair- or stubble-like (fig. 12f); ascomata or pycnidia mostly under 0.4 mm in diameter, irregular in shape or almost spherical; spores massed at the summit, loose or remaining within asci at maturity 4

3. Apothecia broader than stalk; disks pink to brown .... *Baeomyces* key (*Baeomyces*, *Dibaeis*)
3. Apothecia immersed in the tip of the stalk; disks black .... ***Pertusaria dactylina* (Ach.) Nyl.**

4. Stalk summits with a mass of conidia, which are thread-like or moniliform, many-celled ....key to *Gyalideopsis*
4. On bark, wood, or rarely rock; spores (ascospores) often in a dry mass, spherical to ellipsoid, 1- to 2(-4)-celled .... key to *Calicium* and similar lichens

5.(1) Apothecia deeply concave or opening by a small pit-like hole or ostiole; margin sometimes double, with a thalline margin partially or entirely enclosing a well-developed excipulum (figs. 12h; 13d) .... see Key D
5. Apothecia disk- or cup-shaped, or immersed in the thallus or in thalline warts, convex, flat, or concave, not opening by a small pit-like hole or ostiole; margins various (figs. 12 a-c; 13 a-c) 6

6. Apothecia buried in thalline warts, 1 or more per wart; spores very large and thick-walled, 1- to 2-celled, often with walls having 2 layers (fig. 15-l) 7
6. Apothecia disk- or cup-shaped, convex, flat, or concave, not buried in thalline warts, with the hymenium exposed; spores 1- to many-celled, thin- or thick-walled 8

7. Spores 1-celled; apothecia in fruiting warts opening to the surface with one or more small ostioles with the hymenium largely enclosed (fig. 12g) ....*Pertusaria*
7. Spores 2-celled, constricted at the septum and easily breaking in two; apothecia in fruiting warts broadly opened to the surface, pinkish or yellowish, sometimes with the ascus tips showing as tiny glistening dots;

thallus white, thick or thin, sometimes becoming sorediate; medulla K–, C+ red (lecanoric acid); spores 200-400 x 72-135 µm; on moss, dead vegetation, wood, bark, or stones; arctic-alpine .... *Varicellaria rhodocarpa* (Körber) Th. Fr.

8.(6) Apothecia mazaedial: spores lying loose in a powdery mass within the exciple 9

8. Apothecia usually with a well-developed hymenium; spores remaining within the asci until maturity, not forming a loose mass 12

9. Spores 2-celled, each spore enveloped by a cellular layer of pseudoparenchyma; mazaedial surface covered by a greenish pruina; California, rare .... *Texosporium sancti-jacobi* (Tuck.) Nadv.

9. Spores 1- to 4-celled, not covered with pseudoparenchyma; pruina, if present, not greenish 10

10. Spores 4-celled, with unevenly thickened walls; ascomata with 2 chambers, one above the other, the lower one immersed in the substrate, the upper one cylindrical, about 0.5 mm high; thallus UV+ yellow (lichexanthone), thin, continuous; southeast coastal plain .... *Pyrgillus javanicus* Nyl.

10. Spores 1- to 2-celled, rarely submuriform; ascomata with one chamber; thallus UV– or UV+ orange from yellow pigments (lichexanthone absent); mostly northern or western 11

11. Exciple brown to black, well developed; spores ellipsoid, 2-celled or submuriform in one species (*C. notarisii*); thallus thin; apothecia immersed or superficial, not in thalline warts .... *Cyphelium*

11. Exciple pale and weakly developed; spores globose or broadly ellipsoid, 1- to 2-celled; thallus thick; apothecia entirely immersed in thalline warts, disk level with surface .... *Thelomma*

12.(8) Apothecia pale yellow, bright lemon- or yolk-yellow, orange, or red 13

12. Apothecia white, gray, green, olive, brown, pink, or black 37

13. Asci containing numerous spherical spores; apothecia tiny, 0.1-0.3 mm in diameter; rare .... *Strangospora* key

13. Asci containing 1-32 spores; apothecia mostly larger than 0.2 mm in diameter 14

14. Apothecia blood red to cinnabar or pale red 15

14. Apothecia yellow or orange, not red 18

15. Spores polarilocular; apothecia pale red; California

.... *Caloplaca luteominia* (Tuck.) Zahlbr. var. *bolanderi* (Tuck.) Arup

15. Spores thin-walled, not polarilocular 16

16. Apothecia biatorine; spores ellipsoid, 1-celled .... *Pyrrhospora*

16. Apothecia lecanorine; spores fusiform, 2- to 8-celled.. 17

17. Southeastern and southwestern U.S.; thallus pale gray to greenish gray (lacking usnic acid) .. *Haematomma*

17. Western and northern North America; thallus distinctly yellowish (usnic acid in the cortex) .... *Ophioparma*

18.(14) Pigmented tissues (epihymenium or thallus cortex) K+ purple or dark red-purple (anthraquinones)........... 19

18. Pigmented tissues K– or K+ pinkish (lacking anthraquinones) 23

19. Spores polarilocular, with a thickened septum .... *Caloplaca*

19. Spores 1- to several-celled or muriform, not polarilocular 20

20. On dry soil; spores 1- to 2-celled, thin-walled ....*Fulgensia*

20. On rocks, bark, or bryophytes; spores 1- to several-celled or muriform 21

21. On rock, especially limestone; apothecia with thin, disappearing margins; spores 1-celled .... *Protoblastenia*

21. On bark, bryophytes, or decaying vegetation; apothecia with prominent, persistent margins; spores transversely septate or muriform 22

22. Thallus olive to orange, K– or K+ purple; spores with unevenly thickened walls, sometimes forming spiralled locules, or thin-walled and muriform; 2-8 spores per ascus; apothecia usually without pruina ....*Letrouitia*

22. Thallus white to pale gray, K+ yellow; spores thin-walled, muriform, with many cells; one spore per ascus; apothecia usually heavily pruinose ....*Brigantiaea*

23.(18) Apothecia biatorine, lacking algae in the margin 24

23. Apothecia lecanorine (with algae in the margin) or cryptolecanorine (sunken into the thallus, and lacking a recognizable exciple) 29

24. Spores 2- or more celled 25

24. Spores 1-celled 28

25. Apothecial margin fuzzy or tomentose (byssoid); paraphyses branched; asci club-shaped; spores 4-celled, 10-18 x 2.5-5 µm (in species treated here) .... *Byssoloma*

25. Apothecial margin smooth; paraphyses unbranched, slender; asci narrowly cylindrical 26

26. Spores 4- to 8-celled, 16-48 per ascus; Great Lakes to Maine .... *Pachyphiale fagicola* (Hepp) Zwackh

26. Spores 2- to 4-celled, 8 per ascus 27

27. Photobiont *Trentepohlia*; apothecia pale orange to pinkish orange, 0.2-1.5 mm in diameter; spores 2-celled .... *Coenogonium* (syn. *Dimerella*)

27. Photobiont chlorococcoid; apothecia bright yellow-orange; spores 2- to 4-celled .... *Cliostomum vitellinum* Gowan

28.(24) Thallus leprose, bright yellow, UV+ orange (rhizocarpic acid); on rock .... ***Psilolechia lucida* (Ach.) M. Choisy**

28. Thallus smooth to granular, without soredia, greenish yellow, UV– (usnic acid); on bark .... ***Lecanora symmicta* (Ach.) Ach.**

29.(23) Apothecia cryptolecanorine, immersed in thallus, the disk flush with the thallus surface 30

29. Apothecia lecanorine, broadly attached or constricted at the base; thallus pale yellow, yellow-green, or gray 31

30. Spores numerous in each ascus; thallus bright yellow .... *Acarospora* key

30. Spores 8 per ascus; thallus distinctly rusty orange or brownish gray (shade forms) .... ***Ionaspis lacustris* (With.) Lutzoni**

31. Apothecial disks bright yellow or bright yellow-orange 32

31. Apothecial disks pale greenish yellow, pale yellow, or pale pinkish orange 35

32. Thallus and apothecial disks bright yolk-yellow (calycin) 33

32. Thallus or apothecial disks yellow-green or yellow-gray (usnic acid) 34

33. Thallus distinctly lobed; lower surface corticated .... *Candelina*

33. Thallus edge indefinite, never lobed; lower surface without a cortex .... *Candelariella*

34. Apothecial disks not pruinose; apothecial margin flush with disk, becoming thin and disappearing in maturity .... ***Lecanora strobilina* (Sprengel) Kieffer**

34. Apothecial disks yellow pruinose; apothecial margin prominent .... *Lecanora cupressi* Tuck.

35.(31) Spores conspicuously thick-walled, 2-30 µm thick .... *Pertusaria*

35. Spores thin-walled, usually under 2 µm thick 36

36. Spores 10-20 µm long; paraphyses unbranched or slightly branched; apothecia generally less than 1.5 mm in diameter .... *Lecanora* key

36. Spores over 25 µm long; paraphyses highly branched; apothecia frequently over 1.5 mm in diameter .... *Ochrolechia*

37.(12) Apothecia lacking algae in the margin: lecideine, biatorine, or without an exciple; apothecia superficial, or sunken into thallus but retaining a distinct, usually pigmented exciple 38

37. Apothecia with algae in the margin or in thalline tissue surrounding the sunken disk: lecanorine (superficial) or cryptolecanorine (sunken into the thallus), usually without a distinct exciple 107

38. Spores 1-celled, spherical to narrowly ellipsoid (if spores are fusiform with length to width ratio more than 3:1, they may be species that become septate later in development; see, e.g., *Icmadophila*) 39

38. Spores septate (2- or more celled), broadly ellipsoid to thread-shaped 67

39. Apothecia biatorine, with a colorless to pigmented exciple that is soft rather than brittle, with a clearly defined and usually radiating cellular structure throughout, or exciple very reduced and indistinct (if absent, see *Micarea* key) 40

39. Apothecia lecideine, with an exciple that is very dark brown to black or green-black at least at the outer edge, usually somewhat carbonized, occasionally brittle, often with a poorly defined cellular structure, at least on the outer part 63

40. Spores conspicuously thick-walled (more than 2.5 µm thick) 41

40. Spores thin-walled relative to the spore size (usually less than 1.5 µm thick) 42

41. Apothecial disks black; thallus white to pale greenish gray; usually with a blood-red area below the hypothecium; epihymenium green; spores 1or 2 per ascus; thallus cortex K+ yellow (atranorin) .... *Mycoblastus*

41. Apothecial disks dark reddish brown; thallus brown or pale; tissue below hypothecium not red; epihymenium brown; spores 8 per ascus; thallus cortex K– (atranorin absent) .... ***Japewia tornoensis* (Nyl.) Tønsberg**

42. Apothecia pink, convex, over 1 mm in diameter .... *Baeomyces* key
42. Apothecia not pink 43

43. Directly on rock 44
43. On bark, wood, soil, or mosses or dead vegetation 46

44. Apothecia heavily pruinose; spores more than 100 per ascus; on limestone ..... ***Sarcogyne regularis* Körber**
44. Apothecia not pruinose; eight spores per ascus; on non-calcareous rock 45

45. Thallus pale gray, C+ pink (gyrophoric acid); apothecia pinkish or brown, with a rough or ragged margin; asci cylindrical, usually with uniformly K/I+ pale blue walls including the tip (fig. 14-l) .... *Trapelia*
45. Thallus dark, brownish, olive, or dark gray, C–; apothecia dark brown to black, with a smooth margin; ascus club-shaped, the tips staining as pale and dark blue layers in K/I (fig. 14d) .... *Fuscidea*

46.(43) On moss, soil, or dead vegetation 47
46. On bark or wood 53

47. Spores more than 8 per ascus, colorless; apothecia brown, 0.1-0.5 mm in diameter .... *Strangospora* key
47. Spores 8 per ascus 48

48. Hypothecium dark red-brown, brown, or black 49
48. Hypothecium colorless, yellowish, or very pale brown 51

49. Thallus dark brown or olive-brown; asci K/I–; on soil or decayed peat; hypothecium merging with exciple .... *Placynthiella*
49. Thallus pale greenish gray to whitish; asci *Porpidia*-type; on moss or soil; hypothecium distinct from exciple; paraphyses mostly unbranched 50

50. Cephalodia common, thallus-colored or brownish; spores narrowly ellipsoid, 17-22(-24) x 7-9(-10) µm; arctic-alpine, rare .... *Pilophorus dovrensis* (Nyl.) Timdal et al. (syn. *Lecidea pallida*)

50. Cephalodia absent .... *Biatora* key

51.(48) Thallus cortex and medulla C+ pink, KC+ pink (gyrophoric acid); epihymenium shades of olive or green; paraphyses highly branched .... *Trapeliopsis*

51. Thallus cortex or medulla C– or C+ orange, KC– or KC+ yellow to orange (gyrophoric acid absent); epihymenium shades of yellow or brown, or colorless; paraphyses only branched at tips 52

52. Thallus mostly pale green or gray-green; apothecia usually pale to dark yellowish brown, but sometimes dark brown; exciple almost colorless or yellowish internally, sometimes pigmented at the outer edge, with slender radiating hyphae .... *Biatora*

52. Thallus dark brown or olive, thick, mostly continuous or dispersed areolate; apothecia dark brown to almost black; exciple distinct, dark brown, with large cells .... ***Lecidoma demissum*** **(Rutstr.) Gotth. Schneider & Hertel**

53.(46) Spores numerous within the ascus 54

53. Spores 8 per ascus 55

54. Apothecia flat to convex .... *Strangospora* key

54. Apothecia deeply concave; spores 5-7 x 2 µm, halonate; southeastern .... *Ramonia microspora* Vězda

55. Thallus leprose, dark yellow, C+ orange (xanthones); epihymenium K+ deep purple (anthraquinones) .... ***Pyrrhospora quernea*** **(Dickson) Körber**

55. Thallus continuous to areolate or granular, or isidiate; C–, C+ orange, or C+ pink; epihymenium K– 56

56. Spores almost spherical, 5-10(-13) x 4-8 µm; apothecia brown to black; margins PD+ red; very common on conifers in western mountains .... *Lecanora fuscescens* (Sommerf.) Nyl.

56. Spores ellipsoid; apothecial margins PD– or rarely PD+ 57

57. Apothecia greenish to yellow-green; thallus cortex KC+ gold (usnic acid) .... ***Lecanora symmicta*** **(Ach.) Ach.**

57. Apothecia pale beige, pinkish, brown, or black; thallus cortex KC– or KC+ orange or pink (usnic acid absent) 58

58. Paraphyses highly branched and anastomosing 59

58. Paraphyses unbranched except for tips 61

59. Apothecia lacking margins; thallus thin, almost imperceptible, usually C– .... *Micarea*
59. Apothecia with margins, at least when young; thallus well developed, often C+ pink (gyrophoric acid); epihymenium C–, K– 60

60. Hypothecium brown, often merging with exciple; thallus reddish brown to dark brown or gray .... *Placynthiella* key
60. Hypothecium pale to colorless; thallus pale gray to dark greenish gray .... *Trapeliopsis*

61.(58) Thallus a mass of isidia or coarse granules (sometimes squamulose in younger parts); spores narrower than 2.5 µm .... *Phyllopsora*
61. Thallus continuous, areolate to somewhat granular, occasionally sorediate, never squamulose; spores broader than 2.5 µm 62

62. Spores broad (length to width ratio 1.5-2:1); thallus often KC+ orange (somewhat faint); apothecia dark red-brown, 0.2-0.4 mm in diameter; very common on trees in the east .... "*Pyrrhospora*" *varians* (Ach.) R. C. Harris [See Note in *Pyrrhospora* key.]
62. Spores narrow (length to width ratio 2-3:1); thallus KC– or KC+ pink or orange .... *Biatora*

63.(39) Spores 8 per ascus 64
63. Spores many more than 8 per ascus 65

64. Spores brown, walls thickened at equator; prothallus thick, black; arctic-alpine .... ***Orphniospora moriopsis* (A. Massal.) D. Hawksw.**
64. Spores colorless, walls uniformly thin; prothallus present or absent ...key to *Lecidea* and *Lecidea*-like crusts

65. Thallus well developed, areolate; medulla C+ pink (gyrophoric acid); apothecia immersed in thallus between the areoles, disk flush with thallus; epihymenium usually green; spores globose or broadly ellipsoid .... *Sporastatia*
65. Thallus mostly within the substrate, absent from view, lacking gyrophoric acid; apothecia sessile; epihymenium brown or black; spores narrowly ellipsoid 66

66. Apothecial disks rough, umbonate (with sterile carbonized columns of tissue in apothecial disk); apothecia often concave or crushed due to crowding; apothecial margin rough or minutely fissured; exciple dark brown to carbonized throughout .... *Polysporina*

66. Apothecial disks smooth, without umbos; apothecia flat or soon convex; apothecial margin smooth; exciple dark or carbonized only at edge, pale internally .... *Sarcogyne*

67.(38) Spores muriform 68

67. Spores only transversely septate 74

68. Apothecia biatorine, concave, pale and waxy-looking, pink, yellow, or pale orange; epihymenium and hypothecium colorless, paraphyses unbranched; asci narrowly cylindrical .... *Gyalecta* key

68. Apothecia biatorine or lecideine, flat to somewhat convex, brown or black, not waxy; epihymenium pigmented; hypothecium colorless to yellowish, or brown to black; paraphyses unbranched or abundantly branched or difficult to distinguish; asci club-shaped 69

69. Directly on rock; spores halonate or not 70

69. On bark, soil, leaves, or mosses; spores not halonate 71

70. Spores halonate .... *Rhizocarpon*

70. Spores not halonate .... *Buellia* key (*Diplotomma*)

71. On leaves; apothecia 0.2-0.5 mm in diameter, disks pale to dark brown; apothecial margin even with disk .... *Calopadia*

71. On bark, mosses, or soil; apothecia usually more than 0.5 mm in diameter, disks black or almost black; apothecial margin prominent 72

72. Apothecial margin yellow or bright orange, paler than disk; exciple radiate, yellow; epihymenium K+ red or deep purple-red (anthraquinones); on bark in tropical and subtropical regions .... *Letrouitia*

72. Apothecial margin black, the same color as the disk; exciple dark brown at edge, pale internally; epihymenium unchanged in K (anthraquinones absent); on bark, soil, or mosses in northern regions 73
[Note: If on bark in southeastern U.S., see *Graphis* key.]

73. Spores colorless, or pale brown when old, 1-8 per ascus, over 20 μm long .... *Lopadium* key

73. Spores brown, 8 per ascus, less than 30 μm long .... *Buellia* (*Diplotomma*)

74.(67) Apothecia pink to yellowish pink, 1.5-4 mm in diameter; thallus green to greenish white, continuous, sometimes verrucose; growing over well-rotted wood or peat .... ***Icmadophila ericetorum*** **(L.) Zahlbr.**

74. Apothecia white, gray, or pale brown to black (if pinkish, then under 1 mm in diameter); thallus and habitat various 75

75. Spores mostly 4- or more celled 76

75. Spores mostly 2-celled 92

76. Photobiont blue-green; thallus with a conspicuous blue-black prothallus ....*Placynthium*

76. Photobiont green; lacking a blue-black prothallus 77

77. Spores brown ....*Buellia*

77. Spores colorless 78

78. Spores slender and needle-shaped, some often curved or bent; length to width ratio 7:1 or more 79

78. Spores ellipsoid, narrowly ellipsoid, or fusiform, straight or bent; length to width ratio less than 7:1 84

79. Thallus bright yellow; apothecia *Arthonia*-like (figs 12c, 13f); apothecial margin absent; growing on soil or lichens ....*Arthrorhaphis*

79. Thallus greenish, olive-gray, gray, or brownish; apothecia with a persistent margin; on bark, mosses, or dead vegetation 80

80. Apothecia lecideine, carbonaceous or blackened, at least at outer edge; ascus walls and tips K/I– or + pale blue, sometimes containing a small blue-staining "ring structure" in the tip; photobiont *Trentepohlia* 81

80. Apothecia biatorine, lacking carbonaceous tissue; ascus tips K/I+ dark blue or K/I–; photobiont various 82

81. Spores 4- to 7-celled, remaining intact; apothecia typically pruinose at least when young ....*Lecanactis*

81. Spores 6- to 50-celled, often breaking up into segments; apothecia without pruina ....*Bactrospora*

82. Spores 8 per ascus ....*Bacidia* key

82. Spores many per ascus 83

83. Apothecia lead-black or very dark brown; spores 16-55 x 1.2-3.1 µm; thallus edge definite, with a brown prothallus; exciple pale internally, dark brown at edge; ascus *Fuscidea*-type, with a layered appearance when stained with K/I; photobiont Trebouxioid .... ***Ropalospora chlorantha*** **(Tuck.) S. Ekman**

83. Apothecia pink-brown to reddish brown; spores 24-28(-35) x 3.0-3.8(-4.7) µm; thallus indistinct, lacking a prothallus; ascus K/I pale blue, lacking a thickened tip (*Gyalecta*-type); photobiont *Trentepohlia* .... *Pachyphiale fagicola* (Hepp) Zwackh [see also *Gyalecta* key]

84.(78) Thallus thick, areolate to squamulose; apothecia black; on soil or rocks, usually in arid or alpine habitats .... *Toninia*

84. Thallus mostly thin, membranous to leprose or granular; apothecia very pale to black; on bark, moss, or dead vegetation, occasionally on rock or soil 85

85. Exciple absent 86

85. Exciple present 87

86. Asci broad, often balloon-shaped, walls and thickened tips (fig. 14n), K/I– (although "hymenium" may be K/I + blue in some species) ....*Arthonia*

86. Asci club-shaped; walls and tip K/I+ blue .... *Micarea*

87. Exciple lecideine, dark brown to black, carbonaceous at least on outer edge 88

87. Exciple biatorine, usually pale, sometimes dark at the outer edge but not carbonaceous; apothecia not, or lightly, pruinose 89

88. Apothecia usually pruinose or dull; margin not radially fissured ....*Lecanactis* key

88. Apothecia black, shiny, never pruinose; margin broken up by radial fissures; on moss; arctic-alpine .... *Sagiolechia rhexoblephara* (Nyl.) Zahlbr.

89. Apothecial margin prominent or even with disk, fuzzy or tomentose (byssoid), with excipular hyphae extending to surface wth cylindrical cells; apothecia flat when mature; spores 4-celled, 10-18 x 2.5-5 µm (in species treated here) ....*Byssoloma*

89. Apothecial margin prominent and smooth or becoming thin and disappearing in maturity, or absent; excipular cells usually fused with short and rounded to elongate cells; apothecia flat to convex or hemispherical 90

90. Paraphyses highly branched and anastomosing; epihymenium C– or often C+ pink (usually disappearing quickly) .... *Micarea* key

90. Paraphyses unbranched or slightly branched at the tips; epihymenium always C– 91

91. Apothecia flat to convex; asci tips usually K/I+ dark blue .... *Bacidia* key

91. Apothecia deeply concave; ascus walls including tips K/I ± pale blue or – ....*Ramonia* (see *Gyalecta* key)

92.(75) Spores dark brown or dark gray when mature 93

92. Spores colorless 94

93. Spores halonate; ascus tips K/I mostly negative; on rock .... *Rhizocarpon*

93. Spores not halonate; ascus tips K/I+ dark blue, at least in part; on various substrates .... *Buellia* key

94. On rock 95

94. On bark, wood, moss, leaves, or dead vegetation 99

95. Hypothecium yellow-brown; asci balloon-shaped (fig. 14n); exciple absent (fig. 13f); on maritime rocks along both coasts .... *Arthonia phaeobaea* (Norman) Norman

95. Hypothecium dark brown to black; asci club-shaped; exciple well developed, biatorine or lecideine 96

96. Photobiont blue-green; prothallus conspicuous, blue-green; epihymenium emerald green ....*Placynthium*

96. Photobiont green; prothallus conspicuous or not, but not blue-green; epihymenium brown, reddish, to olive or greenish 97

97. Spores small, 6-15 x 2.5-5; paraphyses expanded and pigmented at tips; widespread temperate to boreal .... *Catillaria*

97. Spores large 12-30 x 3-14 98

98. Spores narrow, 3-5 μm wide, not halonate; on rock or soil; epihymenium gray to purplish .... *Toninia*

98. Spores broad, 8-12 μm wide, halonate; on rock; epihymenium brown or greenish, rarely reddish .... *Rhizocarpon*

99.(94) Apothecial margins fuzzy and tomentose (byssoid); thallus sometimes UV+ orange; on bark or leaves on the southeastern coastal plain .... *Byssoloma*

99. Apothecial margin, if present, smooth, not at all fuzzy; thallus always UV–; mostly East Temperate, northern, or western, on various substrates 100

100. Hypothecium dark reddish brown to black .... *Catillaria* key

100. Hypothecium colorless to pale brown 101

101. On leaves of conifers and broadleaf trees and shrubs, rarely on twigs; apothecia tiny, 0.1-0.3 mm in diameter, very pale beige; spores 9.5-14(-16) x 3.5-7 µm, tapered and constricted at the septum; Pacific Northwest to California and southeastern U.S. .... *Fellhanera bouteillei* (Desm.) Vězda

101. On bark, wood, or moss; apothecia (0.1-) 0.2-2 mm in diameter, pink to brown, white, gray, or black ... 102

102. Paraphyses much branched and anastomosing; exciple poorly differentiated .... *Micarea* or *Arthonia* [see couplet 86]

102. Paraphyses mostly unbranched except at the tips; exciple well developed, at least in young apothecia 103

103. Spores 2-4 µm wide 104

103. Spores 4-7 µm wide 106

104. Asci thin-walled, cylindrical, K/I–; photobiont *Trentepohlia;* apothecia white to pale gray, pinkish or pale orange or orange-brown .... *Coenogonium*

104. Asci with a thickened tip, club-shaped, K/I+ blue; photobiont Trebouxioid; apothecia pale to dark brown, black, or pale pinkish to black, mottled 105

105. Apothecia pruinose, very pale pinkish to black or mottled; paraphyses barely expanded and not pigmented at the tips; hypothecium colorless; spores 8-16 x 2.5-3.5 µm; on bark of all kinds; coastal .... ***Cliostomum griffithii* (Sm.) Coppins**

105. Apothecia not pruinose, brown to black, not pinkish; paraphyses expanded or not; hypothecium pale brown .... *Catillaria* key

106.(103) Apothecia black, mostly flat to lightly convex, with a persistent margin; spores all 2-celled, 10-15 x 5-7 µm; on bark of different kinds, especially poplar and cedar; widely distributed, especially in north temperate to boreal region and California .... *Catinaria atropurpurea* (Schaerer) Vězda & Poelt.

106. Apothecia pale to dark brown, often convex; spores 1- to 2-celled .... *Biatora* key

107.(37) Spores 2- or more celled 108
107. Spores 1-celled 120

108. Spores pale to dark brown when mature, mostly 2-celled 109
108. Spores colorless, 2- or more celled 113

109. Spores 4 per ascus, mostly over 35 µm long; cyanobacteria in the continuous crustose thallus, and green algae in lobes forming the apothecial rim; apothecia concave, cup-like, reddish or orange-brown .... ***Solorina spongiosa* (Ach.) Anzi**
109. Spores 8 or more per ascus, mostly under 35 µm long; photobiont green with no cephalodia; apothecia flat or soon convex, black to dark brown 110

110. Spores mostly with unevenly thickened walls at least when young, thallus not forming rosettes; if lobed, then not on rock; otherwise, on bark, soil, mosses, or rock .... *Rinodina* key
110. Spores with thin, uniformly thickened walls 111

111. On soil, mosses, or peat; spores 17-22 µm long .... ***Phaeorrhiza nimbosa* (Fr.) H. Mayerh. & Poelt**
111 On rock, bark, or wood 112

112. On rock; thallus thin, areolate, forming round, often lobed rosettes; spores 8-12 µm long .... *Dimelaena*
112. On bark or wood; spores 10-40 µm long .... *Rinodina* key, couplet 31

113.(108) Spores 1 per ascus, muriform or, rarely, transversely septate; apothecia buried in thallus .... *Phlyctis*
113. Spores 8 per ascus, transversely septate, 2- to 8-celled 114

114. Spores 2-celled, ellipsoid, straight or bean-shaped 115
114. Spores 4- to 8-celled, fusiform 117

115. Spore walls with a thickened septum (polarilocular); apothecial disks brown to almost black, sometimes pruinose .... *Caloplaca*
115. Spore walls evenly thickened, thin-walled; apothecial disks pale to dark brown, frequently pruinose 116

116. Ascus tip *Catillaria*-type (fig. 14c); spores with thin gelatinous perispore .... *Halecania*
116. Ascus tip *Bacidia*-type (fig. 14a); spores lacking a perispore .... *Lecania*

117. Hypothecium brown to black; apothecia white because of a heavy pruina, black beneath; photobiont *Trentepohlia* .... *Dendrographa* key

117.(114) Hypothecium colorless; apothecia lightly pruinose or not pruinose, brown to black; photobiont chlorococcoid 118

118. Spores straight to slightly curved, with rounded ends, not twisted in the ascus; apothecia pale to very dark brown or black, pruinose or not; on trees or rocks .... *Lecania*

118. Spores sinuous, curved and twisted in the ascus; apothecia pinkish brown to red-brown, lightly pruinose or without pruina; on bark in temperate to boreal regions 119

119. Thallus producing long, slender, sometimes branched isidia; thallus K–, PD–; coastal forests of Pacific northwest .... ***Loxosporopsis corallifera* Brodo, Henssen & Imshaug**

119. Thallus lacking isidia; thallus K+ deep yellow, PD+ orange (thamnolic acid); eastern temperate forests to boreal forest .... *Loxospora*

120.(107) Thallus gelatinous, containing cyanobacteria in small packets with gelatinous sheaths; apothecia opening by a pore or disk 121

120. Thallus not gelatinous, most with green photobionts; apothecial disks broad; on various substrates 122

121. Photobiont with a reddish sheath that turns K+ purple (*Gloeocapsa)*; thallus reddish when wet; apothecia opening by a deep, broad, or narrow pore; on streamside or lakeside rocks .... *Pyrenopsis*

121. Photobiont with yellowish sheaths that are K– (chroococcoid); apothecia first opening by a pore, later with a reddish brown disk; spores 12-15(-20) x 5-7.5(-10) µm; widespread on calcareous, often mossy rocks in shady habitats .... *Psorotichia schaereri* (A. Massal.) Arnold

122. Cephalodia conspicuous on the thallus surface, pink to brown, disk-like and more or less lobed or convex and "brain"-like 123

122. Cephalodia absent or relatively inconspicuous (between areoles or squamules); apothecia pink, brown, or black 124

123. Apothecia usually pink, sessile.......... *Placopsis*

123. Apothecia black, usually sunken (cryptolecanorine) .... *Amygdalaria*

124. Apothecia more or less immersed in thallus, disk flush with thallus surface (cryptolecanorine) 125
124. Apothecia superficial: adnate or constricted at base 137

125. Spores numerous in each ascus .... *Acarospora* key
125. Spores 4-8 per ascus 126

126. Thallus distinctly lobed at the margin; ascus tips K/I+ uniformly pale blue (*Aspicilia*-type) 127
126. Thallus margin without distinct lobes; asci of various types 128

127. Thallus and apothecia often pruinose, thin and clearly crustose; apothecia black (beneath the pruina); thallus cortex and medulla K–, PD– .... couplet 13 of *Aspicilia* key
127. Thallus and apothecia without pruina, very thick and almost foliose; apothecia dark brown; thallus cortex and medulla K+ yellow becoming blood red, PD+ yellow (norstictic acid) .... *Lobothallia*

128. Spores (30-)36-65 x 16-36 µm with walls 1.1- 2.5 µm thick .... ***Megaspora verrucosa* (Ach.) Hafellner & V. Wirth**
128. Spores mostly smaller than 35 x 22 µm, thin-walled, rarely over 1.5 µm thick... 129

129. Thallus pale yellow to yellowish white, thick, KC+ gold (usnic acid); epihymenium green .... ***Lecanora marginata* (Schaerer) Hertel & Rambold**
129. Thallus not at all yellowish, KC– (lacking usnic acid); epihymenium shades of yellow, brown, or green, or colorless 130

130. Apothecial disks C+ yellow, heavily pruinose; on siliceous rocks .... ***Lecanora rupicola* (L.) Zahlbr.**
130. Apothecial disks C–, pruinose or not (if pruinose, then on calcareous rock) 131

131. Apothecial disks black; epihymenium often olivaceous to green 132
131. Apothecial disks pink to brown or gray, rarely darkening to black; epihymenium shades of gray, yellow, or brown, lacking green pigments 136

132. Spores with a gelatinous perispore (halonate) sometimes only visible in an ink preparation 133
132. Spores not halonate 134

133. Thallus with cephalodia; spore halo usually distinct and well defined in a water mount without ink; spores 18-34(-45) x 9-18 µm .... *Amygdalaria*

133. Thallus lacking cephalodia; spore halo usually diffuse and only visible in an ink preparation; spores 10-23 x 6-12 µm .... *Porpidia* (species with immersed apothecia)

134. Epihymenium yellow-olive to olive-brown, unchanged or becoming green with nitric acid; ascus K/I – or uniformly pale blue; widespread and common .... *Aspicilia*

134. Epihymenium blue-green to olive-green, changing to wine-red with nitric acid, or unchanged 135

135. Mostly arctic-alpine species; thallus thin, continuous to rimose; ascus tip K/I – or K/I+ blue ....*Ionaspis* key

135. East Temperate; thallus thick, white to pale gray, areolate; ascus with a K/I+ blue tip .... ***Lecanora oreinoides*** **(Körber) Hertel & Rambold**

136.(131) Ascus tips with a IKI+ dark blue "plug" in a light blue tholus (*Porpidia*-type); spores always halonate; on dry rocks; apothecial disks very dark brown or red-brown (especially when wet) .... *Bellemerea*

136. Ascus tips entirely IKI–; spores not, or sometimes vaguely, halonate; on wet or submerged rocks, or on dry rocks; apothecia pinkish to brown .... *Ionaspis* key

137.(124) Spores conspicuously thick-walled (more than 2.5 µm thick) 138

137. Spores thin-walled (usually less than 1.5 µm) 140

138. Apothecia pale, pruinose or not .... *Ochrolechia*

138. Apothecia black, sometimes whitened by pruina 139

139. Epihymenium olive to greenish, K+ more clearly green; spores 30-65 µm long, 8 per ascus; boreal to arctic, on trees, bryophytes, and vegetation .... ***Megaspora verrucosa*** **(Ach.) Hafellner & V. Wirth**

139. Epihymenium brown, K– or K+ violet; spores usually longer than 60 µm, 1-8 per ascus; widespread on various substrates .... *Pertusaria*

140.(137) Spores very numerous in the ascus, 3-6 x 1.5-2.5 µm; asci similar to *Fuscidea*-type; apothecia dark brown to almost black, 0.4-1 mm in diameter; thallus brownish gray, bumpy; on trees; East Temperate and southern California .... *Maronea polyphaea* H. Magn.

[Note: This represents most records of "*M. constans*" by N. Am. authors. The latter is, however, known from the Southern Appalachians.]

140. Spores 1-32 per ascus, larger than 6 x 2.5 µm 141

141. Hymenium purple; hypothecium brown; apothecia pitch black .... ***Tephromela atra* (Hudson) Hafellner**

141. Hymenium colorless; hypothecium colorless to yellowish, rarely brown; apothecia various shades of pink and brown, or black 142

142. Spores with rough, bumpy, sculptured walls (clearly seen under 400X magnification); thallus brownish to green, granular to, more commonly, squamulose; on mossy logs, peat, and soil in boreal to arctic regions 143

142. Spore walls smooth; thallus rarely squamulose, pale to dark; on various substrates 144

143. Photobiont blue-green .... ***Protopannaria pezizoides* (P.M. Jørg.) S. Ekman** (syn. *Pannaria pezizoides*)

143. Photobiont green .... *Psoroma* (see key to *Pannaria*)

144. Spores over 25 µm long; paraphyses highly branched; apothecia pale, usually pinkish to yellow-pink .... *Ochrolechia*

144. Spores mostly 10-20 µm long; paraphyses mostly unbranched; apothecia pale to dark brown or black 145

145. On dry soil or rock (limestone); thallus pale yellowish to chalky white and pruinose, rarely brownish-green, clearly lobed at the margins .... ***Squamarina lentigera* (Weber) Poelt**

145. On rock, wood, bark, dead vegetation, or peat; thallus not yellowish white and chalky pruinose, lobed or not lobed 146

146. Thallus pale greenish gray to pinkish white, C+ pink (gyrophoric acid); apothecial margin often with a frayed or ragged appearance; on rock, especially small stones .... *Trapelia*

146. Thallus pale or dark, C– in North American species (lacking gyrophoric acid); apothecial margins smooth or bumpy but not ragged; on various substrates 147

147. On bark, wood, mosses, or dead vegetation .... *Lecanora*

147. On rock 148

148. Thallus distinctly lobed at the margin 149

148. Thallus not lobed 150

149. Ascus tips K/I± uniformly very pale blue (*Aspicilia*-type); tips of paraphyses septate and often constricted like a string of beads (moniliform) .... *Lobothallia*

149. Asci with K/I+ dark blue tips (*Lecanora*-type); tips of paraphyses not appearing "beaded" .... *Lecanora*

150. Ends of spores pointed; thallus grayish brown to reddish brown, shiny; apothecial disks chocolate brown and shiny .... *Protoparmelia*

150. Ends of spores rounded; thallus rarely brown and shiny; apothecia various colors .... *Lecanora* key

Key G: Squamulose Lichens

1. Thallus gelatinous when wet, containing cyanobacteria 2

1. Thallus dull or shiny, but not gelatinous when wet; containing green algae or cyanobacteria 4

2. Photobiont *Chroococcus*; squamules umbilicate, 1.5-4(-8) mm across, black when dry and rusty red-brown when wet, surface areolate, each areole containing a perithecium-like ascoma; spores 1-celled (often appearing 2-celled), 8-16 spores per ascus; on siliceous often wet rocks; infrequent, widespread montane to boreal .... *Phylliscum demangeonii* (Moug. & Mont.) Nyl.

2. Photobiont *Nostoc* 3

3. Thallus without a cortex .... *Collema* key

3. Thallus with a cellular cortex .... *Leptogium* key

4.(1) Thallus greenish yellow, yellow-green, yellow, or orange 5

4. Thallus pale green, gray, brown, olive, or black, without a yellowish or orange tint 20

5. Thallus greenish yellow or yellowish-green 6

5. Thallus bright yolk-yellow, lemon yellow, or shades of orange 12

6. Directly on rock 7

6. On soil 9

7. Apothecia entirely immersed in thallus, disk flush with thallus; asci containing numerous spores .... *Acarospora* key

7. Apothecia sessile or constricted at base; asci containing 8 spores 8

8. Apothecia constricted at base; disks yellow-orange, lightly pruinose; thallus not forming radiate, lobed rosettes .... ***Rhizoplaca subdiscrepans*** **(Nyl.) R. Sant.**

8. Apothecia sessile; disks pale greenish yellow, not pruinose; thallus usually forming distinctly lobed rosettes .... ***Lecanora muralis*** **(Schreber) Rabenh.**

9.(6) Spores brown, 2-celled .... *Phaeorrhiza*

9. Spores colorless, 1- to 12-celled 10

10. Squamules ascending, with the lower surface easily seen; thallus cortex KC+ yellow, UV– (usnic acid) .... *Cladonia* species, especially ***C. robbinsii*** **A. Evans**

10. Squamules closely appressed or lifting only at the edges; apothecia frequently seen, black, directly on the squamules; thallus cortex KC–, UV+ orange (rhizocarpic acid) 11

11. Squamules flat, often turned up at the edges, sturdy; spores 1-celled; southwestern and south central U.S. .... ***Psora icterica*** **(Mont.) Müll. Arg.**

11. Squamules closely appressed, convex, fragile; spores needle-shaped, 9- to 12-celled; boreal to arctic .... *Arthrorhaphis*

12.(5) Thallus yellow-orange or orange-yellow; thallus cortex K+ dark purplish (anthraquinones) 13

12. Thallus yolk-yellow or lemon yellow; thallus cortex K– (anthraquinones absent) 16

13. Squamules ascending, with white lower surface easily seen; on bark or rock .... *Xanthoria*

13. Squamules ascending or closely appressed; lower surface, if visible, yellow; on wood, soil, or rocks 14

14. Squamules thick, yellow to orange, shiny, ascending, with deep yellow lower surface clearly visible; apothecia black, hemispherical and marginless; spores spherical, colorless, 1-celled, 3-4 μm in diameter; Texas to Mexico, on limestone .... *Xanthopsorella texana* (W. A. Weber) Kalb & Hafellner

14. Squamules mostly thin, appressed, lower surface not easily seen; apothecia orange, clearly rimmed 15

15. On soil in the interior of the continent north to the arctic; spore 1- or 2-celled, with uniformly thin walls .... *Fulgensia*

15. On rock, often in maritime or nearby coastal localities; spores 2-celled, polarilocular .... *Caloplaca*

16.(12) Squamules ascending, with the pale lower surface easily seen; mainly on bark; thallus thin ....*Candelaria*

16. Squamules closely appressed; on rock or soil; thallus thick 17

17. Apothecial disks black, convex; on soil and peat; spores 2- or more celled; rare but widespread, mostly arctic-alpine 18

17. Apothecial disks brown or yellow; spores globose or ellipsoid, 1-celled 19

18. Spores brown, 2-celled, (12-)13-17(-18) x 7-10 µm ....*Catolechia wahlenbergii* (Ach.) Körber

18. Spores colorless, needle-shaped, multi-septate, 25-110 x 2-5 µm ....*Arthrorhaphis*

19. Apothecia sessile, lecanorine; apothecial disks yellow; spores over 10 x 4 µm, 8-32 per ascus ....*Candelariella*

19. Apothecia entirely immersed in thallus, disk level with thallus, apothecial disks dark brown, or reddish brown to orange-brown, spores 4-5 x 2-3 µm, many per ascus ....*Acarospora*

20.(4) Squamules ascending or erect, flat or cylindrical, with the lower surface easily seen 21

20. Squamules closely appressed, or lifting only at the edges 28

21. Thallus dwarf fruticose, consisting of erect cylindrical to somewhat flattened lobes 22

21. Thallus composed of flat, clearly dorsiventral squamules 23

22. Photobiont blue-green; spores broad to globose, 1-celled, more than 8 per ascus .... *Peltula*

22. Photobiont green; spores ellipsoid, muriform, 2 per ascus ....***Endocarpon pulvinatum* Th. Fr.**

23. Photobiont blue-green ....*Pannaria* key

23. Photobiont green 24

24. Squamules elongate and branched; apothecia clustered in groups on the squamule tips; thallus cortex K+ yellow (atranorin); very rare .... ***Cetradonia linearis* (Evans) J.-C. Wei & Ahti** (syn. *Gymnoderma lineare*)

24. Squamules not more than 3x longer than broad; apothecia occurring singly, thallus cortex K– (atranorin absent); relatively common species 25

25. Fruiting bodies are perithecia, appearing as black dots on the lobe surface; often growing in or near water or seepage .... *Dermatocarpon*

25. Fruiting bodies, when present, are apothecia; typically, not in wet habitats 26

26. Squamules lobed or finely divided, apothecia raised on a stalk or stipe; thallus cortex C–, KC– (gyrophoric acid absent as main compound) ....*Cladonia* species (e.g., *C. caespiticia* (Pers.) Flörke)

26. Squamules round, or scalloped (with rounded lobes); apothecia sessile; thallus cortex C+ pink or red, KC+ pink or red (gyrophoric acid) 27

27. Thallus pale gray; squamules thick, distinctly lobed, with edges turned down; apothecia flat or convex, pink to lead gray, with a prominent margin
.................................***Trapeliopsis glaucopholis* (Nyl. *ex* Hasse) Printzen & McCune** [as *T. wallrothii* in LNA]
[Note: This species represents most North American records of *T. wallrothii*.]

27. Thallus olive to greenish brown; squamules thin, round, with edges turned up, apothecia becoming hemispherical, dark brown to black; apothecial margin absent
.... ***Psora nipponica* (Zahlbr.) Gotth. Schneider**

28.(20) Photobiont (and exposed algal layer) blue-green 29

28. Photobiont (and algal layer) green 34

29. Soredia present 30

29. Soredia absent 31

30. On rocks mainly in arid habitats in the western interior; soredia farinose;thallus olive to olive brown, attached by a central holdfast (peltate) .... *Peltula*

30. On bark, rarely rocks or soil, in humid habitats, oceanic and boreal to subarctic; soredia usually coarsely granular; thallus olive-brown or blue-gray to yellowish gray, attached broadly or on one edge, not peltate
....*Pannaria* key

31. Squamules extremely narrow-lobed (0.1- 0.3[-0.4] mm wide), sometimes building into an areolate crust and (or) with abundant isidia; apothecia black, lecideine ....*Placynthium*

31. Squamules mostly broader than 0.3 mm, isidiate or not; apothecia red-brown, lecanorine or biatorine 32

32. Thallus brownish or gray-brown; apothecia sessile or constricted at the base; photobiont *Nostoc* [but packets of cells sometimes must be broken apart to see the short filamentous segments of 3-6 cells]
.... *Pannaria* key (*Pannaria*, *Fuscopannaria*, *Parmeliella*)

32. Thallus olive or olive-brown; apothecia partly or entirely immersed in the thallus; photobiont *Scytonema* or *Chroococcidiopsis*-type 33

33. Spores ellipsoid, 8 per ascus; photobiont filamentous (*Scytonema*-type) .... ***Heppia conchiloba*** **Zahlbr.**

33. Spores globose or broadly ellipsoid, up to 100 per ascus; photobiont in packets of small spherical cells (*Chroococcidiopsis*-type) .... *Peltula*

34.(28) Soredia present 35

34. Soredia absent 38

35. On soil or rock, rarely wood; cortex and medulla K+ persistently yellow, PD+ orange (stictic acid) .... ***Baeomyces rufus*** **(Hudson) Rebent.**

35. On bark, wood, mosses, or lichens; thallus cortex and medulla K–, PD– (stictic acid absent) 36

36. Soralia raised and finally cup-like; thallus olive-brown, with round, tiny squamules (less than 0.5 mm in diameter); apothecia blue-gray, biatorine; spores 4-celled, fusiform; California to Washington .... *Waynea californica* Moberg

36. Soralia on lobe surface or on margin of lower surface, not raised or cup-like 37

37. Thallus pale green; squamules thin with a raised, thickened rim, separate, not overlapping; on mosses and lichens (especially cyanobacterial lichens), also bark and wood in humid habitats, thallus cortex and medulla C–, KC– .... ***Normandina pulchella*** **(Borrer) Nyl.**

37. Thallus olive to brownish or gray-brown; squamules thick, commonly overlapping like shingles; on wood or bark .... *Hypocenomyce* key

38.(34) Ascomata absent; squamules green with raised or thickened rims 39

38. Ascomata present, either perithecia or apothecia 40

39. Squamules less than 2 mm in diameter; on bark, mosses, and lichens (see couplet 37) .... ***Normandina pulchella*** **(Borrer) Nyl.**

39. Squamules 2-5 mm broad, often lobed, closely appressed to soil or peat; arctic-alpine to boreal. Fruiting bodies, when present, pale yellow mushrooms ...... *Lichenomphalia hudsoniana* (H. S. Jenn.) Redhead et al.

40. On bark or wood 41

40. On soil, mosses, dead vegetation, or rocks 45

41. Ascomata perithecia buried in the thallus, only the ostioles showing as dark spots at the surface 42
41. Ascomata apothecia on the thallus surface 43

42. Spores muriform, brown, 2 per ascus; squamules thin, closely appressed, relatively flat edges; perithecial cavity containing algae .... *Endocarpon pallidulum* (Nyl.) Nyl.
42. Spores 1-celled, colorless, 8 per ascus; squamules thick, free at the margins, with the edges turned down; perithecial cavity containing no algae .... ***Placidium arboreum*** **(Schwein.** ***ex*** **E. Michener) Lendemer** (syn. *P. tuckermanii*)

43. Apothecia black .... *Hypocenomyce* key
43. Apothecia reddish-brown 44

44. Apothecia biatorine, 0.3-0.5 mm in diameter, flat to convex, margin becoming thin and disappearing in maturity; spores narrowly ellipsoid, 2-4 µm wide; southeastern U.S. .... *Phyllopsora*
44. Apothecia lecanorine, 0.5-3 mm in diameter, flat to concave, margin prominent; spores ellipsoid, 6.5-12.5 µm wide; boreal to arctic and montane .... ***Psoroma*** (see key to *Pannaria*)

45.(40) Ascomata perithecia buried in the thallus with only the dot-like ostiole showing 46
45. Ascomata apothecia, superficial or sunken into the thallus; or ascomata absent 47

46. Spores muriform, brown, 2 per ascus; perithecial cavity containing algae .... *Endocarpon*
46. Spores 1-celled, colorless, 8 per ascus; perithecial cavity containing no algae .... *Placidium*

47. Apothecia raised on a stalk or stipe .... *Baeomyces*
47. Apothecia not raised on a stalk or stipe 48

48. Apothecia lecanorine or cryptolecanorine (with algae in the margin, or immersed in the thallus) 49
48. Apothecia lecideine or biatorine (lacking algae in the margin) 56

49. Spores up to 100 per ascus .... *Acarospora* key
49. Spores 8 or fewer per ascus 50

50. Spores 2- or more celled 51

50. Spores 1-celled 53

51. Spores dark brown; growing on calcareous soil, mosses, or dead vegetation .... *Phaeorrhiza*

51. Spores colorless; growing directly on non-calcareous rock 52

52. Spore walls evenly thickened; epihymenium K– (anthraquinones absent); in coastal California .... *Lecania* key

52. Spores polarilocular; epihymenium K+ deep purple-red (anthraquinones); in the arid southwest .... ***Caloplaca pellodella* (Nyl.) Hasse**

53. Apothecia entirely immersed in thallus, disks flush with thallus or sunken; apothecial disks black or almost black, heavily pruinose; spores globose or broadly ellipsoid, 15-25 µm wide; on limestone .... ***Circinaria contorta* (Hoffm.) A. Nordin et al.** (syn. *Aspicilia contorta*)

53. Apothecia sessile or somewhat raised; apothecial disks brown to yellowish orange, not or lightly pruinose; spores ellipsoid, 4-10 µm wide; on rock or on soil 54

54. Directly on rocks; thallus shiny .... ***Lecanora muralis* (Schreber) Rabenh.**

54. On soil, mosses, or dead vegetation; thallus dull 55

55. Thallus thick, yellowish white, pruinose; apothecial margin flush with disk, or becoming thin and disappearing in maturity; on soil in relatively dry habitats .... ***Squamarina lentigera* (Weber) Poelt**

55. Thallus thin, brown to gray-green; apothecial margin prominent; on mossy soil and peat in relatively humid habitats .... *Psoroma* (see key to *Pannaria*)

56.(48) At least some spores 2- or more celled 57

56. All spores 1-celled, spore length to width ratio mostly 2.0 or less, or mostly 2.0-3.0 58

57. Spores colorless, 1- to 8-celled, length to width ratio mostly over 3.0 .... *Toninia* key

57. Spores brown, 2-celled, length to width ratio under 3.0 .... *Phaeorrhiza sareptana* (Tomin) H. Mayerh. & Poelt

58. Squamules closely appressed 59

58. Squamules lifting at least at the edges 61

59. On siliceous rock; thallus shiny, with a black prothallus; apothecia lecideine, black; apothecial margin prominent or flush with disk .... *Lecidea*

59. On soil or decaying vegetation; apothecia biatorine or lacking an exciple, reddish brown or orange-brown to black; apothecial margin absent or persistent; spores 12-17 x 7-9 µm 60

60. Apothecia reddish brown or orange brown, undelimited and lacking an exciple, simply transforming the upper cortex of the squamules into a hymenium and appearing like a swelling on the squamules; on calcareous soil in dry sites .... ***Gypsoplaca macrophylla*** **(Zahlbr.) Timdal**

60. Apothecia very dark red-brown to black, round or fusing with neighbors, with a distinct exciple in section although the margins usually excluded in maturity; on soil and vegetation in alpine, subalpine and arctic habitats .... ***Lecidoma demissum*** **(Rutstr.) Gotth. Schneider & Hertel**

61.(58) Apothecial disks lead gray, pale brown, or pink; thallus C+ pink (gyrophoric acid) .... ***Trapeliopsis glaucopholis*** **(Nyl. *ex* Hasse) Printzen & McCune** [as *T. wallrothii* in LNA]

61. Apothecial disks black or almost black, dark brown, or reddish brown to orange-brown; thallus C+ or C– .... *Psora* key (*Psora*, *Psorula*)

KEY H: UMBILICATE AND FAN-SHAPED LICHENS

1. Thallus attached by a holdfast at one edge of the lobe and becoming fan-like 2

1. Thallus attached by a holdfast close to the center of the thallus 5

2. Thallus remaining opaque when wet; with a distinct grass-green algal layer (photobiont a green alga), lower surface white, yellowish, or dark brown; on bark or soil 3

2. Thallus jelly-like and translucent when wet; lacking a distinct algal layer (photobiont cyanobacteria); lower surface black or gray; on rock 4

3. Thallus yellow-green, rather shiny; lobes with a network of depressions and sharp ridges, wrinkled or bumpy (rugose), lower surface about the same color as the upper surface, with a smooth, more or less uniform cortex; on bark .... ***Ramalina sinensis*** **Jatta**

3. Thallus dark gray-green when dry and grass-green when wet, lobes smooth and even, upper surface dull; lower surface distinctly different from the upper surface, entirely without a cortex, webby or cottony, basically white, with conspicuous dark brown veins; on soil .... ***Peltigera venosa*** **(L.) Hoffm.**

4. On dry calcareous rocks in the open; lower surface more or less uniform; thallus black .... ***Lichinella nigritella*** **(Lettau) P. P. Moreno & Egea**

4. Aquatic, on submerged rocks in mountain streams; lower surface with conspicuous veins; thallus mineral gray, brown, or olive, at least when dry .... *Peltigera*

5.(1) Thallus yellow-green; cortex KC+ orange-yellow (usnic acid) 6

5. Thallus gray, brown, or olive, without a yellow tint; cortex KC–, or KC+ red (lacking usnic acid) 7

6. Rhizines abundant, stubby and peg-like; thalli large (up to 15 cm across); apothecia with pale to dark red-brown disks .... ***Omphalora arizonica*** **(Tuck. *ex* Willey) T. H. Nash & Hafellner**

6. Rhizines absent; thalli up to 6 cm across; apothecia various colors, but not clear red-brown .... *Rhizoplaca*

7. Black dots abundant all over upper surface caused by immersed perithecia .... *Dermatocarpon*

7. Black dots absent or rare, and then caused by scatted pycnidia; fruiting bodies, when present, apothecia 8

8. Apothecia brown, sometimes pruinose, broken up into segments by sterile tissue, cryptolecanorine, sunken into thallus; thallus thick, surface areolate or cracked, chalky white to pale gray; spores spherical, 3-4 µm in diameter, many per ascus; rare; on limestone in the arid interior or arctic .... *Glypholecia scabra* (Pers.) Müll. Arg.

8. Apothecia black, usually with concentric or radiating bands of sterile tissue, lecideine, superficial or sunken; thallus thin or thick, very dark brown to gray; spores never spherical, 8 or fewer per ascus; common and widespread, on siliceous rock .... *Umbilicaria* key (*Lasallia, Umbilicaria*)

KEY I: JELLY LICHENS

1. Thallus crustose 2

1. Thallus foliose or almost fruticose 3

2. Thallus granular, reddish when wet; photobiont *Gloeocapsa* .... *Pyrenopsis* key

2. Thallus membranous, olive to black when wet; photobiont *Nostoc* .... *Lempholemma*

3. Aquatic, on submerged rocks; lower surface with conspicuous veins .... *Peltigera*

3. Not aquatic (i.e., not growing on submerged rocks); lower surface without veins 4

4. Lower surface tomentose (hairy or furry) .... *Leptogium* key

4. Lower surface naked, smooth or wrinkled 5

5. Thallus pitch black when dry, attached by a single point with the lobes fanning out; on rocks of different kinds; photobiont *Gloeocapsa* .... *Lichinella* key

5. Thallus shades of gray, brown, or olive, often dark but not pure black, broadly attached, on various substrates; photobiont *Nostoc* 6

6. Upper and lower surfaces with a cortex, usually a single layer of square to roundish cells (fig. 7b); thallus often shiny, gray to reddish brown .... *Leptogium*

6. Upper and lower surfaces lacking a cortex (fig. 7a); thallus always dull; olive to olive- brown or almost black 7

7. Spores transversely septate or muriform; very common lichens .... *Collema* key

7. Spores 1-celled; uncommon, inconspicuous lichens .... *Lempholemma*

Key J: Orange to Yellow or Yellowish Green Foliose Lichens

1. Thallus orange to yellow-orange; cortex K+ dark purple (anthraquinones) .... *Xanthoria* key

1. Thallus deep yolk-yellow, bright yellow, greenish yellow, or yellowish green ("usnic-yellow"), yellowish olive, or yellowish gray; cortex K–, K+ pinkish, or K+ yellow (anthraquinones absent) 2

2. Thallus deep yolk-yellow or bright yellow, K– or K+ pinkish 3

2. Thallus greenish yellow, yellowish green ("usnic-yellow"), yellowish olive, or yellowish gray, K– or K+ yellow 5

3. Thallus very closely attached, almost crustose (but with a lower cortex), lacking rhizines; on rock .... *Candelina*

3. Thallus clearly foliose, with rhizines; on bark or wood, less frequently on rock 4

4. Thallus lobes finely divided, 0.1-0.7 mm wide; medulla white; calycin present, pinastric and vulpinic acids absent .... *Candelaria*

4. Thallus lobes (0.5-)1-4(-7) mm wide; medulla bright yellow; calycin absent, pinastric and vulpinic acids present .... *Vulpicida*

5.(2) Rhizines absent; lower surface with or without tomentum (see fig. 6) 6

5. Rhizines abundant or sparse; lower surface without tomentum 13

6. Thallus attached by a single central point (umbilicate), or not attached to the substrate (vagrant); lower surface without tomentum .... *Rhizoplaca*

6. Thallus not umbilicate or vagrant; lower surface with or without tomentum 7

7. On bark 8

7. On rock, soil, or moss 10

8. Photobiont green; upper surface rather shiny; soredia absent but lobules present; medulla PD+ orange .... ***Lobaria oregana* (Tuck.) Müll. Arg.**

8. Photobiont blue-green; upper surface dull or scabrose; soredia present, lobules absent; medulla PD– or PD+ yellow 9

9. Lobes with a network of depressions and sharp ridges; thallus greenish yellow; lower surface naked, smooth or wrinkled, dark in the center, pale to dark brown close to the margin; medulla PD–, K– .... ***Nephroma occultum* Wetmore**

9. Lobes smooth and even, or with rounded depressions; thallus yellowish olive or pale green; lower surface pale or dark tomentose interrupted by small or large bald spots; medulla PD+ yellow, K+ yellow to orange (stictic acid with some norstictic acid) .... ***Lobaria scrobiculata* (Scop.) D.C.**

10.(7) Thallus of erect, elongated lobes, almost fruticose; lower surface yellow; pycnidia abundant and conspicuous as black dots mostly along the lobe margins .... *Flavocetraria*

10. Thallus prostrate, lobes rounded not erect, closely appressed or loosely attached over entire surface; lower surface pale to dark brown or black; pycnidia absent or sparse and very inconspicuous 11

11. On mossy rock in humid, temperate regions; algal layer dark blue green (photobiont cyanobacteria); soredia present; lower surface with tomentum interrupted by small or large bald spots .... ***Lobaria scrobiculata* (Scop.) D.C.**

11. On the ground or over rocks in alpine or arctic regions; algal layer grass-green (photobiont green algae); soredia absent 12

12. Lobes smooth and even, or with shallow depressions or wrinkles; cephalodia producing broad gray bumps on the thallus surface; pseudocyphellae absent; lower surface with a tomentum; apothecia abundant, produced on the lower surface of lobe margins .... ***Nephroma arcticum* (L.) Torss.**

12. Lobes with a network of depressions and sharp ridges; cephalodia absent; pseudocyphellae usually abundant and conspicuous; lower surface naked, lacking tomentum; apothecia rare, on the upper surface of the lobes .... ***Asahinea chrysantha* (Tuck.) W. L. Culb. & C. F. Culb.**

13.(5) Cilia bulbous, common along lobe margins; small lichens (lobes less than 2 mm wide) of the subtropics and tropics; rare .... *Relicina* (*Bulbothrix* key)

13. Cilia, if present, not bulbous 14

14. On bark or wood 15

14. On rock or soil 30

15. Soredia present 16

15. Soredia absent 24

16. Rhizines forked in regular dichotomies; soredia on upper surface of lobe tips; medulla PD+ distinct yellow, K+ red (salazinic acid) .... ***Hypotyrachyna sinuosa* (Sm.) Hale**

16. Rhizines unbranched; soredia on the lobe margins or on the upper surface; medulla PD– or PD+ red-orange, K– (salazinic acid absent) 17

17. Tiny white dots (pseudocyphellae) present on the upper surface of the lobes; lower surface dark brown to black; medulla C+ red or pink (lecanoric acid) 18

17. Pseudocyphellae absent; lower surface white, brown, or black; medulla C– or C+ red.......... 19

18. Soralia almost entirely marginal; pseudocyphellae sparse and inconspicuous .... ***Flavopunctelia soredica* (Nyl.) Hale**

18. Soralia both on the thallus surface and along the margins; pseudocyphellae abundant and conspicuous .... ***Flavopunctelia flaventior* (Stirton) Hale**

19. Lower surface dark brown to black with a brown naked zone near margins; lobes 1-8 mm broad; medulla PD– or PD+ red-orange, KC+ pink or red 20

19. Lower surface white to pale brown; lobes 0.5-4 mm broad; medulla PD–, KC– 23

20. Lobes 0.5-3 mm broad 21

20. Lobes 3-8 mm broad 22

21. Lobes 0.5-2 mm wide; rhizines abundant to the tips; medulla PD–, UV+ white (divaricatic acid) .... *Parmeliopsis*

21. Lobes 1-3(-5) mm wide; rhizines sparse, and absent near the tips; medulla PD+ red-orange, KC+ pink (protocetraric acid); southern California, on shrubs and cactus .... *Flavoparmelia subcapitata* (Nyl. *ex* Hasse) Hale *ex* DePriest & B. Hale

22. Soredia coarsely granular, entirely laminal; medulla PD+ red orange, KC+ pink, C– (protocetraric acid) .... ***Flavoparmelia caperata* (L.) Hale**

22. Soredia both laminal and marginal; medulla PD–, KC+ red, C+ red (lecanoric acid) .... ***Flavopunctelia soredica* (Nyl.) Hale**

23.(19) Soredia in round soralia on thallus surface; upper surface dull; rhizines abundant, brown; medulla UV+ blue-white (divaricatic acid) .... *Parmeliopsis*

23. Soredia in elongate soralia along the lobe margins; upper surface rather shiny; rhizines sparse, white or very pale tan; medulla UV– (divaricatic acid absent) .... ***Usnocetraria oakesiana* (Tuck.) M.J. Lai & C.J. Wei** (syn. *Allocetraria oakesiana*)

24.(15) Lower surface dark brown or black 25

24. Lower surface white to pale brown or yellow 27

25. Pseudocyphellae abundant and conspicuous; medulla KC+ red, C+ red (lecanoric acid) .... ***Flavopunctelia praesignis* (Nyl.) Hale**

25. Pseudocyphellae absent; medulla KC–, C– 26

26. Thallus closely attached with abundant rhizines almost to the margin; cilia and isidia absent; medulla PD+ red; south Texas, New Mexico, and Missouri .... *Flavoparmelia rutidota* (Hooker f. & Taylor)

26. Thallus loosely attached; rhizines absent from a broad zone at the margin; isidia and cilia present

....*Parmotrema*

27.(24) Medulla bright yellow (pinastric and vulpinic acids); apothecia along the lobe margins .... ***Vulpicida viridis* (Schwein.) J.-E. Mattson & M. J. Lai**

27. Medulla white or pale yellowish orange; apothecia laminal, not marginal 28

28. On southeastern coastal plain; medulla often becoming yellowish to orange in places, K+ yellow, PD+ orange (stictic acid) .... ***Pseudoparmelia uleana* (Müll. Arg.) Elix & T. H. Nash**

28. Western or Appalachian-Great Lakes region; medulla white, PD–, K– 29

29. Pseudocyphellae absent; medulla C–; northwestern and Appalachian-Great Lakes region ....*Ahtiana*

29. Pseudocyphellae usually conspicuous, white; medulla C+ red (lecanoric acid); common in southeastern Arizona ....*Flavopunctelia darrowi* (J.W. Thomson) Hale

30.(14) Soredia present 31

30. Soredia absent 33

31. Soredia on thallus surface; lower surface and rhizines black 32

31. Soredia along the lobe margins; lobes elongated; lower surface and rhizines white to pale brown; medulla PD–, KC– ....***Usnocetraria oakesiana* (Tuck.) M.J. Lai & C.J. Wei** (syn. *Allocetraria oakesiana*)

32. Widespread temperate; lobes rounded; medulla PD+ red, KC+ pink (protocetraric acid) .... ***Flavoparmelia caperata* (L.) Hale**

32. Arctic-alpine; lobes narrow, under 1 mm wide, convex, medulla PD–, KC+ red (alectoronic acid) ....*Arctoparmelia incurva* (Pers.) Hale

33.(30) Hollow pustules resembling isidia present in clumps on the upper surface, sometimes breaking into granular fragments; medulla PD+ red, KC+ pink, K– (protocetraric acid) .... ***Flavoparmelia baltimorensis* (Gyelnik & Fóriss) Hale**

33. Hollow, isidia-like pustules absent; solid isidia present or absent; medulla with various reactions.............34

34. Lobes 8-12 mm wide; medulla PD–, K–; marginal cilia common and abundant; rhizines sparse, absent from a broad zone close to the margin ....*Parmotrema*

34. Lobes less than 4 mm wide; marginal cilia absent; rhizines usually abundant to the thallus edge 35

35. Upper surface dull; lower surface white to dull black; medulla K–, PD–, KC+ red (alectoronic acid); arctic-alpine to boreal .... *Arctoparmelia*

35. Upper surface often shiny; lower surface brown to pitch black, shiny; medulla usually PD+ yellow to red, K+ yellow to red; widespread from arctic to southern temperate .... *Xanthoparmelia*

Key K: Foliose Lichens That Are Not Umbilicate, Jelly-Like, or Yellowish

1. Lichens on the ground (soil or mossy turf) 2
1. Lichens directly on bark, wood, or rock 18

2. Thalli ascending, forming almost erect fruticose tufts; pseudocyphellae conspicuous on the lower surface of the lobes or branches .... *Cetraria* key
2. Thalli prostrate, not erect 3

3. Algal layer blue-green (photobiont cyanobacteria) 4
3. Algal layer green (photobiont green algae) 8

4. Lobes less than 2 mm broad; on mossy or bare soil .... *Massalongia* (*Pannaria* key)
4. Lobes usually more than 2 mm broad 5

5. Lower surface covered by a rather thick, blue-black tomentum; subtropics to eastern coastal plain .... *Coccocarpia*
5. Lower surface with a pale brown to brown-black tomentum, or veined 6

6. Cyphellae present on lower surface; thallus dark brown; rare, arctic-alpine .... *Sticta arctica* Degel.
6. Lacking cyphellae on lower surface 7

7. Lobes mostly under 4 mm broad; apothecia embedded in thallus lobes; lichens mainly of arid localities; spores 1-celled; photobiont *Scytonema* .... ***Heppia conchiloba* Zahlbr.**
7. Lobes mostly greater than 4 mm broad; apothecia marginal; spores septate; photobiont *Nostoc* ..... *Peltigera*

8.(3) Lobes mostly less than 3 mm wide 9
8. Lobes mostly more than 3 mm wide 13

9. Rhizines absent; lobes generally appearing puffed 10

9. Rhizines present, sparse or abundant; lobes solid and flat, not appearing puffed 11

10. Lobes tube-like, hollow .... *Hypogymnia*

10. Lobes solid .... ***Brodoa oroarctica* (Krog) Goward**

11. Thallus lightly to heavily pruinose; lower surface black with a dense mat of squarrose rhizines; lobes 1-3 mm wide .... ***Physconia muscigena* (Ach.) Poelt**

11. Thallus not pruinose, lower surface mostly pale (but dark in the center); lobes under 1.0 mm wide 12

12. Lower surface with sparse, unbranched rhizines; lobes 0.2-0.5(-1) mm wide .... *Phaeophyscia constipata*

12. Lower surface with many rhizines that are branched when mature; lobes 0.4-0.8(-1.0) mm wide .... *Anaptychia bryorum* Poelt

13.(8) Lower surface jet black, shiny; thallus lacking cephalodia; upper surface K+ yellow (atranorin); western arctic-boreal 14

13. Lower surface pale tan to black (or orange in *Solorina crocea*), not shiny; cephalodia present as small warts on the lower or upper surface or as internal patches; upper cortex K– (atranorin absent); arctic to temperate 15

14. Rhizines sparse but present, especially on older parts of thallus; isidia absent; rare, Bering Sea coast of Alaska .... *Cetrelia alaskana* (C. F. Culb. & W. L. Culb.) W. L. Culb. & C. F. Culb.

14. Rhizines entirely absent; isidia present on thallus surface; Alaska to Hudson Bay .... *Asahinea scholanderi* (Llano) W. L. Culb. & C. F. Culb.

15. Upper surface of lobes with a network of ridges and depressions; lower surface uniformly pale brown, sparsely to clearly tomentose .... ***Lobaria linita* (Ach.) Rabenh.**

15. Upper surface of lobes smooth or wrinkled but not forming a network of ridges 16

16. Cephalodia appearing as gray to brown scales on the upper surface; apothecia marginal on small lobes .... *Peltigera*

16. Cephalodia hidden within the thallus, on the lower surface, or forming low bumps on the upper surface 17

17. Thallus lobes strongly crinkled and crisped, ascending; apothecia rare, on the lower side of small marginal lobes .... ***Nephroma expallidum*** **(Nyl.) Nyl.**

17. Thallus more or less smooth, prostrate; apothecia common, immersed in the thallus lobes, often in depressions .... *Solorina*

18.(1) Lobes mostly under 3 mm wide 19

18. Lobes mostly over 3 mm wide 89

19. Lobes inflated and hollow; lower surface lacking rhizines or tomentum .... *Hypogymnia* key (*Hypogymnia, Menegazzia*)

19. Lobes solid; rhizines or tomentum present or absent 20

20. Thallus shades of brown, olive, or black 21

20. Thallus shades of gray 47

21. Algal layer blue green .... *Pannaria* key

21. Algal layer green 22

22. Rhizines and tomentum absent, or thallus so tightly attached that it is hard to tell 23

22. Rhizines or tomentum present, sparse or abundant 28

23. Thallus ascending, very loosely attached at only a few points 24

23. Thallus appressed, closely attached at numerous points 27

24. Lobes strap-shaped, squamule-like, dark olive-green; apothecia spherical; southern Appalachian Mountains, rare and endangered .... ***Cetradonia linearis*** **(Evans) J.-C. Wei & Ahti** (syn. *Gymnoderma lineare*)

24. Lobes almost fruticose, not flat and strap-shaped; apothecia flat; not from southeast 25

25. Branches short, mostly up to 15 mm in length; thallus dark greenish brown to black .... ***Kaernefeltia merrillii*** **(Du Rietz) Thell & Goward**

25. Branches quite long, over 20 mm in length; thallus brown to brownish gray, without a greenish tint 26

26. Lobe margins with long, branched, thallus-colored cilia; lower surface white, webby .... ***Anaptychia crinalis*** **(Schaerer)** (syn. *A. setifera*)

26. Lobe margins often with short, unbranched, black projections but no true cilia; lower surface brown and shiny like the upper surface .... ***Tuckermanopsis subalpina* (Imshaug) Kärnefelt**

27.(23) Lobes convex, appearing inflated but actually solid; on rock .... *Allantoparmelia* key

27. Lobes flat; on bark .... *Hyperphyscia*

28.(22) Cortex K+ yellow (atranorin); thallus mainly brown on the lobe tips, gray in older parts of the thallus 29

28. Cortex K–; thallus uniformly brown, olive, or blackish 30

29. Pseudocyphellae conspicuous, white, round to elliptical; medulla K–, C+ pink (gyrophoric acid) .... ***Punctelia stictica* (Duby) Krog**

29. Pseudocyphellae in a net-like pattern on ridges or along lobe margins; medulla K+ red or brownish, C– (salazinic acid or fumarprotocetraric acid) .... *Parmelia*

30. Lobes distinctly pruinose at least at tips 31

30. Lobes without pruina 32

31. Rhizines dark, usually squarrose; lacking cortical hairs .... *Physconia*

31. Rhizines pale, unbranched at least when young; sometimes with short, colorless cortical hairs ...*Anaptychia*

32. Soredia present 33

32. Soredia absent 38

33. Soredia mainly in patches (soralia) on the thallus surface 34

33. Soredia mainly at the lobe tips or along the margins 36

34. Medulla C+ pink or red .... *Melanelia* key

34. Medulla C– 35

35. Rhizines not visible from above; thallus very dark brown, often shiny; spores colorless, 1-celled .... *Melanelia* key

35. Rhizines often protruding in a fringe around lobes; thallus dark to pale brown or dark greenish gray, not shiny; spores brown, 2-celled .... *Phaeophyscia*

36. Medulla K+ red or C+ reddish .... *Melanelia* key

36. Medulla K–, C– 37

37. Lower surface and rhizines pale brown; rhizines sparse, not extending out from margins and visible from above; lobes chocolate brown to olive-brown, flat, crisped or undulating .... ***Tuckermanopsis chlorophylla*** **(Willd.) Hale**

37. Lower surface mostly black; rhizines mostly black (often with white tips), abundant and extending out from the margins; lobes flat, not undulating or crisped .... *Phaeophyscia*

38.(32) Isidia present 39

38. Isidia absent 41

39. Isidia mainly on the upper surface of the lobes, even on young lobes .... *Melanelia* key

39. Isidia mainly marginal, especially on young lobes 40

40. On rock .... ***Phaeophyscia sciastra*** **(Ach.) Moberg**

40. On wood or bark .... ***Tuckermanella coralligera*** **(W. A. Weber) Essl.** (syn. *Tuckermanopsis coralligera*)

41. Pycnidia prominent, black, along the lobe margins; pseudocyphellae often conspicuous especially along the lobe margins 42

41. Pycnidia absent or immersed in thallus with only a pale or dark ostiole showing, not largely confined to the margins; pseudocyphellae present or absent, not especially marginal 43

42. On bark or wood .... *Tuckermanopsis*

42. On rock .... ***Melanelia hepatizon*** **(Ach.) Thell**

43. Lobes with conspicuous ridges and depressions .... ***Kaernefeltia merrillii*** **(Du Rietz) Thell & Goward**

43. Lobes more or less flat, without ridges and depressions 44

44. Lower surface and rhizines pale brown to white 45

44. Lower surface and rhizines mostly dark brown or black 46

45. Lower surface pale brown, smooth and shiny; on rock in arid habitats; lobules, if present, not strap-shaped; medulla K+ yellow, PD+ orange (stictic acid); spores colorless, 1-celled; southwestern U.S.

....*Xanthoparmelia atticoides* **(Essl.) O. Blanco et al.** (syn. *Neofuscelia atticoides*)

45. Lower surface almost white, fibrous or dull; on rock or bark in shady or arid habitats; lobules, if present, strap-shaped, marginal; medulla, K–, PD–; spores brown, 2-celled ....*Anaptychia*

46. Apothecia very dark brown to black, dull; spores brown, 2-celled; rhizines usually very abundant, often visible from above as a cilia-like fringe and frequently growing on the apothecial margins .... *Phaeophyscia*

46. Apothecia yellowish brown to reddish brown, shiny; spores colorless, 1-celled; rhizines abundant or sparse, not forming a fringe visible from above nor growing on the apothecial margins ....*Melanelia* key

47.(20) Photobiont blue-green ....*Pannaria* key

47. Photobiont green 48

48. Lower surface white, tan, yellow, or orange 49

48. Lower surface dark brown to black at least in center (sometimes paler near lobe tips, or with scattered or extensive white blotches) 61

49. Lobes thick and convex, appearing inflated (although solid); rhizines absent; PD+ yellow; on rock ....*Allantoparmelia* key (*Allantoparmelia*, *Lobothallia*)

49. Lobes thin, convex or flat, not appearing inflated; rhizines present or absent; on various substrates 50

50. Rhizines and tomentum absent 51

50. Rhizines or tomentum present, sometimes sparse 57

51. Thallus very closely attached to substrate over the entire thallus surface (almost crustose in appearance) ....*Hyperphyscia*

51. Thallus loosely attached and ascending 52

52. Marginal cilia absent 53

52. Marginal cilia present and conspicuous 56

53. On rock 54

53. On trees 55

54. Lobes convex, 0.1-0.3(-0.5) mm wide; on dry limestone or sandstone, southcentral U.S.

.... ***Speerschneidera euploca*** **(Tuck.) Trevisan**

54. Lobes flat, 0.8-1.3 mm wide, strap-shaped; on wet rock walls in forests of the Southern Appalachian Mountains .... ***Cetradonia linearis*** **(Evans) J.-C. Wei & Ahti** (syn. *Gymnoderma lineare*)

55. Lobes smooth to wrinkled, often strongly convex; eastern or southern .... *Pseudevernia*

55. Lobes generally wrinkled and ridged, flat or concave; humid forests on the west coast .... *Platismatia*

56.(52) Soredia produced on the lower surface of the lobe tips; cortex K+ yellow (atranorin) .... ***Heterodermia leucomela*** **(L.) Poelt.**

[Note: Linda in Arcadia (2012) recommends conserving the spelling "*leucomelos*."]

56. Soredia absent; cortex K– .... ***Anaptychia crinalis*** **(Schaerer) Vězda** (syn.*A. setifera*)

57.(50) Lower surface with a short, pale tomentum, sometimes very sparse and limited to the low areas between raised bald spots; cephalodia visible on the lower surface as small bumps that are dark blue-green inside .... *Lobaria*

57. Lower surface with distinct rhizines; cephalodia absent 58

58. Soredia present .... *Physcia* key (*Heterodermia, Physcia, Physciella*)

58. Soredia absent 59

59. Black cilia with bulbous bases fringing the lobe margins .... *Bulbothrix*

59. Cilia, if present, not black and bulbous 60

60. Thallus cortex PD+ orange, K+ deep yellow (thamnolic acid); apothecia pale brown, without pruina; spores colorless, 1-celled .... *Imshaugia*

60. Thallus cortex PD– or pale yellow, K+ yellow (atranorin); apothecia, if present, dark brown to black, often pruinose; spores brown, 2-celled .... *Physcia* key (*Heterodermia, Physcia*)

61.(48) Distinct rhizines absent or very sparse 62

61. Distinct rhizines present and abundant at least in central parts of thallus 67

62. Lower surface with a spongy black (rarely dark brown) hypothallus consisting of intricately interconnected fibers (fig. 6g,h) .... *Anzia*

62. Lower surface naked, without a spongy hypothallus 63

63. Lower surface pitted with many tiny perforations; thallus small, usually forming rosettes less than 2.5 cm across .... "*Cavernularia*" (see *Hypogymnia* key)

63. Lower surface not perforated or pitted; thalli larger than 3 cm long or broad 64

64. On alpine or arctic rocks; lobes thick, convex .... *Allantoparmelia* key (*Allantoparmelia*, *Brodoa*)

64. On bark or wood 65

65. Closely appressed over most of the thallus surface .... *Dirinaria*

65. Very loosely attached by relatively few points 66

66. Lobes flat to concave, often wrinkled; pycnidia along the lobe margins; medulla C–; in humid or boreal forests along the west coast .... *Platismatia*

66. Lobes convex; pycnidia buried in lobe tips, not along the margins; medulla C+ red (lecanoric acid); montane and interior sites .... *Pseudevernia*

67.(61) Pseudocyphellae present 68

67. Pseudocyphellae absent 69

68. Pseudocyphellae dot-like, most easily seen on young lobes; medulla K– .... *Punctelia*

68. Pseudocyphellae net-like on reticulate ridges or along lobe margins; most species with medulla K+ red (salazinic acid) .... *Parmelia*

69. Cilia present on lobe margins or in the axils of the lobes (sometimes sparse!) 70

69. Cilia absent 74

70. True cilia arising from the margins or axils of the lobes .... *Parmelia* key (*Bulbothrix*, *Myelochroa*, *Parmelina*, "*Parmelinopsis*," *Relicina*)

70. "Cilia" actually a fringe of rhizines extending beyond the lobe margins, not arising from the lobe margins themselves 71

71. Rhizines long and forked, uniformly black .... *Hypotrachyna*

71. Rhizines unbranched or squarrose, or frayed to brush-like at the tips, often with white tips 72

72. Cortex K– .... *Phaeophyscia*

72. Cortex K+ yellow 73

73. Lobe margins abundantly squamulose .... ***Heterodermia squamulosa* (Degel.) W. L. Culb.**

73. Lobe tips sorediate, not squamulose .... *Heterodermia casarettiana* (A. Massal.) Trevisan

74.(69) Rhizines forked in regular dichotomies; thallus without pruina .... *Hypotrachyna*

74. Rhizines unbranched, squarrose, or brush-like, rarely forked; thallus with or without pruina 75

75. Thallus very loosely attached, strongly wrinkled, especially on the lower surface; on conifers in the west .... ***Esslingeriana idahoensis* (Essl.) Hale & M. J. Lai**

75. Thallus closely attached over most of its surface; smooth, or only slightly wrinkled, on both surfaces 76

76. Cortex K– (without atranorin); thallus usually dark greenish gray 77

76. Cortex K+ yellow (atranorin); thallus usually pale gray 79

77. Thallus without pruina .... *Phaeophyscia*

77. Thallus pruinose at least at the lobe tips 78

78. Rhizines squarrose (like bottle brushes); mostly northern .... *Physconia*

78. Rhizines unbranched or rarely forked; northern to subtropical .... *Pyxine* (*Physcia* key)

79.(76) Thallus pruinose at least at lobe tips .... *Physcia* key (*Physcia*, *Pyxine*)

79. Thallus without pruina 80

80. Medulla pale yellow, at least close to the algal layer .... *Myelochroa* (*Parmelia* key)
[Note: Rare non-pruinose specimens of *Pyxine eschweileri* or adnate *Esslingeriana* will key out here.]

80. Medulla white 81

81. Soredia absent; squamulose lobules abundant on lobe margins .... ***Heterodermia squamulosa* (Degel.) W. L. Culb.**

81. Soredia present 82

82. Soredia on or close to the lobe margins; medulla C– 83

82. Soredia laminal or on the lobe tips 84

83. Medulla PD–, K+ yellow; rhizines sparse and distinct .... ***Physcia sorediosa* (Vainio) Lynge**

83. Medulla PD+ orange, K–; rhizines abundant, forming an intricate mat .... ***Pyxine eschweileri* (Tuck.) Vainio** (non-pruinose morph)

84. Lobes tips often curled into tubes; medulla C+ pink (gyrophoric acid) .... ***Hypotrachyna revoluta* (Flörke) Hale**

84. Lobes more or less flat; medulla C– 85

85. Lobes 0.5-2 mm wide 86

85. Lobes 2-4 mm wide 88

86. On rocks; medulla PD+ red-orange (protocetraric acid); southeastern U.S. .... *Canoparmelia alabamensis* (Hale & McCull.) Elix (syn. *Paraparmelia alabamensis*)

86. On bark or wood; medulla PD– or PD+ orange 87

87. Rhizines black; southeastern coastal plain .... ***Pyxine eschweileri* (Tuck.) Vainio**

87. Rhizines pale tan to brown, not black; western and northern .... ***Parmeliopsis hyperopta* (Ach.) Arnold**

88. Medulla UV+ white (divaricatic acid); lower surface and rhizines dark brown .... ***Canoparmelia texana* (Tuck.) Elix & Hale**

88. Medulla UV–; lower surface and rhizines black .... ***Myelochroa aurulenta* (Tuck.) Elix & Hale** (rare specimens with a white medulla)

89.(18) Lobes inflated and hollow .... *Hypogymnia*

89. Lobes solid 90

90. Thallus brown, brownish green, olive, or black when dry 91

90. Thallus pale gray to greenish gray when dry 105

91. Algal layer blue-green 92

91. Algal layer green (with a lower, secondary, blue-green layer in *Solorina*) 97

92. Lower surface smooth or with a short tomentum, usually pale brown (or gray to bluish in *Pannaria*), lacking veins; mostly with a lower cortex; without discrete rhizines 93

92. Lower surface cottony or webby, often with branching or interconnecting veins, lacking a lower cortex; white to brown or black, with discrete or tufted rhizines .. *Peltigera* key (*Peltigera*, *Erioderma*, *Leioderma*)

93. Small round holes or pits (cyphellae) on the lower surface (plate 15) ....*Sticta*

93. Cyphellae absent 94

94. Tiny, white or yellow, raised spots (pseudocyphellae) on the lower surface (plate 14) .....*Pseudocyphellaria*

94. Pseudocyphellae absent 95

95. Tomentum on lower surface mostly gray to bluish; apothecia on upper surface of lobes ....*Pannaria*

95. Tomentum on lower surface, or lower surface itself, pale brown 96

96. Thallus olive-gray to brownish gray sorediate; apothecia infrequent, laminal ....*Lobaria*

96. Thallus brown, sorediate or not; apothecia frequent, on lower surface of lobe margins; medulla PD– ....*Nephroma*

97.(91) Lower surface tomentose or veined at least in part; cephalodia present (warts or secondary algal layers containing cyanobacteria) 98

97. Lower surface smooth or wrinkled, not at all tomentose; cephalodia absent 100

98. Lower surface more or less veined or webby, without a cortex; apothecia immersed in thallus, often in depressions ....*Solorina*

98. Lower surface uniform in color, pale brown, partly or entirely tomentose, often with raised, smooth, naked areas, with a cortex; apothecia superficial or raised 99

99. Pseudocyphellae (yellow or white raised spots) on lower surface of lobes ....*Pseudocyphellaria*

99. Pseudocyphellae lacking on lower or upper surface ....*Lobaria*

100. Thallus loosely attached, ascending 101

100. Thallus more or less closely appressed to substrate, except at the lobe tips 102

101. Thallus greenish black or dark olive-brown; apothecia black or very dark brown; pycnidia immersed in thallus; pseudocyphellae inconspicuous or absent .... ***Kaernefeltia merrillii* (Du Rietz) Thell & Goward**

101. Thallus brown to olive brown; apothecia red-brown; pycnidia black, prominent; pseudocyphellae often conspicuous .... *Tuckermanopsis*

102. Lobe margins fringed with black rhizines extending out from below and appearing like cilia; rhizines forming an interwoven mat below .... ***Phaeophyscia hispidula* (Ach.) Essl.**

102. Lobe margins without cilia-like rhizines; rhizines separate and distinct 103

103. Lower surface and rhizines black; upper surface usually gray in part, K+ yellow (test the parts that remain gray); lobe surface usually with a network of ridges and depressions, appearing like hammered metal 104

103. Lower surface and rhizines pale to dark brown (rhizines rarely black); upper surface uniformly brown or olive, K–; lobes smooth or rough, but rarely with a network of ridges .... *Melanelia* key

104. Pseudocyphellae milk white, very conspicuous, round to elongated; medulla K–, C+ pink (gyrophoric acid) .... ***Punctelia stictica* (Duby) Krog**

104. Pseudocyphellae pale, net-like or irregular in shape, occasionally round; medulla K+ red, C– (salazinic acid) .... *Parmelia*

105.(90) Lower surface tomentose, cottony, or veined, at least in part, rarely naked; discrete rhizines sometimes present; photobiont blue-green or green 106

105. Lower surface smooth, wrinkled, or rough, but not tomentose or cottony; discrete rhizines usually well developed; photobiont green 114

106. Pseudocyphellae present on the upper or lower thallus surface 107

106. Pseudocyphellae absent 108

107. Pseudocyphellae on the upper surface of the lobes; medulla C+ red .... *Punctelia*

107. Pseudocyphellae on the lower surface of the lobes; medulla C– .... *Pseudocyphellaria*

108. Soredia present 109

108. Soredia absent 111

109. Lobes 2-5 mm wide; soredia marginal, coarsely granular; medulla PD+ orange (pannarin)

....***Pannaria conoplea* (Ach.) Bory**

109. Lobes 5-20 mm wide 110

110. Lower surface vaguely veined, cottony (not tomentose); soredia mostly marginal
....***Peltigera collina* (Ach.) Schrader**

110. Lower surface uniformly brown, tomentose (or with scattered naked areas); soredia on both the margins and upper surface of the lobes ....*Lobaria*

111.(108) Algal layer blue-green 112

111. Algal layer green 113

112. Lower surface with a thick or thin gray to blue-black tomentum ......*Pannaria* key (*Coccocarpia*, *Pannaria*)

112. Lower surface with distinct or indistinct veins ....*Peltigera*

113. Cephalodia in the form of brown or gray warts or lobed squamules on the upper surface of the thallus; lower surface webby or cottony, white at the lobe edge and dark brown to black in the center, sometimes with veins ....*Peltigera*

113. Cephalodia in the form of small round warts or galls on the lower surface of the thallus; lower surface without veins, abundantly or sparsely tomentose ....*Lobaria*

114.(105) Lower surface entirely white or very pale brown; rhizines white or pale brown 115

114. Lower surface pale to dark brown or black, sometimes blotched with white over small or large areas, but in such cases, always brown to black in the oldest, central area; rhizines black or brown.............................116

115. White dots (pseudocyphellae) present on upper surface of lobes ....*Punctelia*

115. Pseudocyphellae absent ....***Physcia biziana* (A. Massal.) Zahlbr.**

116. Upper cortex K–, UV–; lobes fringed with black cilia-like rhizines .....***Phaeophyscia hispidula* (Ach.) Essl.**

116. Upper cortex K+ yellow, UV–, rarely K–, UV+ yellow; with or without cilia or cilia-like rhizines 117

117. Apothecia common, black; spores brown, 2-celled; closely adnate tropical species on bark ....*Dirinaria*

117. Apothecia, when present, brown; spores colorless, 1-celled ....*Parmelia* key

Identification Keys to Species

ABSCONDITELLA

1. Spores 2-celled, 9-14 x 2.5-4(-5) μm; forming a thin crust on *Sphagnum* moss; apothecia pale yellowish to pinkish .... *Absconditella sphagnorum* Vězda & Poelt

1. Spores 4-celled, 12-15 x 4.5-6.5 μm; on shaded logs and rotting wood, rarely on moss; apothecia 0.1-0.3 mm in diameter, white to pale yellowish, with persistent margins; thallus thin and membranous, often forming a shiny alga-like film; widespread temperate to boreal, but very inconspicuous .... *Absconditella lignicola* Vězda & Pišút

ACAROSPORA (including *Caeruleum*, *Myriospora*, and *Pleopsidium*) Based in large part on Knudsen (2007a); Knudsen, Lendemer & Harris (2011); and Harris & Ladd (2005).

[Note: Spot tests for gyrophoric acid with C should be done on a section of the thallus observed under a stereo or compound microscope. Mount sections under a fragment of a cover slip, remove the water with blotting paper, add a small quantity of C with a dropper or capillary pipette, and observe the reaction in the cortex.]

1. Thallus bright yellow, at least beneath pruina; UV+ orange (rhizocarpic acid) 2
1. Thallus brown, rusty orange, white, or pink, UV– 10

2. On soil 3
2. On rock or parasitic on lichens on rock 4

3. Areoles and squamules imbricate; spores subglobose, 2-4 x 2-2.5 μm; exclusively on soil .... *Acarospora schleicheri* (Ach.) A. Massal.
3. Areoles and squamules contiguous; spores ellipsoid, (3-)4-5 x 2-3 μm; usually also found growing on rocks nearby .... *Acarospora socialis* H. Magn.

4. Parasitic on lobate *Caloplaca*, at least initially; thallus usually heavily white pruinose; spores subglobose, 3-4 x 2-3 μm .... *Acarospora stapfiana* (Müll. Arg.) Hue
4. Growing directly on rock, not parasitic; spores ellipsoid 5

5. Thallus rimose-areolate with a clearly lobed margin; apothecia yellow, usually 1 per areole; asci K/I+ blue in lower part of ascus tip (tholus); containing fatty acids by TLC 6

5. Thallus areolate to subsquamulose, thallus margin usually not clearly defined and not lobed at the edge, although the squamules themselves can sometimes be lobed; apothecia reddish brown, one or more per areole; ascus tip entirely K/I–; containing no fatty acids 7

6. Lobes mostly flat to slightly convex, usually contiguous, dull and often somewhat rough; center of thallus areolate; apothecia flat to slightly convex, up to 1 mm in diameter; spores narrowly ellipsoid, 2.0-2.5:1; common in the west .... ***Pleopsidium flavum*** **(Bellardi) Körber**

6. Lobes strongly convex, often verrucose, smooth and shiny; peripheral lobes typically distinct and separate, forming small rosettes; apothecia convex when mature, up to 3 mm in diameter; spores ellipsoid, 1.5-2.0:1; rare, in southern California, arctic-alpine, under rock overhangs .... *Pleopsidium chlorophanum* (Wahlenb.) Zopf

7. Thallus medulla K+ red (norstictic acid); areoles dull yellow, with or without pruina; central to southwestern U.S. .... *Acarospora heufleriana* Körber

7. Thallus medulla K– (norstictic acid absent) 8

8. Thallus medulla C+ pink (best examined under microscope); dispersed areolate, not pruinose; mainly Ozarks to Texas and Arizona .... ***Acarospora tuckerae*** **K. Knudsen** [Plate 91 as "*A. contigua*" in LNA]

8. Thallus medulla C– 9

9. Thallus areolate, not lifting at edges to become subsquamulose; spores narrowly ellipsoid, 3-4(-6) x 2-3 µm; apothecia typically rimmed; mainly Arizona to South Carolina, north to Nebraska .... *Acarospora chrysops* (Tuck.) H. Magn.

9. Thallus areolate, but areoles soon lifted at edges to become subsquamulose; spores broadly ellipsoid, (3-)4-5 x 2-3 µm; apothecia variable, 1-many per areole, rimmed or not; very common, mainly California to southern Texas .... *Acarospora socialis* H. Magn.

10.(1) Thallus rusty orange, dispersed areolate; areoles containing 2-8 minute apothecia; on metal-rich, siliceous rocks; spores 100+ per ascus, 3-3.5 x 1.2-1.6 µm; eastern arctic to northeastern North America .... *Acarospora sinopica* (Wahlenb.) Körber

10. Thallus brown, whitish, or pinkish 11

11. Spores (8-)16-80 per ascus, ellipsoid to globose, more than 8 µm long; thallus cortex and medulla K–, C–, containing no lichen substances 12

11. Spores more than 100 per ascus, tiny, grain-like, less than 8 µm long; thallus cortex and medulla K– or K+ red, C– or C+ pink, sometimes containing lichen substances 14

12. On soil; thallus whitish scabrose, dispersed areolate, 1 apothecium per areole; spores globose, 9-13(-15) µm in diameter, 8-48 per ascus; southern California .... *Acarospora thelococcoides* (Nyl.) Zahlbr.

12. On rock; thallus brown, areolate to subsquamulose, usually 1 apothecium per areole; spores ellipsoid; widespread 13

13. On calcareous rocks; spores 40-80 per ascus, 8-12 x 4-5 µm; infrequent, widely distributed .... *Acarospora macrospora* (Hepp) A. Massal. *ex* Bagl.

13. On siliceous rocks; spores 16-24(-40) per ascus, 10-12 x 5-7 µm; southwestern U. S., Wisconsin, and Alberta .... *Acarospora oligospora* (Nyl.) Arnold

14.(11) On soil or eroding sandstone; southwestern 15

14. On stable rock, or parasitic on rock lichens 16

15. Thallus whitish, squamulose, squamules up to 5 mm across, medulla C–, K+ red or K– (usually with norstictic acid); spores subglobose, (3-)4-6 x 3-4 µm .... *Acarospora nodulosa* (Dufour) Hue

15. Thallus brown, areolate-verruculose, areoles mostly under 1 mm across; medulla C+ pink, K– (gyrophoric acid); spores ellipsoid, 4-5(-7) x 1-2(-2.5) µm .... *Acarospora obpallens* (Nyl. *ex* Hasse) Zahlbr.

16. Thallus whitish to gray, sometimes due to thick pruina 17

16. Thallus dark to pale brown, pruinose or not 20

17. Parasitic on *Caloplaca trachyphylla* .... *Acarospora stapfiana* (Müll. Arg.) Hue (see couplet 4)

17. Not parasitic, growing directly on rock 18

18. Spores globose to subglobose, 7-9 x 5-7 µm; thallus dispersed areolate; southeastern or midwestern .... *Acarospora sphaerosperma* R. C. Harris & Knudsen

18. Spores narrower than 4 µm, cylindrical, ellipsoid, or subglobose 19

19. Spores ellipsoid when mature, (3-)4-7 x 2.5-4 µm; thallus areolate to verruculose, occasionally dispersed; on rocks of different kinds, western .... ***Acarospora strigata* (Nyl.) Jatta**

19. Spores cylindrical when mature, 4-8 x 1.3-3 µm; thallus dispersed areolate; on limestone, widely distributed .... **Acarospora glaucocarpa (Ach.) Körber**

20.(16) On calcareous rock (producing bubbles with strong acid) 21

20. On siliceous rock (not bubbling with strong acid) 26

21. Thallus cortex or medulla C+ pink (gyrophoric acid); thallus without pruina, thick, areolate to subsquamulose, forming round, slightly lobed patches; areoles and squamules up to 4 mm in diameter; western; rare .... *Acarospora rosulata* (Th. Fr.) H. Magn. [= A. bullata Anzi in Sonoran flora]

21. Thallus cortex and medulla C– (containing no lichen substances) 22

22. Ascus tips (tholus) K/I+ dark blue; thallus often reduced to very small, dispersed, pale brown areoles; southern California (scattered elsewhere, Alaska, B.C., Manitoba, MN), rare .... *Caeruleum heppii* (Nägeli *ex* Körber) K. Knudsen & L. Arcadia (syn. *Acarospora heppii*)

22. Ascus tips K/I– or uniformly very pale blue 23

[Note: The hymenium is K/I+ dark blue in both choices.]

23. Apothecia dark brown to black, punctiform to more or less expanded, but generally not lecanorate, often rough; thallus dark red-brown to black, usually glossy, never pruinose, with areoles 0.2-0.6 mm in diameter; western, mainly southern California .... *Acarospora elevata* H. Magn.

23. Apothecia reddish brown, sometimes pruinose, disk punctiform or expanded; thallus pale to dark brown, not brown-black 24

24. Thallus rarely with pruina; areoles small, 0.3-0.7(-1) mm in diameter; apothecia all lecanorate, almost filling the areoles; eastern Canada .... *Acarospora canadensis* H. Magn.

[Note: *Acarospora janae* (couplet 32) can be very similar, but contains gyrophoric acid.]

24. Thallus typically pruinose, at least in part; areoles small to very large; apothecia lecanorate or not; western or widely distributed 25

25. Upper surface of algal layer very uneven and layer frequently interrupted by medullary tissue; thallus very variable in appearance, dispersed areolate to squamulose, often reduced to a rim around the expanded apothecium giving it the appearance of a lecanorine apothecium; very widely distributed and common on calcareous rock .... **Acarospora glaucocarpa (Ach.) Körber**

[Note: If lecanorate, see also *A. badiofusca*, which rarely occurs on calcareous rocks, but has blackish margins and is never pruinose. See also *A. strigata* (couplet 19), which rarely can lack pruina.]

25. Upper surface of algal layer fairly even, not jagged, infrequently interrupted by medullary tissue; thallus dispersed areolate to subsquamulose, rarely reduced to a thalline margin around the apothecia; rarely on calcareous rock .... *Acarospora americana* H. Magn. (see couplet 43)

26.(20) Thalli forming round patches that are clearly showing short lobes at the margin, areolate in the older parts, not pruinose 27

26. Thalli areolate to subsquamulose or squamulose throughout, not forming lobed patches even though the squamules themselves may be slightly lobed, pruinose or not pruinose 28

27. Cortex C+ pink (gyrophoric acid); western .... *Acarospora rosulata* (Th. Fr.) H. Magn. (= *A. bullata* Anzi in Knudsen [2007a])

27. Cortex and medulla C– (no substances); mainly eastern, from arctic to coastal maritimes .... *Acarospora molybdina* (Wahlenb.) Trevisan

28. Thallus K+ red (norstictic acid), dispersed areolate; 2-4 apothecia per areole; east coast and Pacific Northwest .... *Myriospora smaragdula* (Wahlenb.) Arcadia & Knudsen (syn. *Acarospora smaragdula*, *Silobia smaragdula*)

28. Thallus K– (lacking norstictic acid); areolate to squamulose or subsquamulose, typically one apothecium per areole but sometimes more than one 29

29. Cortex or medulla C+ pink (containing gyrophoric acid) (test on thallus sections under a scope); thallus not pruinose 30

29. Cortex and medulla C– (containing no substances); thallus with or without pruina 36

30. Thallus squamulose with convex, often overlapping, shiny squamules with varying shades of brown (often variegated), 0.3-1.5 mm in diameter, that are somewhat raised on a black stipe; spores narrowly ellipsoid; cortex C+ strong red; throughout California, less common in Rocky Mountains and Arizona .... *Acarospora thamnina* (Tuck.) Herre

30. Thallus rimose-areolate to subsquamulose, areoles without stipes; cortex C+ pink to red, sometimes weak 31

31. Thallus entirely dispersed areolate, areoles mostly under 1.5 mm in diameter; apothecia one per areole, pruinose or not 32

31. Thallus rimose areolate to dispersed or subsquamulose, areoles commonly reaching 1.5-3 mm in diameter, one to several apothecia per areole, not pruinose 34

32. Apothecia red-brown, not pruinose; areoles smooth; widely distributed, especially in east; on siliceous rock .... *Acarospora janae* K. Knudsen

32. Apothecia usually pruinose or scabrid 33

33. Areoles often with fissures, tiny pits, and depressions, under 0.7 mm in diameter; on soil or crumbling sandstone; southern California and Arizona, rare in east ..... *Acarospora obpallens* (Nyl. *ex* Hasse.) Zahlbr.

33. Areoles lacking tiny pits or fissures, 0.2-1.5 mm in diameter; on sandstone, mainly in Ozark region .... *Acarospora nicolai* B. de Lesd.

34. Thallus medium to dark red-brown, shiny, usually growing in small round patches; 1- several apothecia per areole .... *Acarospora rosulata* (Th.Fr.) H. Magn. (see also couplet 27)

34. Thallus pale to yellow-brown, dull 35

35. Areoles and squamules convex, often fissured; apothecial disks usually very dark, one per areole, very rough, almost gyrose, with sterile tissue breaking up the apothecial surface and sometimes forming ridges; reported from southwest, Ontario, and arctic regions .... *Acarospora peliscypha* Th. Fr.

35. Areoles and squamules rather flat or concave, not fissured; apothecia pale to dark brown, typically several per areole, mostly punctiform but sometimes filling areole, smooth; common and widespread .... ***Acarospora fuscata* (Schrad.) Th. Fr.**

36.(30) Ascus tip K/I+ blue; thallus scanty, consisting of pale brown, sometimes pruinose areoles; areoles less than 0.6 mm in diameter, dispersed or continguous .... *Caeruleum heppii* (Nägeli *ex* Körber) K. Knudsen & L. Arcadia (see also couplet 22)

36. Ascus tip K/I– ; thallus scanty or well developed 37

37. Thallus reduced, often largely within the rock (endolithic), or areolate, pale to yellowish brown or dark brown, surface dull; algal layer discontinuous, broken up by bundles of medullary hyphae; spores very small, 3-4 x 1-1.5(-2) µm, ellipsoid; California coast .... *Myriospora hassei* (Herre) K. Knudsen & L. Arcadia (syn. *Acarospora hassei, Silobia hassei*)

37. Thallus well developed, areolate to squamulose, pale to dark brown or black; algal layer distinct, continuous or discontinuous; spores 3-6(-7) x 1-2.2(-3) µm 38

38. Thallus areoles or squamules forming stipes below; apothecia one to several per areole 39

38. Thallus areoles or squamules not stipitate 40

39. Thallus squamulose, with contiguous to commonly overlapping areoles up to 2 mm in diameter; algal layer very uneven, discontinuous; southern Arizona to southern California ....*Acarospora obnubila* H. Magn.

39. Thallus dispersed to contiguous areolate, areoles 0.5-1.0 mm in diameter; algal layer continuous; southeastern piedmont region .... *Acarospora piedmontensis* K. Knudsen

40. Apothecia one per areole, often expanded with a thin or thick raised rim resembling a lecanorine apothecium and much broader than surrounding areoles, 0.8-2.0 mm in diameter, disc dark reddish brown often with a black margin; spores narrowly ellipsoid; mid-point of paraphyses 2.0-3 µm in diameter; widely distributed from montane west to arctic and Great Lakes region ....*Acarospora badiofusca* (Nyl.) Th. Fr.

40. Apothecia one to several per areoles, with or without rims but not expanded and appearing lecanorine, 0.1-0.7 mm in diameter, and fertile areoles not substantially broader than sterile areoles; spores broad or narrow; mid-point of paraphyses 1.0- 2.1 µm in diameter 41

41. Thallus very dark brown to black, glossy, without pruina; areolate with scattered convex areoles, 0.3-1.5 mm in diameter; apothecia commonly one per areole, often rimmed; western North America .... *Acarospora elevata* H. Magn.

41. Thallus pale to dark brown, rarely black, dull, pruinose or not; apothecia one to several per areole; areoles and squamules flat 42

42. Areoles 0.2-1.0(-1.5) mm in diameter, dark brown, not pruinose; hymenium 65-90 µm high; apothecia often rough with sterile tissue forming bumps and ridges, not rimmed; southwestern and central to southeastern U.S. ....*Acarospora veronensis* A. Massal.

42. Areoles 0.3-3 mm in diameter, with or without pruina; hymenium more than 100 µm high; apothecia smooth, without bumps or ridges 43

43. Areoles dark brown, frequently lightly to heavily pruinose, 0.5-2.5 mm in diameter, not rimmed in black; hymenium (90-)100-120(-150) µm high; apothecia frequently with raised rims, especially in eastern populations; widespread in North America ....*Acarospora americana* H. Magn.

43. Areoles never pruinose, 0.3-1 mm in diameter, typically pale brown, roundish, dispersed (but can be contiguous), with black margins; hymenium 100-200 µm high; apothecia without raised rims; mainly eastern North America centered in the Ozarks, also Alaska and Arizona .... *Acarospora dispersa* H. Magn.

AHTIANA

1. Thallus closely appressed, surface strongly rugose; apothecia and pycnidia on thallus surface, not marginal; lower surface white or almost white; western montane ...... ***Ahtiana sphaerosporella* (Müll. Arg.) Goward**

1. Thallus loosely attached or ascending, surface rather smooth or with shallow depressions, not rugose; apothecia developing along lobe margins, pycnidia marginal or laminal 2

2. Thallus dull, sometimes pruinose, lobes ascending; lower surface pale yellow, sharply wrinkled; lobe margins sometimes toothed or lobulate; northwestern .... ***Ahtiana pallidula* (Tuck. *ex* Riddle) Goward & Thell**

2. Thallus usually shiny, never pruinose, lobes flat, at least at periphery; lower surface pale brown, smooth; lobe margins never lobulate; Appalachian-Great Lakes region .... ***Ahtiana aurescens* (Tuck.) Thell & Randlane**

ALECTORIA

1. Thallus gray, olive, or shades of brown to almost black; if yellowish gray, then cortex KC+ pink, not gold; pseudocyphellae conspicuous or inconspicuous, usually level with the surface or slightly to deeply depressed .... *Bryoria* key

1. Thallus pale greenish yellow or yellowish green; cortex KC+ gold (usnic acid); pseudocyphellae conspicuous, usually slightly raised 2

2. Thallus forming erect clumps, or sometimes prostrate, on the ground; tips of branches usually becoming black or greenish black .... ***Alectoria ochroleuca* (Hoffm.) A. Massal.**

2. Thallus shrubby to pendent, on trees or shrubs, rarely with blackened branch tips 3

3. Thallus forming bushy clumps, usually less than 10 cm long 4

3. Thallus pendent to slightly pendent, usually 8-20 cm long when mature 5

4. Thorny isidia and spinules developing in elongate pseudocyphellae and fissures; apothecia very rare; cortex K+ bright yellow, PD+ orange (thamnolic acid) or K–, PD– (squamatic acid); medulla KC– .... ***Alectoria imshaugii* Brodo & D. Hawksw.**

4. Isidia and spinules absent; apothecia usually present, brown; cortex K–, PD–; medulla KC+ red (alectoronic acid) .... ***Alectoria lata* (Taylor) Lindsay**

5. Branches very slender and pale, with tips curled up and granular or sorediate; medulla KC– .... ***Ramalina thrausta* (Ach.) Nyl.**

5. Branches slender or thick, not curled up at the tips; without soredia; medulla KC+ red (alectoronic acid), or infrequently KC– (usnic acid alone) .... ***Alectoria sarmentosa* (Ach.) Ach.**

ALLANTOPARMELIA (including *Brodoa*)

1. Lobes long and divergent, 0.3-2 mm wide, never coalescing or appearing crustose in older parts of the thallus, gray to dark brown ; medulla K–, KC+ pink, C–, PD– or rarely PD+ red (physodic acid sometimes with fumarprotocetraric acid); common, arctic-alpine .... ***Brodoa oroarctica* (Krog) Goward**

1. Lobes long or short, contiguous at the margin of the thallus, not divergent, often appearing crustose in older parts of thallus 2

2. Medulla C+ red, PD– (olivetoric acid); thallus dark red-brown to black; lobes very short and narrow, 0.2-0.4 mm wide; older parts often verrucose and lobulate; infrequent .... *Allantoparmelia almquistii* (Vainio) Essl.

[Note: See also *Pseudephebe minuscula*, which can be superficially similar but contains no lichen substances.]

2. Medulla C– ; thallus lobes 0.3-1.5 mm wide 3

3. Medulla PD+ deep yellow, K–, KC+ red (alectorialic and barbatolic acids); apothecia always superficial, not immersed, up to 7 mm across; spores 7.5-10 x 5-7 µm .... ***Allantoparmelia alpicola* (Th. Fr.) Essl.**

3. Medulla PD–, K– or PD+ yellow, K+red, KC– (lacking alectorialic acid, sometimes with norstictic acid); apothecia at first immersed, later superficial, under 2.5 mm across; spores 10-14 x 6-10 µm ...... *Lobothallia*

AMYGDALARIA

1. Sorediate; medulla C+ pink (gyrophoric acid); widespread in cool, humid areas .... ***Amygdalaria panaeola* (Ach.) Hertel & Brodo**

1. Not producing soredia 2

2. Apothecia forming between areoles, not immersed; medulla and cortex C+ pink, usually K– (gyrophoric acid, rarely with stictic acid); Alaska and western mountains .... *Amygdalaria elegantior* (H. Magn.) Hertel & Brodo

2. Apothecia immersed in thallus areoles, appearing cryptolecanorine but with a brown-black exciple present, thin at sides and thick below hypothecium 3

3. Medulla K+ yellow, PD+ orange (stictic acid); cortex C+ pink or C– (gyrophoric acid present or absent); apothecia sunken into large, verrucose areoles; Pacific Northwest, but common only in Haida Gwaii, B.C. .... *Amygdalaria subdissentiens* (Nyl.) Inoue & Brodo

3. Medulla K–, PD–; cortex and sometimes medulla C+ pink (gyrophoric acid); areoles usually flat, but sometimes convex, apothecia frequently one per areole, giving a pseudo-lecanorine appearance; Pacific Northwest .... *Amygdalaria pelobotryon* (Wahlenb.) Norman

ANAPTYCHIA See Esslinger (2007).

1. Thallus lobes with soredia on the margins and tips, heavily pruinose and often with colorless cortical hairs; on rock, western interior .... ***Anaptychia elbursiana* (Szatala) Poelt** (syn. *Physconia thomsonii*)

1. Thallus without soredia 2

2. Thallus almost fruticose, with long, slender, branching, ascending lobes with long marginal cilia and no rhizines; Great Lakes to northeast and Pacific coast; on both rocks and tree bark .... ***Anaptychia crinalis* (Schaerer) Vězda** (syn. *A. setifera, A. kaspica*)

2. Thallus clearly foliose, not ascending, lacking cilia, but with rhizines 3

3. Lobes pruinose, often fuzzy with cortical hairs, 0.3-1 mm wide; on rock; mainly in the arid western interior .... *Anaptychia ulotrichoides* (Vainio) Vainio

3. Lobes largely lacking pruina and cortical hairs; conspicuously lobulate 4

4. Appalachian-Great Lakes-Ozarks; on bark or rock, usually in the shade; lobes 0.7-1.5(-2) mm wide, sometimes scabrose on the lobe tips; abundant lobules remain prostrate, not ascending; rhizines at first weakly branched but later becoming squarrose .... ***Anaptychia palmulata* (Michx.) Vainio**
[Note: Compare with *Physconia subpallida*, which is usually heavily pruinose.]

4. Arctic and western alpine, on the ground or among mosses on rocks; lobes elongate 0.4-0.8(-1.0) mm wide, with very narrow ascending lobules 0.15-0.3 mm wide; rhizines pale to dark, unbranched at first, but becoming branched later .... *Anaptychia bryorum* Poelt
[Note: Compare with *Phaeophyscia constipata*, which has unbranched rhizines.]

ANZIA

1. Thallus lacking soredia or isidia; apothecia usually abundant; common, East Temperate .... ***Anzia colpodes* (Ach.) Stizenb.**
1. Thallus with isidia or soredia; apothecia absent; rare 2

2. Thallus lobes long and slender, convex, dichotomously branched, sorediate at lobe tips; lacking rhizines; southern Appalachian Mountains .... *Anzia americana* Yoshim. & Sharp
2. Thallus lobes short or long, flat, irregularly branched, granular isidiate along the margins; sparse black, unbranched rhizines sometimes emerging from thick black hypothallus; southeastern coastal plain .... *Anzia ornata* (Zahlbr.) Asahina

ARTHONIA (including *Arthothelium*)

[Note: Some spore measurements from Fink (1935); Harris (1990).]

1. Spores muriform; ascomata lobed to star-like or almost round 2
1. Spores only transversely septate; ascomata round to script-like 4

2. Spores (15-)17-24(-26) x 7-9.5(-10.5) µm, 5-8 transverse septa .... *Arthonia ruana* A. Massal. (syn. *Arthothelium ruanum*)
2. Spores 24-36 x 10-15 µm, up to 9 transverse septa 3

3. Ascomata 0.5-1.5 mm long, generally flat, naked; northern; photobiont containing orange oil globules (*Trentepohlia*) .... *Arthothelium spectabile* (Flotow) A. Massal.
3. Ascomata 0.3-0.6 mm long, like branched dotted lines, emerging from bark with remnants of bark on surface giving it a pruinose appearance; southeastern; photobiont chlorococcoid, cells 8-14 µm in diameter, without orange oil globules .... *Arthonia susa* R. C. Harris & Lendemer (= "*A. taediosum*" of American authors)
[Note: *Arthonia albovirescens* Nyl. is a very similar coastal plain species having spores with up to 15 transverse septa.]

4. Ascomata round or slightly irregular in shape, with the appearance of a biatorine apothecium 5
4. Ascomata elongate or branched lirellae 14

5. On rocks 6
5. On bark 7

6. On maritime siliceous rocks; thallus brown, membranous; spores (3-)4(-6)-celled, 15-23 x 5.5-8 µm; on both Atlantic and Pacific coasts ....*Arthonia phaeobaea* (Norman) Norman

6. On inland calcareous rocks; thallus very thin to endolithic; spores 2–celled, 11-19 x 4-7.5 µm; arctic to temperate ....*Arthonia lapidicola* (Taylor) Branth & Rostrup

7. Ascomata brown to black, not at all pruinose, flat to convex 8

7. Ascomata bluish gray to white due to a heavy pruina, convex or flat 11

8. Spores mostly 3- to 4(-5)-celled, 10.5-13 x 4.2-5.0 µm, slightly tapered; on bark of deciduous trees; East Temperate...*Arthonia diffusa* Nyl. (syn. *A. willeyi*)

[Note: The rare *Arthonia diffusella* Fink in Hedr. has strongly tapered spores, (2-)3-celled, 12-15 µm long, and is on wood.]

8. Spores mostly 2-celled 9

9. Spores (8-)10-12 x 3.5-5 µm, remaining colorless; growing almost exclusively on trembling aspen trees, widespread .... *Arthonia patellulata* Nyl.

9. Spores 11-17 x 4.3-7 µm, typically gray to pale brown when mature; on trees of all kinds, Great Lakes to the Maritimes 10

10. Ascomata tiny, flat, dot-like or slightly elongate, black; spores up to 7 µm broad; tissues often turning violet in K, not wine red .... *Arthonia didyma* Körber

10. Ascomata hemispherical, 0.2-0.5 mm in diameter, round to irregular in shape; spores up to 5.5 µm broad; tissues always turning wine red and then purple in K ....*Arthonia vinosa* Leighton

11.(7) California, common; spores 14-17 x 6-8 µm, 4- to 6-celled, equal-sized cells; thallus C+ red or C– (sometimes containing arthoniaic acid) ....*Arthonia pruinata* (Pers.) Steud. *ex* A.L. Sm.

11. Northeastern; spores 4-6 µm wide; thallus and ascomata C– 12

12. Ascomata blue-gray pruinose; thallus yellowish (usnic acid), granular to leprose in small patches; spores 4-celled, cells equal in size, 15-20 x 4-6 µm; on a variety of trees in the northeast .... ***Chrysothrix caesia* (Flotow) Ertz & Tehler** (syn. *Arthonia caesia*)

12. Ascomata white pruinose; thallus whitish or brownish (lacking usnic acid), granular or extremely thin and barely perceptible; spores 4- to 6-celled 13

13. Spores 4- to 5(-6)-celled, with upper cell usually noticeably larger than others, 12-16(-23) x 4.5-6(-8) µm; ascomata convex, marginless, pruinose; in very old forests on various trees .... *Arthonia byssacea* (Weig.) Almqu.

13. Spores 4-celled approximately equal in length, tapering, (18-)22-26(-30) x 6-7(-8.5) µm; ascomata appear to have a white margin, with a black pruinose disk; on oak bark, mainly southeastern, north to Massachusetts along the coast .... *Schismatomma glaucescens* (Nyl. ex Willey) R. C. Harris (syn. *Arthonia glaucescens*)

14.(4) Spores 2-celled, brownish .... *Arthonia didyma* Körber (see couplet 10)

14. Spores 4- to 8-celled.. 15

15. Ascomata long and narrow with thick carbonized walls, opening by a narrow slit; thallus white in delimited patches; spores 13-18 (-20) µm long, 4-celled .... *Arthonia atra* (Pers.) A. Schneid. (syn. *Opegrapha atra*)

15. Ascocarps lacking a carbonized wall, not opening by a slit; thallus various; spores 15-35 µm long 16

16. Spores with 4 equal-sized cells, cigar-shaped; ascomata without pruina .... ***Arthonia radiata* (Pers.) Ach.**

16. Spores 4- to 8-celled, slightly to strongly tapered; ascomata pruinose or not 17

17. Spores with cells all about the same size, 4- to 5-celled; thallus whitish, thin, granulose to leprose; ascomata branched, often with a thin white pruina and (or) whitish at the edges; northeastern coast .... *Arthonia leucopellaea* (Ach.) Almqu.

17. Spores with end cell larger than the others; not northeastern 18

18. Spores 15-28 µm long 19

18. Spores 26-35 x 10-15 µm 21

19. Ascomata not pruinose, irregularly branched, flat or immersed (red pigments absent); spores 17-20(-25) x 6-10 µm, 6-celled, usually immature or absent; thallus lacking algae; common in southeast, rarer farther north .... *Arthonia quintaria* Nyl.

19. Ascomata pruinose; thallus and (or) ascomata pink or red; spores 4- to 6-celled, 15-23 x 5.5-8.0 µm 20

20. Thallus greenish, sometimes dusted with red pruina, PD+ yellow (psoromic acid); ascomata ellipsoid to Y-shaped, convex, red pruinose on the margins or over the entire surface (anthraquinones); spores (4-)5-celled; on Sabal palm leaf bases............. ***Arthonia rubrocinta* G. Merr.** [Plate 115 in LNA, as "*A. cinnabarina*."]

20. Thallus usually pink, PD–; ascomata irregular to elongate, under 0.5 m long, not branched, margins black with a white pruina on the disk surface; spores 4- to 5(-6)-celled; on bark on various kinds .... *Arthonia cinnabarina* (DC.) Wallr. [*not* Plate 115; see above]

21.(18) Ascomata flat and superficial, black to dark reddish brown; spores 5- to 7-celled; coastal Pacific Northwest .... *Arthonia ilicina* Taylor

21. Ascomata usually immersed and narrowly or widely open, appearing like irregular cracks in the bark, dark red to reddish brown; spores 4- to 5-celled; southeast .... *Arthonia rubella* (Fée) Nyl.

ARTHRORHAPHIS

1. Colorless oxalate crystals absent from medulla; thallus areoles and squamules small, mostly under 0.5 mm in diameter, often dissolving into granular soredia; spores (45-)50-65(-110) µm long .... ***Arthrorhaphis citrinella* (Ach.) Poelt**

1. Colorless oxalate crystals present in thallus medulla; thallus composed of large, convex areoles or verrucae, (0.2-)0.5-1.5 mm in diameter, smooth to rough, occasionally dissolving into soredia; spores (25-)45-52 µm long .... *Arthrorhaphis alpina* (Schaerer) R. Sant.

ASPICILIA (including *Aspilidea, Circinaria, Megaspora,* and *Teuvoa*) Based on Norden & Owe-Larrson (2007) for southwestern species and McCune & Rosentreter (2007) for vagrant species.

1. Thallus at least partially fruticose (one species mostly crustose: see *Aspicilia reptans* below), on soil, unattached or lightly attached to soil or vegetation, with clumps of tangled, terete branches; apothecia rare 2

1. Thallus entirely crustose, on rock, moss, peat, soil, wood, or bark; apothecia usually abundant 6

2. Branches with white depressed pseudocyphellae abundant and conspicuous on branch surface; common in the intermontane arid high steppes, partially attached to soil .... *Circinaria hispida* (Mereschk.) A. Nordin, S. Savić & Tibell (syn. *Aspicilia hispida*) [Note: Plate 122 in LNA shows *Circinaria rogeri,* not *C. hispida* or *C. fruticulosa*.]

2. Branches lacking white pseudocyphellae, or pseudophellae confined to the branch tips 3

3. Thallus forming compact rounded clumps, not attached to soil; pseudocyphellae on branch tips; locally common in southern and central Idaho, rare elsewhere ... ***Circinaria rogeri* (Sohrabi) Sohrabi** (syn. *Aspicilia rogeri; Aspicilia or Circinaria fruticulosa* of N. Am. authors) [Plate 122 in LNA]

3. Thallus crustose to fruticose and loosely branched, prostrate and attached to soil; pseudocyphellae absent 4

4. Medulla K+ red (norstictic acid); rare, central California .... *Aspicilia californica* Rosentreter
4. Medulla K– 5

5. Entirely fruticose, branches narrow; spores 11-26 µm in diameter; Oregon and Washington to Montana and Wyoming .... *Aspicilia filiformis* Rosentreter
5. Mostly crustose, developing fruticose branches in places along thallus margin; spores 10-12 µm in diameter; widespread in cold, arid habitats in central and western North America .... *Aspicilia reptans* (Looman) Wetmore

6.(1) Growing on wood, bark, peat, or soil; medulla and cortex PD–, K–, C–; no lichen substances 7
6. Growing on rock 9

7. Spores thick-walled, 30-65 x 16-36 µm; tips of paraphyses not expanded or moniliform; widespread, especially in west and arctic, rare in temperate Ontario; on various substrates other than rock .... ***Megaspora verrucosa* (Ach.) Hafellner & V. Wirth**
7. Spores thin-walled, 13-30 x 8-16 µm; tips of paraphyses moniliform 8

8. Spores (13-)16-19(-22) x (10-)11-14(-16) µm; epihymenium dark olive-brown to brownish; lacking a prothallus; conidia 5-8 µm long; on *Juniperus*, in interior arid parts of Colorado Plateau and Great Plains .... *Teuvoa junipericola* Sohrabi & S. Leavitt
8. Spores (16-)18-25(-31) x (8-)10-15(-17) µm; epihymenium blue-green; often with a blue-green prothallus; conidia 15-30 µm long; on conifers in shaded, often moist habitats; California .... *Aspicilia cyanescens* Owe-Larss. & A. Nordin

9.(6) Spores 2-6 per ascus, almost spherical (but see couplet 11), 13-28 µm wide 10
9. Spores mostly 8 per ascus, mostly ellipsoid, 6-16 µm wide 13

10. Thallus areolate with areoles dispersed or occasionally contiguous, flat to strongly convex or pyramidal, dark olive to dark olive-gray, often with white pseudocyphellae .... ***Circinaria contorta* (Hoffm.) A. Nordinet al.** (syn. *Aspicilia contorta*)

10. Thallus usually continuous, smooth or rimose-areolate, or areolate, white to gray, rarely olive; lacking pseudocyphellae 11

11. Thallus gray, rarely brownish gray; spores 13-16 µm wide, globose only when young; mostly northwestern ....*Circinaria caesiocinerea* (Nyl. ex Malbr.) A. Nordin et al. *s. lat.* (see couplet 22)

11. Thallus brown to brownish gray; spores 13-28 µm wide, mostly globose; southwestern to central 12

12. Algal layer continuous; areoles mostly 0.4-1.5 mm in diameter; apothecia 0.2-0.7(-1.4) mm in diameter; aspicilin usually present; strictly southwestern .... *Circinaria arida* Owe-Larss. et al.

12. Algal layer in patches interrupted by hyphal tissue; areoles mostly 1.0-2.5 mm in diameter; apothecia 0.4-1.4(-1.8) mm; aspicilin absent; southwest to the Great Plains .... *Circinaria elmorei* (E.D. Rudolph) Owe-Larss. et al.

13.(9) Thallus with short or long lobes forming at the periphery; on calcareous rock 14

13. Thallus not at all lobed at the periphery 15

14. Thallus with long, narrow lobes, contiguous or separate, otherwise areolate to dispersed areolate, variegated pale to dark gray or brownish gray; apothecia filling small areoles, up to 0.5 mm in diameter, with prominent, often black margins; spores 12-20 x 7-12 µm; arctic .... *Aspicilia perradiata* (Nyl.) Hue (syn. *A. disserpens*)

14. Thallus with very short contiguous, fimbriate lobes, chalky white, at first continuous rimose-areolate, later sometimes eroding and becoming dispersed areolate; apothecia buried in thallus, usually pruinose, margins rarely prominent but often becoming dark gray; spores 14-24 x 10-16 µm [fide Thomson (1997)]; western mountains and arctic ....***Aspicilia candida* (Anzi) Hue**

15. Thallus chalky white; on calcareous rock; all spot tests negative despite the occasional presence of stictic acid ....***Aspicilia candida* (Anzi) Hue**

15. Thallus creamy white or pale to dark gray to brown or brownish gray, rarely chalky white; on non-calcareous rocks 16

16. Paraphyses tips largely non-septate, not bead-like (moniliform) at tips 17

16. Paraphyses tips septate, with several rounded cells at the tip, resembling a string of beads (moniliform); ascus tips K/I+ uniformly pale blue or K/I– 19

17. Asci stained uniformly K/I+ blue or with a dark blue coating .... *Aspilidea myrinii* (Fr.) Hafellner

17. Asci with dark blue-staining tholus or tube-structure in tips with K/I; paraphyses expanded at the tips 18

18. Spores not halonate; ascus with a K/I + dark blue tip and a clear axial body (*Lecanora*-type); medulla PD–, K– .... ***Lecanora oreinoides* (Körber) Hertel & Rambold**

18. Spores halonate (in ink preparation); ascus with a dark blue tube structure (*Porpidia*-type) in K/I; medulla PD+ yellow and K+ red, or PD– and K– .... *Bellemerea*

19.(16) Medulla and (or) cortex K–, PD– 20

19. Medulla and (or) cortex K+ red or yellow, PD+ yellow or orange (norstictic or stictic acid) 27

20. Northern species, not found in southwest 21

20. Southwestern species; spores 17-28 x 9-16 µm 23

21. Thallus brownish gray to gray-brown, rimose-areolate to verruculose or rugose; spores rarely more than 20 µm long; northeastern to central, common

........................*Aspicilia* cfr. *verrucigera* Hue [stictic acid deficient, brownish population; probably a distinct species]

21. Thallus gray; spores usually longer than 20 µm 22

22. Thallus areolate; conidia 6-12 µm; often containing aspicilin; on dry rocks, widespread but mainly western

............................*Circinaria caesiocinerea* (Nyl. *ex* Malbr.) A. Nordin et al. *s. lat.* (syn. *Aspicilia caesiocinerea*)

22. Thallus smooth to rimose-areolate; conidia 15-25 µm; lacking lichen substances; in moist habitats, western mountains .... *Aspicilia supertegens* Arnold

23.(20) Thallus pale gray to whitish; apothecia aften showing a darker margin; southern to central California

....*Aspicilia fumosa* Owe-Larss. & A. Nordin

23. Thallus medium to dark gray or brown; apothecial margin absent or paler than thallus 24

24. Thallus dark brown to gray-brown, often mottled; conidia 7-15 µm long; southern to central California

.... *Aspicilia phaea* Owe-Larss. & A. Nordin

24. Thallus mainly gray; conidia 14-30 µm long 25

25. Common in Arizona, New Mexico, and Colorado; areolae flat, 0.4-1.5(-3.0) mm in diameter

.... *Aspicilia americana* de Lesd.

25. California 26

26. Thallus with convex, well-defined areolae, 0.4-1.0(-2.0) mm in diameter
....*Aspicilia confusa* Owe-Larss. & A. Nordin

26. Thallus rimose-areolate, areoles in older parts of thallus flat, irregular in size
....*Aspicilia cyanescens* Owe-Larss. & A. Nordin

27.(19) Medulla and (or) cortex K+ red, PD+ yellow (norstictic acid alone) 28

27. Medulla and (or) cortex K+ yellow, PD+ orange (stictic acid, sometimes with a trace of norstictic acid)
....31

28. Tips of paraphyses not expanded or moniliform; medulla under apothecia IKI+ reddish purple, sometimes blue; hymenium IKI (1.5%) persistent dark blue, not turning red-orange (euamyloid); asci K/I+ uniform pale but distinct blue or with a dark blue coating; arctic-alpine in west; infrequent
....*Aspilidea myrinii* (Fr.) Hafellner

28. Tips of paraphyses with a series of 3-4 rounded cells (moniliform); medulla IKI–; hymenium IKI (1.5%) first turning dark blue, rapidly changing to reddish orange (hemi-amyloid); asci entirely K/I– or faintly blue
....29

29. Hymenium 90-115 µm high; spores 12-17(-22) x 6-12 µm; thallus pale to dark gray, rarely with brownish tint; widespread and very common ....***Aspicilia cinerea* (L.) Körber** *s. lat.*

29. Hymenium 120-280 µm high; spores (15-)19-28(-33) x 10-16(-20) µm; thallus gray or brown; southwestern 30

30. Thallus red-brown to yellowish brown, shiny, becoming grayer or greener in shade, areoles convex to flat; conidia 16-35 µm long; apothecia 0.4-1.0(-1.6) mm in diameter, often with a whitish rim; throughout California ....*Aspicilia cuprea* Owe-Larss. & A. Nordin

30. Thallus olive-brown to olive, mottled with white; areoles flat; conidia 10-13 µm long; apothecia 0.2-0.8(-1.1), often with a whitish rim; common in Arizona, rare in southern California
.... *Aspicilia olivaceobrunnea* Owe-Larss. & A. Nordin

31.(27) Thallus pale gray to yellowish white; apothecia 0.1-0.4(-0.8) mm in diameter, white pruinose; spores (14-)19-28(-33) x (10-)12-16 µm; southwestern; prothallus often seen
.... *Aspicilia pacifica* Owe-Larss. & A. Nordin

31. Thallus pale or dark gray to brownish gray; apothecia lacking pruina or rarely lightly pruinose; spores (12-)15-18(-22) x (7-)8-10 µm; mostly northeastern to north central; prothallus present or absent 32

32. Thallus continuous, thin, smooth to rimose, sometimes rimose-areolate; spores ellipsoid to narrowly ellipsoid; widespread, especially in the northeast, on wet or dry rocks .... *Aspicilia laevata* (Ach.) Arnold

32. Thallus thick, becoming lumpy with curved, worm-like areoles or verrucae; spores ellipsoid to broadly ellipsoid; widespread temperate to boreal, on dry rocks .... *Aspicilia verrucigera* Hue

BACIDIA (including *Arthrosporum*, *Bacidina*, *Bilimbia*, *Herteliana*, *Ropalospora*, and *Scoliciosporum*)

1. On rocks and occasionally wood, rarely tree bases 2

1. On bark or mosses, very rarely on rock or wood 8

2. In periodically submerged habitats such as streams, lake shores, and seashores 3

2. In dry habitats; epihymenium greenish, brownish, or colorless; thallus without a prothallus 4

3. Maritime rocks; epihymenium brown to grayish red; thallus thick, pinkish brown, lacking a prothallus; spores (2-)4-celled, slightly curved, 20-25 x 3.3-4.0 µm; apothecia pinkish brown to dark brown, margin usually excluded; Washington to Alaska... *Herteliana alaskensis* (Nyl.) S. Ekman

3. Freshwater habitats;; epihymenium brown; thallus thin, greenish, usually with a white prothallus; spores needle-shaped, 4(-8)-celled, 24-43 x 2-3 µm; apothecia brown to black most with a thin margin; probably widespread temperate to boreal .... *Bacidina inundata* (Fr.) Vězda

4. Spores fusiform, length/width ratio < 6:1, not spirally twisted; on limestone 5

4. Spores needle-shaped, length/width ratio > 6:1, straight or spirally twisted; mostly on siliceous rocks 6

5. Apothecia pale to dark brown; spores 18-30(-40) x 5-8 µm, 4- to 6-celled ....***Bilimbia sabuletorum* (Schreber) Arnold** (syn. *Myxobilimbia sabuletorum*, *"Bacidia" sabuletorum*)

5 Apothecia black; spores 11-20 x 2.5-4.5 µm, 4-celled; central to eastern temperate ....*Bacidia coprodes* (Körber) Lettau (syn. *B. granosa*)

[Note : This is the lichen called *B. trachona* by many North American authors.]

6. Spores 35-50 x 5-7 µm, 8-celled, strongly tapered [Thomson 1997]; thallus thick, brown to gray-brown; arctic-alpine in Alaska and northeast .... *Ropalospora lugubris* (Sommerf.) Poelt

6. Spores 19-32 x 1.2-3 µm; thallus thin, inconspicuous, gray-brown to greenish gray; temperate 7

7. Spores spirally twisted, 4(-8)-celled; apothecia red-brown to black, margins disappearing; hypothecium entirely colorless; widespread temperate, especially in east .... *Scoliciosporum umbrinum* (Ach.) Arnold

7. Spores straight to slightly curved, 1- to 4(-8)-celled; apothecia mottled dark brown to black with thin, persistent margins; hypothecium reddish brown above and colorless below; widespread in east but easily overlooked .... *Bacidina egenula* (Nyl.) Vězda

8.(1) Spores 30-50 per ascus, 16-55 x 1.5-3.0 µm; thallus edge definite, with a brown prothallus; exciple pale internally, dark brown at edge; ascus *Fuscidea*-type, with a layered appearance when stained with K/I .... ***Ropalospora chlorantha* (Tuck.) S. Ekman**

8. Spores 8-16 per ascus; thallus edge usually indefinite; prothallus present or absent; exciple pigmented in various patterns, or sometimes colorless; ascus usually *Bacidia*-type, with a more or less uniformly dark blue tholus in K/I 9

9. Most spores strongly curved or "S"-shaped 10

9. Spores straight or curved, not twisted in ascus 11

10. Spores 8 per ascus, strongly tapered, spirally twisted in the ascus; apothecia brown to black, lacking margins; hypothecium colorless to pale brown .... *Scoliciosporum* (see also couplet 18)

10. Spores 8-16 per ascus, not tapered, (7-)9-13(-18) x 3.5-5 µm; apothecia black with thin black margins; hypothecium reddish brown .... *Arthrosporum populorum* A. Massal. (see also couplet 20)

11. Apothecial disks black or almost black when dry; epihymenium reddish violet to brown or green 12

11. Apothecial disks pale brown, or reddish brown to orange-brown; epihymenium reddish violet to brown, yellowish or colorless, without green pigments 21

12. Hypothecium or subhymenium dark orange-brown to red-brown 13

12. Hypothecium colorless to pale yellowish brown 16

13. Spores 5-8 µm wide, fusiform; apothecia strongly convex to hemispherical, margins disappearing; spores fusiform, under 40 µm long, 4- to 6-celled. On moss on tree bases, rarely directly on bark .... ***Bilimbia sabuletorum* (Schreber) Arnold** (syn. *Myxobilimbia sabuletorum, Bacidia sabuletorum*)

13. Spores 1.5-5.0 µm wide, needle-shaped; apothecia mostly flat 14

14. Hypothecium broad, merging with exciple; spores mostly 45-75 x 1.5-3.0 µm, straight or curved; apothecia 0.6-1.5 mm in diameter with prominent margins; on bark, rarely on mosses .... ***Bacidia schweinitzii* (Fr. *ex* E. Michener)**

14. Hypothecium distinct from exciple; spores mostly 20-50 µm; apothecia less than 1.0 mm in diameter 15

15. On soil, decaying vegetation, and mosses, rarely on bark; pigmented part of hypothecium narrow, distinct from exciple; spores 21-45 x 1.5-3.0 µm, straight; widespread except in southeast ....*Bacidia bagliettoana* (A. Massal. & De Not.) Jatta

15. On bark, usually deciduous trees; hypothecium dark red-brown; exciple dark within, pale at edge; spores 20-50(-64) x 2.0-5.0, straight to very slightly bent; apothecia 0.3-0.7 mm in diameter; western and Great Lakes region .... *Bacidia subincompta* (Nyl.) Arnold

16.(12) Photobiont *Trentepohlia*; apothecial disk very rough and irregular; asci K/I mostly negative or light blue, thick walled, without a conspicuously thickened tip (tholus), sometimes with a dark blue lining on the upper half and (or) a dark blue "ring structure" around the tiny apical chamber; multicellular spores often fragile and breaking into few-celled fragments; paraphyses branched and anastomosing .... *Bactrospora*

16. Photobiont Trebouxioid or chlorococcoid; apothecial disks smooth; asci with a conspicuous tholus, *Bacidia*- or *Lecanora*-type ascus tips; spores not breaking into small fragments; paraphyses unbranched or branched 17

17. Spores long and sinuous, 9- to 29-celled, (34-)45-80(-108) x 2.0-3.5 µm; epihymenium reddish violet to reddish brown, K+ purple; apothecia 0.5-1.0 mm in diameter with even, persistent margins; Great Lakes eastward and Rockies to west coast .... *Bacidia laurocerasi* (Delise ex Duby) Zahlbr.

17. Spores straight or curved, 4- to 8-celled, under 40 µm long; epihymenium usually showing some green pigments, K–; apothecial margins persistent or not 18

18. Spores comet-shaped, usually tapering, 5- to 8-celled; thallus dark green to brownish green, granular; apothecia convex and soon marginless; spores 18-35(-40) x 3-5 µm ; eastern .... ***Scoliciosporum chlorococcum* (Stenh.) Vězda**

18. Spores ellipsoid, rod- or club-shaped, mostly straight and not tapering, 4- to 8-celled; thallus gray-green to pale gray, areolate; apothecia mostly remaining flat with a persistent margin 19

19. Spores 11-37(-45) x 1.6-3.7 µm; Pan-Temperate at low elevations .... *Bacidia circumspecta* (Nyl. *ex* Vainio) Malme

19. Spores more than 3.5 µm wide 20

20. Spores (8-)9-13(-18) x 3.5-5 µm, ellipsoid to fusiform, some bent, 8-16 per ascus
.... *Arthrosporum populorum* A. Massal.

20. Spores 13-24(-28) x 4-5(-6) µm, fusiform, 8 per ascus
.... *Lecania naegelii* (Hepp) Diederich & van den Boom

21.(11) Hypothecium and exciple below the hypothecium light yellow-brown to dark orange-brown or red-brown
.... 22

21. Hypothecium and exciple below the hypothecium colorless, pale yellowish, or pale brown 24

22. Brown pigments in apothecial tissues K–; apothecia not pruinose, often strongly convex to hemispherical; spores (2-)4- to 6(-8)-celled, 18-30(-40) x 5-8 µm; widespread temperate to boreal
............................***Bilimbia sabuletorum* (Schreber) Arnold** (syn. *Myxobilimbia sabuletorum*, "*Bacidia*" *sabuletorum*)

22. Brown pigments in apothecial tissues K+ purple-red; apothecia commonly pruinose at least on margins when young; spores 4- to 12-celled, 31-74 x 2-5 µm; East Temperate 23

23. Thallus continuous, bumpy to warty, not granular; hymenium 12-25% of apothecial height
.... *Bacidia polychroa* (Th. Fr.) Körber

23. Thallus granular, granules 35-100 µm in diameter; hymenium 20-40% of apothecial height
.... *Bacidia diffracta* S. Ekman

24.(21) Spores fusiform, 4(-6)-celled, 4-8 µm wide 25

24. Spores needle-shaped, many-celled, mostly less than 4 µm wide 29

25. Asci K/I–; apothecia 0.1-0.3 mm in diameter, white to pale yellowish, with persistent margins; thallus thin and membranous, often forming a shiny alga-like film; spores 4-celled, 12-15 x 4.5-6.5 µm; on shaded logs and rotting wood, rarely on moss; widespread temperate to boreal, but very inconspicuous
.... *Absconditella lignicola* Vězda & Pišút

25. Asci with a K/I+ blue tholus or tholus structures 26

26. Exciple not evident; paraphyses branched throughout .... *Micarea*

26. Exciple clearly present; paraphyses branched only at tips if branched 27

27. Growing on bark, wood, or peat ....*Lecania naegelii* (Hepp) Diederich & van den Boom (see also couplet 20)

27. Growing on bryophytes, less frequently on tree bases 28

28. Apothecia white or pale gray; spores 15-20 x 4-6 µm
.... *Mycobilimbia carneoalbida* (Müll. Arg.) S. Ekman & Printzen

28. Apothecia dark brown; spores (14-)16-26(-30) x 5.5-8 µm
.... ***Mycobilimbia tetramera* (De Not.) Vitik. et al.**

29.(24) Thallus consisting of large, round granules; apothecia orange-brown, sometimes pruinose on the margins when young; all apothecial tissues negative or colors intensifying with K
.... ***Bacidia rubella* (Hoffm.) A. Massal.**

29. Thallus thin or thick, continuous or cracked, not consisting of large round granules; apothecia pruinose or not; brown pigments in apothecial tissues K+ purple 30

30. Exciple with radiating clusters of crystals; apothecia yellow-brown to purplish brown, rarely red-brown, usually pruinose in part; outermost 4-8 cell layers of exciple very large and distinct from inner cells, which are much narrower; East Temperate .... *Bacidia suffusa* (Fr.) A. Schneider

30. Exciple normally without crystals; apothecia mainly orange-brown, not pruinose or pruinose; only outermost 1-2 cell layers of exciple have enlarged cells 31

31. Brown pigment of the epihymenium deposited as distinct caps on the tips of the paraphyses (seen best when the hymenium is squashed in K); spores 4- to 16-celled, 32-67(-73) x 2.5-4.5 µm; on deciduous trees and shrubs; southeastern coastal plain and along the Pacific coast
.... *Bacidia heterochroa* (Müll. Arg.) Zahlbr.

31. Brown pigment of the epihymenium distributed uniformly in the upper hymenial jelly; spores 8- to 29-celled, (34-)45-80(-108) x 2.0-3.5 µm; on conifers and deciduous trees; Great Lakes to New England, and Rockies to west coast .... *Bacidia laurocerasi* (Delise *ex* Duby) Zahlbr.

BACTROSPORA See Harris (1990), Egea & Torrente (1993), Egea et al. (2004), Ponzetti & McCune (2006).

1. Spores less than 50 µm long, up to 10 cells per spore; common, in southeast
.... *Bactrospora carolinensis* (Ellis & Everh.) R. C. Harris (syn. *B. mesospora*)

1. Spores more than 50 µm long, more than 10 cells per spore 2

2. Exciple open (incomplete) below; spores (45-)50-65(-75) x 3.5-4 µm; asci 55-65(-90?) µm long; mainly in Maritime provinces, but also California .... *Bactrospora brodoi* Egea & Torrente

2. Exciple closed (complete) or open below; spores 57-85 x 3-4 µm; asci 60-100(-135) µm long; western 3

3. Spores very narrow, 2.5-3.0 µm wide; exciple closed at base; Washington, in Cascade Mountains .... *Bactrospora cascadensis* Ponzetti & McCune

3. Spores 3.5-4.0 µm wide; exciple closed or open at base; mainly California, but also B.C. .... *Bactrospora patellarioides* (Nyl.) Almq.

BAEOMYCES (including *Dibaeis*)

1. Apothecia pink; thallus almost white to green 2

1. Apothecia brown; thallus pale green to gray green or brownish 4

2. Apothecia almost spherical, turban-like, on distinct stalks 2-6 mm high; on soil; medulla PD+ bright yellow, K– or K+ yellowish (baeomycesic and squamatic acids) .... ***Dibaeis baeomyces* (L. f.) Rambold & Hertel** (syn. *Baeomyces roseus*)

2. Apothecia flat to slightly convex, sessile or on extremely short, inconspicuous stalks 3

3. Thallus white to greenish; growing on peat, soil, or mosses; medulla PD+ orange, K+ yellow (thamnolic acid); common, mainly boreal to arctic .... ***Icmadophila ericetorum* (L.) Zahlbr.**

3. Thallus green, membranous; on rocks in shaded habitats; medulla PD+ bright yellow, K– or K+ yellowish (baeomycesic and squamatic acids); rare, Ozarks .... *Dibaeis absoluta* (Tuck.) Kalb & Gierl

4.(1) Thallus thick, distinctly lobed at the margin; medulla PD+ orange, K+ yellow (stictic acid) .... ***Baeomyces placophyllus* Ach.**

4. Thallus thin, edge indefinite or definite, not lobed at the margin 5

5. Medulla PD+ orange, K+ yellow (stictic acid); very common and widespread in boreal zone .... ***Baeomyces rufus* (Huds.) Rebent.**

5. Medulla PD+ yellow, K+ red (norstictic acid); rare, arctic-alpine .... *Baeomyces carneus* (Retz.) Flörke

BELLEMEREA (including *Aspilidea*)

1. Medulla K+ red (norstictic acid) 2

1. Medulla K– 3

2. Spores not halonate; ascus tips K/I + uniformly pale blue; arctic-alpine, especially in the west .... *Aspilidea myrinii* (Fr.) Hafellner

2. Spores halonate; ascus tips with a K/I+ dark blue tube structure (*Porpidia*-type); western montane .... ***Bellemerea alpina* (Sommerf.) Clauzade & Cl. Roux**

3. Thallus rusty red; rare, arctic-alpine .... *Bellemerea diamarta* (Ach.) Hafellner & Cl. Roux

3. Thallus pale gray to brownish gray 4

4. Spores (12-)14-21 x 6-13 µm; western arctic-alpine.. *Bellemerea sanguinea* (Kremp.) Hafellner & Cl. Roux

4. Spores (7-)8-16(-19) x (4-)5-9(-10) µm; arctic-alpine to North Temperate .... *Bellemerea cinereorufescens* (Ach.) Clauzade & Cl. Roux

BIATORA (including *Bilimbia, Cliostomum, Catillaria, Frutidella, Micarea (p.p.),* and *Mycobilimbia*) Based on Printzen & Tønsberg (2004), Printzen & Tønsberg (1999), and Printzen (2005). [See also: Notes on similar genera, *Myochroidea, Helocarpon, Japewiella, Xyleborus,* in Printzen, Spribille, & Tønsberg (2008).]

1. Thallus sorediate, often sterile 2

1. Thallus esorediate, fertile 10

2. Thallus and soralia KC–, C– 3

2. Thallus and soralia KC+ orange, C+ orange or KC+ red, C+ rose-red 4

3. Soralia PD+ red (argopsin); soralia gray to greenish gray, coalescing into a partially leprose crust in maturity; on tree bases, twigs, and moss; humid parts of the Pacific Northwest, Great Lakes region, and northeast .... *Biatora efflorescens* (Hedl.) Räsänen

3. Soralia PD– 7

4. Thallus and soralia KC+ orange, usually C+ orange, xanthones present; soralia usually pale green, flat to weakly convex, punctiform or irregular, soon becoming confluent; apothecia grayish yellow to dark brownish gray, often with a dark blue tinge; Appalachians and Maritimes .... *Biatora pontica* Printzen & Tønsberg

4. Thallus and soralia C+ rose-red, gyrophoric acid present, soralia mostly thick, convex and elevated above thallus level, apothecia without dark blue tinge 5

5. Thallus and soralia PD+ orange red (argopsin); soralia small, yellow-green, discrete or coalescing; on bark .... *Biatora printzenii* Tønsberg

5. Thallus and soralia PD– (argopsin absent) 6

6. Soralia mostly rounded, raised, and well delimited; thallus membranous, inconspicuous; apothecial sections C+ rose-red, spores (7-)9–11(-14) × 3–4 µm; on bark, southern Appalachians .... *Biatora appalachensis* Printzen & Tønsberg

6. Soralia mostly confluent, usually poorly delimited; thallus well developed between soralia; apothecial sections C–, spores 11.9–15.6 × 3.8–5.9 µm; on moss over bark or rock, Great Lakes region .... *Biatora chrysantha* (Zahlbr) Printzen

7.(3) Soralia convex and well defined, punctiform, remaining discrete, yellowish green; thallus containing no lichen substances; apothecia rare, pale to dark brown; spores 1(-2)-celled, ellipsoid, 11-15(-17) x 2.5-3.5 (-4.5) µm; on bark, North Temperate .... *Lecania croatica* (Zahlbr.) Kotlov

7. Soralia in irregular, often diffuse soralia, sometimes becoming confluent or leprose; spores 2- to 6-celled 8

8. Spores 4(-6)-celled, 15-25 x 4-7 µm; apothecia red-brown to pale brown; pycnidia rare, pale; usually on bryophytes or mossy bark; mostly boreal, Yukon to Maritimes; containing no lichen substances .... *Mycobilimbia epixanthoides* (Nyl.) Vitik. et al.

8. Spores mainly 2-celled, 8-14(-17) x 2.5-4(-5) µm 9

9. Apothecia pink to pale yellow (containing usnic acid); pycnidia black, common and conspicuous; thallus yellowish white, patchy soralia becoming confluent and leprose; containing atranorin and caperatic acid; Maritime Canada in old-growth forests on bark .... *Cliostomum leprosum* (syn. *C. luteolum*)

9. Apothecia whitish to pale gray or brownish gray; pycnidia white, inconspicuous; thallus consisting of fine, soredia-like, green goniocysts giving it a leprose appearance; containing micareic acid; widespread, boreal to temperate .... *Micarea prasina* Fr. (see couplet 22)

10.(1) Thallus C+ orange, KC+ orange, UV+ white (thiophanic acid, sphaerophorin), granular with corticate granules 0.15-0.2 mm in diameter; apothecia bluish-black, hemispherical; spores 1-celled; on soil and mosses, arctic-alpine .... *Frutidella caesioatra* (Schaerer) Kalb (see *Micarea* key)

10. Thallus and apothecial sections C–, KC– or C+ pink, KC+ pink, without xanthones; apothecia pale to dark brown or almost black, not bluish-black 11

11. Spores predominantly 1-celled, occasionally 2-celled 12

11. Spores predominantly with 2 or more cells 21

12. Hypothecium dark, red-brown to dark brown 13

12. Hypothecium colorless or almost so (slightly yellowish or very pale brown) to yellow-orange.................. 16

13. Asci *Porpidia*-type, with a K/I+ blue tube structure; apothecia typically dark red-brown to black, rarely pale brown 14

13. Asci *Biatora*-type ; apothecia typically pale reddish brown or yellow-brown .... *Biatora subduplex* (form with dark brown hypothecium) (see also couplet 20)

14. Tips of paraphyses expanded to 4.5- 6 µm in diameter; paraphyses thick and septate; clumps of black granules absent from hymenium and hypothecium or extremely rare; spores 10-13(-16) x 3.5-4.5(-5.5) µm, rarely 2-celled; thallus coarsely granular to areolate or continuous, rarely membranous, very pale greenish gray (almost white in old collections); apothecia dark red-brown, with convex apothecia soon losing their disk-colored margins, or infrequently flat with thin black margins .... "*Lecidea*" *berengeriana* (A. Massal.) Nyl. (syn. *Mycobilimbia berengeriana*)

14. Tips of paraphyses not expanded greater than 3-3.5 µm wide; paraphyses slender or indistinct; sparse or abundant clumps of black or purple-black granules sometimes present in hymenium and (or) hypothecium, turning blue or blue-green in K 15

15. Spores ellipsoid, 11-16(-18) x (4.5-)5.0-7.5 µm, with warty wall (visible under oil, appearing granular at 400X), spores occasionally 2-celled; apothecia pale brown to black, typically with persistent black margins that become thin and finally disappear; black, K+ blue-green granules found in most specimens; thallus membranous to well developed and continuous (not granular), dark greenish gray or brownish gray; on moss, bark, and plant remains near or over calcareous soil or rock .... *Bryobilimbia hypnorum* (Lib.) Fryday et al. (syn. *Mycobilimbia hypnorum*)

15. Spores narrowly ellipsoid, (9.0-)10.0-14.0(-16) x 3.5-5.0(-5.5) µm, with smooth cell walls; 2-celled spores extremely rare; apothecia usually dark red-brown, infrequently yellowish brown, with convex apothecia soon losing their disk-colored margins, or sometimes flat with thin black margins, rarely with black, persistent margins; black granules infrequent; thallus finely or coarsely granular, rarely membranous, very pale greenish gray (almost white in old collections); not associated with calcareous habitats .... *Bryobilimbia sanguineoatra* (Wulfen) Fryday et al.

16. Apothecia tiny, 0.15-0.3 mm in diam., white or gray, less frequently pale orange-brown; thallus almost endophloeodal (within the bark); spores (8-)10-12(-17) x 2.0-3.5 µm .... "*Biatora*" *albohyalina* (Nyl.) Bagl. & Carestia (syn. *Lecidea albohyalina*)

16. Apothecia mostly 0.3-0.7(-1.4) mm in diam., light brown to grayish brown; thallus always evident, rugose to verruculose; spores over 3.5 µm wide 17

17. Spores narrowly ellipsoid or bacilliform, usually 4–5 times as long as broad; corticolous on deciduous trees; mainly Appalachian, into the Maritimes .... *Biatora longispora* (Degel.) Lendemer & Printzen

17. Spores narrowly ellipsoid to ellipsoid, 3–4 times as long as broad; on bryophytes, peat, or twigs of shrubs near the ground, or directly on bark of trees 18

18. Thallus medulla PD+ orange red (argopsin) (best seen on thallus sections at 100-200X) 19

18. Thallus PD–, without major secondary substances 20

19. On soil and mosses; thallus whitish, thick, distinctly areolate, areoles 0.25–1 mm in diam.; spores 1(-4)-celled, ellipsoid, (10-)13-17.5(-20) x (3.5-)4.5-5.2(-6) µm; (sub)arctic ....*Biatora cuprea* (Sommerf.) Fr.

19. On bark; thallus thin, continuous to rimose or minutely areolate, areoles 0.15–0.35 mm in diam.; spores 1-celled, narrowly ellipsoid, (8.5-)12.5-15(-19) x (3.0-)3.5-4.5(-5.5); boreal coniferous and mixed coniferous forests .... *Biatora pycnidiata* Printzen & Tønsberg

20. Exciple colorless outside, yellowish, orange- or dark brown within or near the hymenium; spores 9.5–15.0 × 3.5–5.0 µm; thallus whitish gray; (sub)arctic-(sub)alpine ....*Biatora subduplex* (Nyl.) Printzen

20. Exciple and hypothecium colorless to brownish yellow, but without markedly contrasting pigmentation; spores 12.5–19 × 4-6(-7) µm; thallus greenish gray; in woodlands .... ***Biatora vernalis* (L.) Fr.**

21.(11) Spores predominantly 4-celled or more .... *Bacidia* key

21. Spores predominantly 2-celled 22

22. Thallus consisting of fine, soredia-like, green goniocysts giving it a leprose appearance; spores (7-) 8-14 x 2.5-4(-5) µm, ellipsoid; on bark and wood .... *Micarea prasina* Fr.

22. Thallus thin or thick, verruculose to rugose, never with powdery green goniocysts; spores 8-17 x 2-7 µm, ellipsoid to fusiform; on various substrates 23

23. Paraphyses capitate, with pigmented tips 24

23. Paraphyses not conspicuously expanded at tips, not pigmented 25

24. Epihymenium greenish, intensifying with K .... *Biatora globulosa* (Flörke) Fr. (syn. *Catillaria globulosa*)

24. Epihymenium brownish, not changing with K .... see *Catillaria* key

25. Asci K/I–; forming a thin crust on *Sphagnum* moss .... *Absconditella*

25. Asci with a K/I+ blue tholus; not on *Sphagnum* 26

26. Apothecia with persistent, prominent margins; disks lightly pruinose, varying from pale pinkish gray to black; black pycnidia conspicuous; on bark or wood .... ***Cliostomum griffithii* (Sm.) Coppins**

26. Apothecia soon marginless, margins never prominent; disks epruinose, pale to dark brown; pycnidia inconspicuous; on bryophytes, peat, or bark 27

27. Hypothecium colorless; spores 9.5-11(-17) x 3.5-5 µm; apothecia pale brown to reddish brown; on bryophytes, peat, and bark .... *Mycobilimbia pilularis* (Körber) Hafellner & Türk (syn. *Catillaria sphaerioides*)

27. Hypothecium dark brown; spores (10-)12.0-16(-19) x (4.5-)5.0—7.5 µm; apothecia black to very dark brown; on bryophytes .... *Bryobilimbia hypnorum* (Lib.) Fryday et al. (syn. *Mycobilimbia hypnorum*)

BRIGANTIAEA

1. Growing on soil or vegetation, arctic-alpine; spores (50-)74-80(-115) x (24-)34-45(-55) µm; apothecia dark violet with an orange pruina, 0.8-3 mm in diameter; uncommon .... *Brigantiaea fuscolutea* (Dickson) R. Sant.

1. Growing on bark or wood 2

2. Apothecial margins black, disks dark, covered by orange-red pruina; thallus with patches of granular soredia; spores 47-80 x 20-35 µm; apothecia 0.5-1.3 mm in diameter; rare, in northwestern forests .... *Brigantiaea praetermissa* Hafellner & St. Clair

2. Apothecial margins bright orange; thallus typically without soredia; spores 50-80(-105) x 20-30(-40) µm; apothecia bright orange with a yellow pruina, 0.7-1.6(-2) mm in diameter; very common in southeastern coastal plain .... ***Brigantiaea leucoxantha* (Sprengel) R. Sant. & Hafellner**

BRYORIA (including *Bryocaulon, Gowardia, Nodobryoria*, and *Sulcaria*) Based in part on Velmala et al. [2014] and Myllys et al. (2014).

1. Thallus forming erect or prostrate clumps on soil, heath, or sometimes rock (rarely on shrubs); northern boreal to arctic-alpine 2
1. Thallus bushy to pendent, on trees and shrubs, rarely on rock 6

2. Thallus brown with a distinct reddish tint 3
2. Thallus gray to black to very dark brown, not especially reddish 4

3. Pseudocyphellae dot-like, level to slightly raised, white, C+ red (olivetoric acid) .... ***Bryocaulon divergens*** **(Ach.) Kärnefelt**
3. Pseudocyphellae round to elliptical, depressed, white, C– .... *Cetraria* key

4. Thallus dull, gray to yellowish gray at the base with blackened branch tips, occasionally entirely black; pseudocyphellae raised, white; cortex PD+ yellow, KC+ pink (use the filter paper test) .... ***Gowardia nigricans*** **(Ach.) P. Halonen et al.** (syn. *Alectoria nigricans*)
4. Thallus uniform brown to brown black or olivaceous brown; pseudocyphellae, if present, not raised; cortex PD– or PD+ red, KC– 5

5. Thallus erect, mostly shiny, uniformly dark brown to black, not olivaceous, with smooth branches even in diameter; pseudocyphellae level with the surface, brown and very inconspicuous (seen as dull, fusiform areas); soralia absent; outer cortex PD–; medulla and inner cortex PD+ red (fumarprotocetraric acid; diffusing onto filter paper only where the branch is broken) .... ***Bryoria nitidula*** **(Th. Fr.) Brodo & D. Hawksw.**
5. Thallus prostrate and sprawling, brown to brown-black, often olivaceous, main branches typically twisted and flattened at the axils with deep depressions and grooves; pseudocyphellae absent; soralia sparse, sometimes absent, PD+ red, but cortex is PD– .... ***Bryoria fuscescens*** **(Gyeln.) Brodo & D. Hawksw.** *s. lat.* (syn. *B. chalybeiformis*)

6.(1) Thallus forming rounded or irregular bushy tufts or clumps with divergent branching 7
6. Thallus pendent or almost pendent when mature 11

7. Branches without soredia or isidia 8
7. Branches with scattered soralia 9

8. Thallus red-brown; apothecia shiny red-brown, decorated with spiny cilia on the margins; epihymenium K– .... ***Nodobryoria abbreviata* (Müll. Arg.) Common & Brodo**

8. Thallus brown at base, greenish black at tips; apothecia greenish black; epihymenium K+ purple .... ***Kaernefeltia californica* (Tuck.) Thell & Goward**

9. Thallus and soralia PD–; soralia round, often greenish, wider than the branches, northern boreal, often mixed with *Bryoria lanestris* .... *Bryoria simplicior* (Vainio) Brodo & D. Hawksw.

9. Thallus and (or) soralia PD+ orange-yellow or red 10

10. Thallus shiny, uniformly brown, with fissural soralia that contain tiny, thorn-like isidia; outer cortex and medulla PD+ red, K–, KC– (fumarprotocetraric acid) .... ***Bryoria furcellata* (Fr.) Brodo & D. Hawksw.**

10. Thallus dull, gray to pale or dark brown, with round, warty soralia that contain no isidia; cortex PD+ orange-yellow, K+ bright yellow, KC+ pink (alectorialic acid) .... ***Bryoria nadvornikiana* (Gyelnik) Brodo & D. Hawksw.**

11.(6) Branches with soralia 12

11. Branches without soralia 16

12. Soredia bright yellow; thallus with some stout main branches and slender, perpendicular side branches; medulla and soralia PD– .... ***Bryoria fremontii* (Tuck.) Brodo & D. Hawksw.** (syn. *B. tortuosa*)

12. Soredia white to pale green (sometimes flecked with black) 13

13. Pseudocyphellae present, rather long, but sometimes hard to see; cortex PD+ orange-yellow, K+ yellow, KC+ pink (alectorialic acid); main branches with short or long, perpendicular side branches .... ***Bryoria nadvornikiana* (Gyelnik) Brodo & D. Hawksw.**

13. Pseudocyphellae absent or rare (but see couplet 15); cortex PD– or PD+ red; soredia PD+ red (fumarprotocetraric acid); perpendicular side branches present or absent 14

14. Thallus very dark brown, almost black; branches very slender (less than 0.2 mm in diameter), brittle; soralia typically abundant, often broader than the branch; cortex PD– .... ***Bryoria lanestris* (Ach.) Brodo & D. Hawksw.** (Regarded as conspecific with *B. fuscescens* by Velmala et al. [2014].)

14. Thallus dark to medium brown to olive; branches 0.2-0.6 mm in diameter, not very brittle; soralia relatively sparse, usually not broader than the branch; cortex PD+ red or PD– 15

15. Thallus dark to medium brown, typically dull, often paler at the base; angles between the branches (axils) acute, not rounded; soralia tuberculate or fissural; cortex PD+ red or PD–
.... ***Bryoria fuscescens*** **(Gyelnik) Brodo & D. Hawksw.**
[Note: Infrequent specimens with pale fissural pseudocyphellae may be *B. vrangiana* (Gyeln.) Brodo & D. Hawksw.]

15. Thallus olive to olive-brown, shiny, not paler at base; branch axils broad and rounded; soralia always fissural; cortex PD– .... ***Bryoria glabra*** **(Motyka) Brodo & D. Hawksw.**

16.(11) Branches with deep, longitudinal grooves (special pseudocyphellae), often quite long; branches commonly twisted; cortex K+ yellow, PD+ yellowish or brownish, KC– or KC+ yellow (atranorin); very rare, California and Oregon ....***Sulcaria badia*** **Brodo & D. Hawksw.**

16. Branches without deep grooves, with or without pseudocyphellae 17

17. Pseudocyphellae absent; medulla PD– 18

17. Pseudocyphellae present, usually conspicuous; cortex and medulla PD+ red or yellow, or PD– 19

18. Thallus uniformly red-brown, with slender, grooved and pitted, brittle branches 0.1-0.2 mm in diameter; apothecia fairly common, with red-brown, non-pruinose disks; cortex with jigsaw puzzle-shaped cells as viewed at 100-400X magnification ....***Nodobryoria oregana*** **(Tuck.) Common & Brodo**

18. Thallus dark reddish brown to yellowish brown; thallus with stout, mostly smooth twisted main branches (dented only at the axils), 0.4-1.5(-4) mm in diameter, and slender, perpendicular side branches; apothecia uncommon, with yellow-pruinose disks; cortex with straight, long, parallel cells
.... ***Bryoria fremontii*** **(Tuck.) Brodo & D. Hawksw.**

19. Pseudocyphellae yellow, twisting around the branch in spirals; thallus reddish brown or often yellowish (vulpinic acid) .... ***Bryoria fremontii*** **(Tuck.) Brodo & D. Hawksw.** (syn. *B. tortuosa*)

19. Pseudocyphellae white or pale brown, dot-like, fissural, or long and twisting; thallus never yellowish 20

20. Pseudocyphellae often 2-4 mm long, twisting around the branches; perpendicular lateral branches common; thallus usually variegated, dull, pale reddish brown; rare, coastal, northern California to Oregon 21

20. Pseudocyphellae dot-like to fissural, under 1.5 mm long, ± lateral perpendicular branches 22

21. Thallus cortex PD+ yellow, K–, KC+ pink (alectorialic and barbatolic acids)
...............*Sulcaria spiralifera* (Brodo & D. Hawksw.) Myllys et al., Chemotype 1 (syn. *Bryoria pseudocapillaris*)

21. Thallus cortex PD+ yellow, K+ reddish (norstictic acid and atranorin) ....*Sulcaria spiralifera* (Brodo & D. Hawksw.) Myllys et al., Chemotype 2 (syn. *Bryoria spiralifera*)

22. Thallus not brittle, pale to dark brown or red-brown; outer cortex PD– 23

22. Thallus usually brittle, pale gray to gray brown or dark brown; outer cortex and sometimes medulla PD+ yellow, or PD– 25

23. Thallus distinctly red-brown; pseudocyphellae dot-like, slightly raised; inner cortex and medulla PD–, C+ red (olivetoric acid); rare, coastal Pacific Northwest ....*Bryocaulon pseudosatoanum* (Asahina) Kärnefelt

23. Thallus brown, without a reddish tint; pseudocyphellae elongate (fusiform or linear); inner cortex and medulla usually PD+ red (fumarprotocetraric acid; seen through the cortex rendered transparent in the filter paper test), C–; common 24

24. Pseudocyphellae conspicuous, pale, slightly raised; branching in irregular dichotomies; branches uneven in diameter; Appalachian-Great Lakes and Pacific Northwest .... *Bryoria trichodes* (Michaux) Brodo & D. Hawksw. (syn. *B. trichodes* subsp. *trichodes*)

24. Pseudocyphellae dark, slightly depressed, often inconspicuous; branching in even dichotomies; branches very uniform in diameter; along both coasts, south to California in the west ....***Bryoria americana* (Motyka) Holein** (syn. *B. trichodes* subsp. *americana*)

25.(22) Cortex PD–, C+ pink (gyrophoric acid), often with a pale olive tinge; frequent on west coast, rare on east coast ....*Bryoria friabilis* Brodo & D. Hawksw.

25. Cortex PD+ yellow, C+ pink, or C– (gyrophoric acid absent) 26

26. Cortex K–, KC– (psoromic acid); thallus pale to dark brown or olive-brown; widespread boreal to North Temperate, but uncommon .... *Bryoria kockiana* Velmala et al. (= N. Am. records of *B. implexa s. str.*)

26 Cortex K+ yellow or reddish brown 27

27. Cortex K+ bright yellow, KC+ red (alectorialic and barbatolic acids); thallus usually pale gray to brownish gray; mostly in coastal or humid localities ....***Bryoria pikei* Brodo & D. Hawksw.** (= N. Am. records of *B. capillaris*)

27. Cortex K+ brownish to red, KC– 28

28. Pacific Northwest to Rocky Mountains; containing norstictic acid; thallus usually dark brown, less frequently pale; common .... ***Bryoria pseudofuscescens* (Gyelnik) Brodo & D. Hawksw.**

28. Maritime provinces of northeast; containing salazinic acid; thallus often with olive tinge; very rare ....*Bryoria salazinica* Brodo & D. Hawksw.

BUELLIA (including *Amandinea, Catolechia, Diploicia, Diplotomma,* and *Tetramelas*) Based in part on Harris (1995), Marbach (2000), and Giralt and Clerc (2011).

[Note: The classification of some species of *Amandinea* as given below is disputed by Bungartz et al. (2007), who say they probably do not belong in the same genus as *A. coniops*, the type of the genus.]

1. Spores 4-celled to muriform 2

1. Spores 2-celled 8

2. Spores 4-celled (rarely with one subdivided cell), hymenium without oil drops; on bark, wood, or rock 3

2. Spores muriform to few-celled muriform 5

3. Thallus very pale to dark yellow, C+ orange (arthothelin); spores 19-28 x 7.5-10.5 µm; on conifer bark or wood, western subalpine .....*Tetramelas triphragmoides* (Anzi) A. Nordin & Tibell (syn. *Buellia triphragmoides*)

3. Thallus white to pale gray, C– 4

4. On calcareous rock; apothecia convex when mature; spores mostly 16-20 x 6.5-8.5 µm; mainly central plains, scattered elsewhere .... *Diplotomma venustum* Körber (= "*D. epipolium*" of most authors)

4. On bark or wood; apothecia usually flat; spores mostly 14-21 x 5.5-9.0 µm; Oregon, California, and Arizona .... *Buellia triseptata* A. Nordin

5.(2) Apothecia pruinose, convex; hymenium lacking oil drops; spores few-celled muriform, 4- to 6(-8)-celled, (11-)15-20(-30) x (5.5-)8-10(-17) µm; on rocks, bark, or wood; thallus C–, K–; widespread temperate .... *Diplotomma alboatrum* (Hoffm.) Flotow (syn. *Buellia alboatra*)

5. Apothecia without pruina, usually flat; hymenium filled with tiny oil drops; on wood and bark; western 6

6. Thallus pale yellow, C+ orange, K– (diploicin); spores many-celled muriform, 40-50(-65) X 15-21 µm; along the west coast .... *Buellia oidalea* (Tuck.) Tuck.

6. Thallus gray to white, C–, K+ yellow (atranorin) 7

7. Spores 19-24 x (9-)10-13(-15) µm, fewer than 12 cells per spore; western mountains

.... *Buellia penichra* (Tuck.) Hasse (syn. *Diplotomma penichrum*)

7. Spores (21-)25-33(-37) x 11.5-14(-15) µm, more than 12 cells per spore; mainly along the west coast at low elevations, California to B.C. .... *Buellia muriformis* A. Nordin & Tønsberg

8.(1) Thallus yellow or with a yellow tint 9

8. Thallus white to pale gray to greenish gray or brown 11

9. Thallus squamulose, bright lemon-yellow; cortex C–, KC– (containing pulvinic acid-type pigments); on soil and peat; spores (12-)13-17(-18) x 7-10 µm; rare but widespread arctic-alpine .... *Catolechia wahlenbergii* (Ach.) Körber

9. Thallus crustose, pale or greenish yellow; cortex C+ orange, KC+ orange, and (or) UV+ yellow or orange (xanthones) 10

10. On maritime rocks, California coast; spores 11-17 x 6-10 µm .... ***Buellia halonia* (Ach.) Tuck.**

10. On wood or bark, widespread in the west; spores 16-28 x 8-10 .... *Tetramelas chloroleucus* (Körber) A. Nordin (syn. *Buellia chloroleuca*; = *B. zahlbruckneri* of many authors) (see couplet 29)

11.(8) Thallus brown, thick or thin 12

11. Thallus white to pale gray or greenish gray, or disappearing 14

12. Thallus rimose-areolate to verrucose, apothecia superficial or somewhat sunken, black to dark brown; spores 13.5-16(-18) x -7.0-9.5 µm, septum thickened in young spores; on coastal rocks, B.C. to Alaska and Newfoundland .... *Amandinea coniops* (Wahlenb.) M. Choisy *ex* Scheid. & H. Mayrhofer

12. Thallus areolate to subsquamulose; apothecia black; spores 9-15 x 5-8 µm 13

13. Prothallus black and conspicuous around and between areoles; apothecia sunken between areoles, sometimes emergent and adnate; mostly southwestern interior, montane, on siliceous rock, never parasitic .... *Buellia tyrolensis* Körber (syn. *B. novomexicana*)

13. Prothallus absent, thallus thick, becoming subsquamulose; apothecia adnate, not sunken; widespread, especially in central and western U.S.; on siliceous rock or weathered wood; initially parasitic on a variety of lichens, later independent .... *Buellia badia* (Fr.) A. Massal.

14.(11) Thallus thick, distinctly lobed 15

14. Thallus thick or thin to inconspicuous, not at all lobed 16

15. Growing on bark, wood, or rock, California, mostly coastal; spores with angular locules (*Physcia*-type); thallus pale gray, lobes short, pruinose, and sorediate .... ***Diploicia canescens* (Dickson) A. Massal.**

15. Growing on soil, prairies and scattered arctic localities; spores with uniformly thickened walls; thallus white, lobes elongate, scabrose, but not sorediate .... *Buellia elegans* Poelt

16. Mature apothecia entirely, or almost entirely immersed in the thallus, the disk level with the thallus surface or slightly depressed; thallus continuous or rimose-areolate to areolate, often with a black prothallus; on non-calcareous rock 17

16. Mature apothecia superficial, not sunken into thallus; thallus well developed or thin and disappearing, lacking a black prothallus; on bark, wood, rock, or soil and peat 19

17. Exciple brown, without green pigments; thallus continuous in patches, becoming rimose or rimose-areolate, K+ red (norstictic and connorstictic acids); prothallus thin, only at thallus margin; eastern .... *Buellia maculata* Bungartz (syn. *Buellia stigmaea* Tuck., *non Buellia stigmaea* Körber)

17. Exciple containing green pigments (HNO_3+ red-violet) 18

18. Thallus rimose-areolate to areolate, areoles often more than 0.4 mm across, yellowish gray; prothallus often not developing or thin; medulla K+ yellow to orange or red (stictic or norstictic acid), IKI+ blue; mostly southern U.S., North Carolina to California .... ***Buellia spuria* (Schaer.) Anzi**

18. Thallus minutely areolate, areoles less than 0.4 mm across, white or pale gray; black prothallus well developed between and around areoles and thallus; medulla K– (2'-O-methylperlatolic acid and [or] confluentic acid), IKI–; along Pacific coast, rare in east .... *Buellia stellulata* (Taylor) Mudd

19.(16) On soil, mosses, and peat, arctic-alpine; thallus thick, verrucose to subsquamulose, K+ yellow (atranorin); spores 15-25 x 7-10 µm .... *Tetramelas papillatus* (Sommerf.) Kalb (syn. *Buellia papillata*)

19. On rock, bark, or wood 20

20. On siliceous rock 21

20. Mostly on bark or wood, occasionally on rock 22

21. Thallus composed of scattered or contiguous areoles, sometimes white pruinose; spores (9-)12-16(-19) x (4-)6.5-9 µm; California and the southern Rockies .... *Buellia dispersa* A. Massal. (syn. *B. retrovertens*)

21. Thallus almost endolithic, inconspicuous; spores narrow, 10-14 x (3-)4-6; central to southwestern U.S. .... *Buellia abstracta* (Nyl.) H. Olivier [= *B. sequax sensu* Bungartz et al. 2007]

22. Spore walls rough and sculptured, obvious under 400X magnification; spores broad, 15-18 x 10-12 µm; thallus UV– (lacking lichen substances); southeast .... *Amandinea langloisii* Imshaug *ex* Marbach

22. Spore walls appearing smooth under 400X magnification 23

23. Spores mostly less than 16 µm long 24

23. Spores mostly 16-32 µm long 27

24. Spores 12-24(-32) per ascus, 8.5-12(-14) x 4.0-5.5 µm, exciple pale within and black at the outer edge; East Temperate, common on deciduous trees .... *Amandinea polyspora* (Willey) E. Lay & P. May

24. Spores 8 per ascus 25

25. Apothecial sections and thallus K+ deep yellow to red (norstictic acid); apothecia 0.5-1.0 mm in diameter; spores (10.5-)12-17 x 5-8 µm; East Temperate and Pacific Northwest coast .... **Buellia stillingiana J. Steiner**

[Note: This species was called *Buellia erubescens* Arnold by Bungartz et al.(2007). However, the species called "*Buellia erubescens*" by authors in the past (and on page 187 of LNA) is actually a different species, *Tetramelas chloroleucus* (Körber) A. Nordin (Bungartz et al. 2007) [see couplet 29]. To avoid further confusion, the well-understood name *B. stillingiana* is being used here as it has by other recent authors (e.g., Lendemer et al. 2013).]

25. Apothecial sections and thallus K– (norstictic acid absent); apothecia usually less than 0.5 mm in diameter 26

26. Spores (7-)11-16 x (4-)5-8 µm; very common and widespread on wood, but also sometimes on bark, rock or plant detritus .... ***Amandinea punctata*** **(Hoffm.) Coppins & Scheid.**

26. Spores (5.5-)7.0-9.5(-10.0) x 2.5-3.5(-4.5) µm; frequent in northeast, usually on conifer bark and wood, but also on maples .... *Buellia schaereri* De Not.

27.(23) Apothecia tiny, 0.25-0.4 mm in diameter; thallus PD+ red, K– or + brownish (fumarprotocetraric acid); spores 18-30 x 8-12 µm; frequently on hemlock; eastern temperate and California .... *Buellia dialyta* (Nyl.) Tuck. (syn. *Chrismofulvea dialyta*)

27. Apothecia more than 0.3 mm in diameter; thallus PD–, K- or PD+ yellow, K+ red 28

28. Hymenium lacking oil droplets; spores with or without a thickened septum 29

28. Hymenium containing abundant oil droplets; thallus K–, K+ yellow or K+ red; spores often with a thickened septum 31

29. Thallus and apothecial sections K–, usually C+ orange or KC+ orange (xanthones); spores 16-28 x 8-10 µm, often pointed and slightly curved with a thickened septum when young; western, from Arizona and California to the arctic
.... *Tetramelas chloroleucus* (Körber) A. Nordin (syn. *Buellia chloroleuca*; = *B. zahlbruckneri* of many authors)

29. Thallus or apothecial sections K+ yellow changing to blood red (norstictic acid), C–, KC– 30

30. Spores with a thickened septum, 18-22 x 7-8.5(-10) µm; southeastern U.S. and Baja California, infrequent except in Florida .... *Buellia imshaugiana* R.C. Harris

30. Spores lacking a thickened septum, 16-23 x 6-9 µm; southeastern, common
.... *Buellia curtisii* (Tuck.) Imshaug

31. Epihymenium greenish above, K+ purplish; thallus K+ red (norstictic acid); spores 14-22 x 6-8 µm; common in Florida .... *Buellia bahiana* Malme (syn. *Hafellia bahiana*)

31. Epihymenium brownish above, K– or K+ red 32

32. Spores mostly 18-26 x 6-13 µm; thallus K+ yellow or K– (atranorin, no norstictic acid); especially boreal regions, scattered in west.......... ***Buellia disciformis* (Fr.) Mudd** (syn. *Hafellia disciformis*)

32. Spores 22-32 x (9-)11-15 µm 33

33. Thallus K+ yellowish or K– (lacking norstictic acid); Great Lakes region east to New Brunswick
.... *Buellia arnoldii* Servit (syn. *Hafellia arnoldii*)

33. Thallus K+ red (norstictic acid); southeastern
.... *Buellia callispora* (C. Knight) J. Steiner (syn. *Hafellia callispora*)

BULBOTHRIX (including *Relicina*) Based in part on Benneti and Elix (2012).

[Note: All these species are found in the southeastern coastal plain.]

1. Thallus yellowish green ("usnic–yellow"), containing usnic acid; lower surface black 2

1. Thallus gray, lacking usnic acid; lower surface brown or black 3

2. Isidia present on the thallus surface; apothecia rare; medulla K+ red, PD– (norstictic acid); on rock or bark, coastal plain .... *Relicina abstrusa* (Hale & Kurok.) Hale

2. Isidia absent; apothecia abundant; medulla K– or brownish, PD+ red (fumarprotocetraric acid); southern Florida ....*Relicina eximbricata* (Gyelnik) Hale

3. Isidia absent; apothecia common; lower surface dark brown to black ....**_Bulbothrix confoederata_ (W. L. Culb.) Hale**

3. Isidia abundant on the thallus surface 4

4. Lower surface and rhizines mostly black, but can be brown at edge; apothecia sometimes abundant; medulla C+ red (lecanoric acid present as main compound) .... **_Bulbothrix laevigatula_ (Nyl.) Hale**

4. Lower surface and rhizines pale beige to brown throughout; apothecia rare; medulla C+ pink or C–, K+ red or K– 5

5. Medulla C+ pink, K– (gyrophoric acid present as main compound); lobes under 2 mm wide, more or less truncate .... **_Bulbothrix scortella_ (Nyl.) Hale** [= most N. Am. records of *B. goebelii* and Plate 151 in LNA.]

5. Medulla C–, K+ red (salazinic acid); lobes (1-)2-4 mm wide, rounded, often crenate ....*Bulbothrix isidiza* (Nyl.) Hale

BYSSOLOMA Based in part on Lücking (2008).

1. Thallus UV+ yellow (xanthones); hypothecium colorless to light brown; exclusively on bark ....**_Byssoloma meadii_ (Tuck. _ex_ Willey) S. Ekman**

1. Thallus UV–; hypothecium dark brown or purplish; on leaves or bark 2

2. Apothecia pitch black; epihymenium blackish brown to purplish gray or blue-green; spores 10-17 x 3-5 µm, length to width ratio 3-3.5:1 .... *Byssoloma subdiscordans* (Nyl.) P. James

2. Apothecia pale yellowish to dark purplish brown or blue-gray; epihymenium colorless, indistinct; spores 10-18 x 2.5-3.5 µm, length to width ratio 4-5:1 .... *Byssoloma leucoblepharum* (Nyl.) Vainio

CALICIUM (including *Chaenotheca, Chaenothecopsis, Microcalicium, Mycocalicium, Phaeocalicium, Sphinctrina*, and *Stenocybe*) See also Tibell (1999).

1. Spores ellipsoid to fusiform, 1- to 4-celled, brown to greenish 2

1. Spores spherical, 1-celled, colorless to brown 19

2. Spores (2-)3- to 4-celled; thallus not lichenized 3

2. Spores 1- to 2-celled; thallus lichenized or not 4

3. Stalks unbranched, 0.8-1.5 mm tall; spores 18-36 x 5-8.5(-11) µm; common on the trunks and branches of fir (*Abies*) .... *Stenocybe major* (Nyl.) Körber

3. Stalks commonly branched, under 0.7 mm tall; spores 10-17(-20) x 4-6 µm; common on alder (*Alnus*) .... *Stenocybe pullatula* (Ach.) Stein

4. Spores 1-celled 5

4. Spores 2-celled 7

5. Thallus granular, bright yellow to yellow-green; spores broad to subglobose, 4.0-5.5(-9) x 3.5-5.2 µm, pale to dark brown; capitulum and upper part of stalks yellow pruinose; on bark or wood in boreal region .... *Chaenotheca chrysocephala* (Ach.) Th. Fr.

5. Thallus not evident, not lichenized; spores ellipsoid, dark brown 6

6. On old snags in forests; capitula cylindrical, not flattened; spores 6.0-8.0(-10.5) x 3.0-4.0 µm .... *Mycocalicium subtile* (Pers.) Szatala

6. On alder twigs and branches; capitula strongly flattened and fan-like; spores 10-15 x 3.7-5.7 µm .... *Phaeocalicium compressulum* (Szatala) A.F.W. Schmidt

7.(4) Spores narrow, 5.5-9 x 2-2.5 µm; capitula without pruina, black or greenish 8

7. Spores broader than 3.5 µm; capitula with or without pruina 9

8. Spores brown, (5.5-)7-9 µm long; thallus saprophytic on wood, not lichenized; widespread boreal .... *Chaenothecopsis debilis* (Turner & Borrer *ex* Sm.) Tibell

8. Spores blue-green, 5-7 µm long; parasitic on lichens (*Psilolechia lucida*); mainly northeastern .... *Microcalicium arenarium* (Hampe *ex* A.Massal.) Tibell

9. Thallus not lichenized; parasitic on living fungi, shrubs, or trees; thallus not visible; asci remaining intact when mature; spores 10-15 x 3.5-6 µm, smooth 10

9. Thallus lichenized, usually superficial although sometimes within the substrate; on wood or bark; asci disintegrating leaving the spores in a powdery mass; spores ornamented with ridges or warts; usually with algae in the thallus 12

10. Growing on the upper surface of small polypore bracket fungi, *Trichaptum biforme* (syn. *Hirschioporus pargamenus*), usually toward the outer edge .... *Phaeocalicium polyporaeum* (Nyl.) Tibell

10. Growing on the bark of shrubs or trees 11

11. On living or dead twigs and branches of poplars .... *Phaeocalicium populneum* (Brond. *ex* Duby) A. F. W. Schmidt

11. On the 2- or 3-year-old branches of staghorn sumac .... *Phaeocalicium curtisii* (Tuck.) Tibell

12.(9) Thallus yellowish green, granular; capitulum margin and base without pruina .... ***Calicium viride*** **Pers.**

12. Thallus whitish, gray, or yellowish white, or within substrate and almost absent from view, sometimes forming a whitish stain 13

13. Capitulum with a yellow pruina 14

13. Capitulum with a white or rusty pruina, or not pruinose 15

14. Yellow pruina on the margin and base of capitulum; thallus indistinct, within substrate; common in Great Lakes region, scattered elsewhere .... ***Calicium trabinellum*** **(Ach.) Ach.**

14. Yellow pruina on the upper surface of the capitulum; thallus gray to white, well developed; rare, western .... *Calicium adspersum* Pers.

15. Capitulum brown to rusty brown, almost globose when young; spores (7-)8-12 x 4-6 (-7) µm .... *Calicium salicinum* Pers.

15. Capitulum black, but sometimes white pruinose on the rim or underside, usually flaring gradually from the stalk 16

16. Spores (9-)11-13(-18) x (4.5-)6-7.5(-9.5) µm; outer layer of stalk IKI+ deep blue; thallus yellowish white, granular to leprose or disappearing; stalks thick, unbranched capitulum often with a white pruinose rim or disk; common along the coast in the Pacific Northwest, occasional elsewhere ...... *Calicium lenticulare* Ach.

16. Spores 7-12 x 3.5-6 µm; outer layer of stalk IKI–; thallus areolate to granular, or within the wood and almost imperceptible; stalks thick or slender; capitulum sometimes white pruinose below; widespread boreal species 17

17. Spores 3.5-5.0 µm wide, well constricted; asci club-shaped ....*Calicium parvum* Tibell

17. Spores 5.0-7.0 µm wide, barely or not constricted; asci cylindrical 18

18. Spores 12-15 µm long; capitulum lacking pruina .... *Calicium abietinum* Pers.

18. Spores 9-13 µm long; typically white pruinose on the edge and lower side of capitulum .... *Calicium glaucellum* Ach.

19.(1) Parasitic on lichens; stalk extremely short or absent (not much longer than the capitulum, or shorter), light to dark brown; spores almost spherical, over 5 µm in diameter 20

19. Not parasitic on lichens; stalks much longer than capitulum, black or almost black; spores less than 5 µm in diameter 21

20. Parasitic on species of *Pertusaria*; spores 5-7 µm in diameter; widespread .... *Sphinctrina turbinata* (Pers.:Fr.) De Not.

20. Parasitic on minutely granular thallus of *Protoparmelia hypotremella*; spores 6-10 µm in diameter; Great Lakes region to the northeast, rare west of Great Lakes .... *Sphinctrina anglica* Nyl.

21. Stalk and capitulum covered with yellow or yellow-green pruina; thallus leprose, with bright yellow-green powdery soredia; frequently on soil over roots of upturned trees .... ***Chaenotheca furfuracea* (L.) Tibell**
[Compare with *Chaenotheca chrysocephala*; see couplet 5]

21. Stalk and capitulum without pruina, brown to black; thallus type various 22

22. Thallus within substrate, absent from view, or sometimes consisting of a very thin greenish powder; spores 3.3-4.5 µm in diameter .... ***Chaenotheca brunneola* (Ach.) Müll. Arg.**

22. Thallus well developed, verrucose to entirely sorediate, gray to greenish or yellow-green, with scattered yellow to orange patches that turn K+ pink to red; spores 5-7 µm in diameter .... *Chaenotheca ferruginea* (Turner *ex* Sm.) Mig.

CALOPADIA Based in part on Lücking (2008).

[Note: All species are from southeastern coastal plain, especially Florida.]

1. Apothecia black; spores 60-80 µm long; hypothecium dark brown, K+ purplish; exclusively on bark .... *Calopadia lecanorella (*Nyl.) Kalb & Vězda

1. Apothecia brown; on leaves or bark 2

2. Apothecia reddish or orange-brown; hypothecium pale brown; spores 60-85 µm long .... ***Calopadia fusca* (Müll. Arg.) Vězda**

2. Apothecia grayish brown, without reddish tinge; hypothecium dark brown to greenish; spores 55-85 µm long .... *Calopadia puiggarii* (Müll. Arg.) Vězda

CALOPLACA

[Note: Arup et al. (2013) have proposed a revised taxonomy for *Caloplaca* based on genetic studies. Since only some of the species in this key were included in the study, the new combinations made in that paper are not adopted here but are included as synonyms, with their authors, pending further studies. Species with an asterisk (*) are retained in *Caloplaca* in the strict sense.]

1. Thallus subfruticose, composed of a tangle of cylindrical branches 2
1. Thallus crustose, or squamulose 3

2. Branches less than 0.3 mm wide, brittle and delicate, entirely orange, not forming verrucose clumps; medulla IKI–; maritime rocks, California to B.C. .... ***Caloplaca coralloides* (Tuck.) Hulting** (syn. *Polycauliona coralloides* (Tuck.) Hue)
2. Branched up to 0.5 mm wide, tough and woody, dark gray or brown at base, forming clumps that appear verrucose from above, with globose verrucae at the tips of the branches; medulla IKI+ blue; on soil, interior western montane, not maritime .... *Caloplaca cladodes* (Tuck.) Zahlbr. (syn. *Pachypeltis cladodes* (Tuck.) Søchting et al.)

3. Thallus within substrate and absent from view, or very thin and indistinct 4
3. Thallus clearly visible and well developed 20

4. Parasitic; apothecia deep rusty orange .... 5 (See also couplet 70)
4. Not parasitic 6

5. Apothecial margin the same color as the disk; growing on a variety of saxicolous lichens; widespread arctic-alpine .... ***Caloplaca epithallina* Lynge**
5. Apothecial margin brown to black; growing on *Candelariella*; central plains and into the southwest and western Great Lakes region .... *Caloplaca grimmiae* (Nyl.) H. Olivier

6. Growing on bark, wood, moss, plant remains or humus 7
6. Growing on rock 15

7. Growing on bark, wood 8
7. Growing on moss or humus .... see couplet 42

8. In maritime habitats along the west coast, especially on beach logs
.... ***Caloplaca inconspecta* Arup** (syn. *Polycauliona inconspecta* (Arup) Arup et al.)

8. In inland localities throughout the continent, on bark and wood 9

9. Apothecial margin (including base) containing no algae, essentially biatorine, ranging from yellow to "waxy" gray to almost black; north temperate to subarctic ....*Caloplaca borealis* (Vainio) Poelt

9. Apothecial margin lecanorine, containing some algae, at least at base (an expanded proper exciple [parathecium] sometimes becomes well developed laterally and pushes the thalline amphithecium to the base); widely distributed 10

10. Apothecia rusty red-orange; spores 12-18(-20) x 6-10(-11) µm
.... *Caloplaca ferruginea* (Hudson) Th. Fr. (syn. *Blastenia ferruginea* (Hudson) A. Massal.)

10. Apothecia dark orange to yellowish orange; spores 9-13 x 4-7 µm 11

11. Spore septum usually more than 1/3 the length of the spore, 3.5-5.5(-6.5) µm 12

11. Spore septum about 1/3 the length of the spore or less, 2.8-4.5 µm 13

12. Thallus gray to yellowish, often consisting of small, scattered, yellow to orange areoles; apothecia orange, usually thick, and often with a double margin: an inner, disk-colored ring surrounded by a thick, thallus-colored (gray to yellow) outer margin; very common on poplar trees and other deciduous trees, sometimes wood ....*Caloplaca pyracea* (Ach.) Th. Fr. (syn. *Athallia pyracea* (Ach.) Arup et al.)

12. Thallus entirely within the substrate; apothecial disks orange-yellow to orange; infrequent on wood or bark (mostly on rock)
.... ***Caloplaca holocarpa* (Hoffm. *ex* Ach.) A. E. Wade** (syn. *Athallia holocarpa* (Hoffm.) Arup et al.)

13. On wood or sometimes conifer bark or rock; thallus within the substrate; apothecia 0.4-1.6 mm in diameter; disk dark brownish orange, with a prominent paler margin; arctic-alpine or on exposed coasts
.... *Caloplaca fraudans* (Th. Fr.) H. Olivier

13. On bark or sometimes wood; thallus dark gray, thin and inconspicuous, occasionally producing concave soralia with gray soredia; apothecia yellow to yellow-orange, or if dark, then with a gray margin 14

14. Apothecial margin orange; north-central to northeastern; on hardwood trees ....*Caloplaca ahtii* Søchting

14. Apothecial margin gray, thin or thick; southwest to north central; mostly on conifers

.... *Caloplaca pinicola* H. Magn.

15.(6) Apothecia biatorine; hymenium containing abundant oil drops

.... ***Caloplaca luteominia* (Tuck.) Zahlbr.** (syn. *Polycauliona luteominia* (Tuck.) Arup et al.)

15. Apothecia lecanorine; hymenium clear, without oil drops 16

16. On shoreline rocks along the west coast; apothecial disks yellow orange; spores with a broad septum (over 2.5 µm) .... ***Caloplaca inconspecta* Arup** (syn. *Polycauliona inconspecta* (Arup) Arup et al.)

16. On non-maritime rocks, eastern or western 17

17. Spore septum more than 1/3 the length of spores (usually over 3.0 µm); apothecial disks bright orange

.... ***Caloplaca holocarpa* (Hoffm. *ex* Ach.) A. E. Wade** (syn. *Athallia holocarpa* (Hoffm.) Arup et al.)

[Note: Forms with a very thin yellowish thallus can be called *C. vitellinula* (Nyl.) H. Olivier.]

17. Spores with a very narrow septum (under 2.5 µm); apothecial disks usually dark orange or brownish orange

.... 18

18. On calcareous rock and concrete; apothecial margin even with disk; spores 13.5-18 x (5-)6-8 µm; epihymenium C-; east of the Mississippi

.... ***Caloplaca feracissima* H. Magn.** (syn. *Xanthocarpia feracissima* (H. Magn.) Frödén et al.)

[Note: See also *C. crenulatella* (Nyl.) H. Olivier (syn. *Xanthocarpia crenulatella* (Nyl.) Frödén et al.); west of the Mississippi? Still under study.]

18. On non-calcareous rock such as granite or gneiss; apothecial margin prominent or not 19

19. Spores 10-15(-17) x 3.5-5.5 µm; apothecia rusty orange to olive brown; epihymenium C+ purple; Great Lakes to Maritimes and Rockies

.... ***Caloplaca arenaria* (Pers.) Müll. Arg.** (syn. *Rufoplaca arenaria* (Pers.) Arup et al.)

19. Spores 8-11.5(-13) x 3.5-4.5 µm; apothecia yellow-orange to dark orange; epihymenium C+ red; mainly southwestern .... *Caloplaca approximata* (Lynge) H. Magn.

20.(3) Thallus distinctly lobed at the margin 21

20. Thallus not lobed at the margin (but sometimes with lobed squamules) 32

21. Thallus sorediate 22

21. Not sorediate, but sometimes granular or isidiate in part 24

22. Round to irregular soralia containing granular soredia forming on upper surface of thallus; lobes convex, elongate, sometimes pruinose on lobe tips; growing on limestone, usually on dry overhangs; northeastern and Colorado .... *Caloplaca cirrochroa* (Ach.) Th. Fr. (syn. *Leproplaca cirrochroa* (Ach.) Arup et al.)

22. Soralia mostly on the lobe margins; western 23

23. Soralia forming on the lower side of the squamule margins; soredia fine, yellow to yellow-orange; thallus appearing squamulose or forming rosettes, rarely becoming almost leprose; terminal lobes 0.2-0.5 mm long, often pruinose; on non-calcareous rocks, mainly California to B.C., rare elsewhere .... *Caloplaca stellata* Wetmore & Kärnefelt (syn. *Polycauliona stellata* (Wetmore & Kärnefelt) Arup et al.)

23. Soralia on older lobes; soredia coarsely granular, orange; terminal lobes 0.5-2.0 mm long, not pruinose; on calcareous rocks (also see couplet 28); widespread in west .... *Caloplaca decipiens* (Arnold) Blomb. & Forss. (syn. *Calogaya decipiens* (Arnold) Arup et al.) (see couplet 28)

24.(21) On maritime rocks in the salt spray zone 25

24. On non-maritime rocks, not subjected to salt spray; central and western .... 27 [See also *Xanthoria* key]

25. Older parts of thallus dissolving into granular isidia or granules, with a ring of narrow lobes at the periphery up to 5 mm long; apothecia usually sparse or absent; on east and west coasts .... ***Caloplaca verruculifera* (Vainio) Zahlbr.** (syn. *Polycauliona verruculifera* (Vainio) Arup et al.)

25. Older parts of thallus not dissolving into granules; lobes less than 3 mm long; apothecia abundant in central parts of thallus 26

26. On northeastern coast; thalli composed of tiny rosettes less than 10(-15) mm wide; terminal lobes 0.5-1.5 mm long; spore septum 3.5-5.0 µm .... *Caloplaca scopularis* (Nyl.) Lettau (syn. *Athallia scopularis* (Nyl.) Arup et al.)

26. On west coast, California to Oregon; thallus rosettes up to 25 mm wide; terminal lobes very convex, 1.5-3 mm long; spore septum 2.5-3.5 µm .... *Caloplaca brattiae* W. A. Weber (syn. *Polycauliona brattiae* (W. A. Weber) Arup et al.)

27.(24) Thallus brownish to pinkish orange, pruinose, with thin, flat lobes; apothecia common, with thallus-colored margins and dark orange, non-pruinose disks; central plains .... *Caloplaca galactophylla* (Tuck.) Zahlbr. (syn. *Squamulea galactophylla* (Tuck.) Arup et al.)

27. Thallus orange or red-orange, not pruinose 28

28. Thallus forming abundant corticate granules on lobes in the center of the rosettes; terminal lobes 0.5-2 mm long (otherwise similar to *C. verruculifera* above); widespread in west .... *Caloplaca decipiens* (Arnold) Blomb. & Forss. (syn. *Calogaya decipiens* (Arnold) Arup et al.) (see couplet 23)

28. Thallus lacking corticate granules 29

29. Thallus with thick, convex lobes 30

29. Thallus thin, with rather flat lobes, often broadest at the tips 31

30. Thallus with long terminal lobes, up to 5 mm; apothecia originating on older parts of thallus; eastern California to western plains in open, arid habitats .... ***Caloplaca trachyphylla* (Tuck.) Zahlbr.** (syn. *Xanthomendoza trachyphylla* (Tuck.) Frödén et al.)

30. Thallus forming short lobed rosettes, terminal lobes up to 1.5 mm; apothecia originating near lobe tips, at first immersed; widespread in west extending into the Arctic, rare in the east .... *Caloplaca saxicola* (Hoffm.) Nordin

[Note: See also *C. subsoluta* (couplets 69 and 73), which can sometimes have lobed areoles, but the thallus and apothecial margins are paler than the disks, and the apothecia are frequently immersed, at least when young.]

31. Lobes about as long as they are broad, more squamulose than lobate; with a distinct tissue made up of very small round cells (pseudoparenchyma) under the hypothecium extending to the surface laterally (fig. 13c); common in the southwest from southwestern Texas to California .... *Caloplaca squamosa* (B. de Lesd.) Zahlbr. (syn. *Squamulea squamosa* (B. de Lesd.) Arup et al.)

31. Lobes longer than broad at the thallus periphery; lacking a distinct tissue made up of pseudoparenchyma under the hypothecium; found only in southern California .... ***Caloplaca ignea* Arup** (syn. *Polycauliona ignea* (Arup) Arup et al.)

32.(20) Thallus lacking apothecia, with soredia or isidia 33

32. Thallus usually fertile, with apothecia; soredia and (or) isidia present or absent 34

33. Thallus white to gray or brown, K– .... see couplet 46

33. Thallus yellow or orange, K+ deep red-purple .... see couplet 56

34. Apothecial disks brown to black, K–; epihymenium K– or K+ purple; spores 13-19 x 7-10 µm 35

34. Apothecial disks yellow to orange, K+ deep red-purple; epihymenium K+ red to red-violet (anthraquinones) 40

35. Growing on bark 36

35. Growing on rock 38

36. Epihymenium dark blue-green to blackish purple, K+ purple to blue; algae present in margin; Pacific Northwest .... *Caloplaca atrosanguinea* (G. Merr.) I. M. Lamb

36. Epihymenium brown, K+ reddish or K– 37

37. Apothecia pruinose, with algae in margins; epihymenium K–; on bark, eastern forests .... *Caloplaca camptidia* (Tuck.) Zahlbr.

37. Apothecia lacking pruina and lacking algae in margins; epihymenium often K+ reddish-purple; on bark and especially wood; East Temperate .... *Caloplaca pollinii* (A. Massal.) Jatta

38.(35) Thallus sorediate, thick, areolate, the soredia developing from an eroding cortex; epihymenium K+ violet; spore septum 1.0-2.0 µm; on calcareous rock and cement, central states north to Saskatchewan .... *Caloplaca pratensis* Wetmore

38. Thallus without soredia; epihymenium K– or K+ weakly purple; spore septum 1.5-3(-4) µm; on calcareous or non-calcareous rocks; mostly western 39

39. Thallus subsquamulose to raised areolate; apothecia dark brown to black; mostly western mountains to California .... *Caloplaca albovariegata* (de Lesd.) Wetmore

39. Thallus continuous to rimose-areolate; apothecia brown; mostly western, but east to Ohio .... *Caloplaca atroalba* (Tuck.) Zahlbr.

40.(34) Thallus pale to dark gray or pale brown, not yellowish 41

40. Thallus distinctly yellow, or rusty to bright orange 55

41. Growing on moss, plant remains, and humus, arctic-alpine and prairies 42

41. Growing on bark or rock 46

42. Apothecia with a gray to blue-gray thalline margin; disk yellow-orange to orange or sometimes becoming olive to almost black, frequently pruinose; spores 11-14 x 6.5-9.5 µm, septum 4-6 µm .... *Caloplaca stillicidiorum* (Vahl) Lynge*

42. Apothecia with a yellow to orange margin, thick or thin, sometimes excluded; apothecia rarely pruinose

....43

43. Apothecia yellow-orange to orange or darkening to olive and finally black; apothecia remaining flat, margins well developed and persistent; spores 12-17(-20) x 7-10.5 µm 44

43. Apothecia rusty orange to brownish orange or red-brown, very convex with thin or excluded margins 45

44. Apothecia 0.3-0.6 mm in diameter, yellow orange to dark orange or darkening to olive black, margins disk-colored or paler; thallus membranous and often not evident; spore septum (2.5-)3-6 µm ....*Caloplaca tiroliensis* Zahlbr. (syn. *Parvoplaca tiroliensis* (Zahlbr.) Arup et al.)

44. Apothecia 0.5-1.5(-2.0) mm in diameter, orange, with margins of the same color; spore septum (1.5-)2.0-3.5 µm ....*Caloplaca jungermanniae* (Vahl) Th. Fr. (syn. *Bryophora jungermaniae* (Vahl) Søchting et al.)

45. Spores 8 per ascus, 15-20 x 8.0-11.5 µm; thallus white to pale gray, well developed ....*Caloplaca sinapisperma* (Lam. & DC.) Maheu & A. Gillet (syn. *Bryophora sinapisperma* (Lam. & DC.) Søchting et al.)

45. Spores 4 per ascus, 21-26(-28) x (11-)13-16(-20) µm; thallus pale gray, membranous ....*Caloplaca tetraspora* (Nyl.) H. Olivier (syn. *Bryophora tetraspora* (Nyl.) Søchting et al.)

46.(41) Growing on bark; thallus pale brown or gray to blue-gray, K– 47

46. Growing directly on rock 51

47. Thallus pale brown, not gray; apothecia with yellow-orange margins; thallus with very irregular, concave soralia containing fine greenish to greenish yellow soredia; central states .... *Caloplaca ulcerosa* Coppins & James [Note: Compare with *C. ahtii* (couplet 14).]

47. Thallus gray; apothecia with gray apothecial margins 48

48. Thallus with coarse, almost isidioid blue-gray soredia mostly on areole margins; mainly in shaded habitats, mainly northeastern, rare on west coast ....*Caloplaca chlorina* (Flotow) H. Olivier*

48. Thallus with or without soredia; on sunny bark, widespread 49

49. Spore isthmus 2.0-3.5(-4.0) µm; thallus sometimes sorediate; usually on conifers ....*Caloplaca pinicola* H. Magn.* (see also couplet 14)

49. Spore isthmus 3.5-7.0 µm; thallus not sorediate; usually on deciduous trees, especially poplars.................50

50. Thallus thin, continuous; apothecia not pruinose ....***Caloplaca cerina* (Ehrh. *ex* Hedwig) Th. Fr.***

50. Thallus well developed, sometimes subsquamulose; apothecia yellow-pruinose ....*Caloplaca ulmorum* (Fink) Fink*

[Note: This species is often regarded as a synonym of *C. cerina.*]

51.(46) Thallus pale beige, thin; apothecia biatorine with margins the same color as the disk (orange or red); coastal localities, California to Vancouver Island .... ***Caloplaca luteominia* (Tuck.) Zahlbr.** (syn. *Polycauliona luteominia* (Tuck.) Arup et al.)

51. Thallus white to gray; apothecial margins lecanorine 52

52. Apothecial margins orange; thallus thick, areolate, with coarse, dark gray soredia developing on the margins and cracks of the areoles; on calcareous rocks and concrete, central states ....*Caloplaca soralifera* Vondrák & Hrouzek

52. Apothecial margins gray to black 53

53. Thallus thin to thick, coarsely sorediate to almost isidiate, soredia blue-gray; northeastern, but also known from coastal British Columbia ....*Caloplaca chlorina* (Flotow) H. Olivier* (see also couplet 48)

53. Thallus without soredia 54

54. Thallus pale gray, continuous to areolate, not squamulose or lobed; apothecia with a dark, narrow ring of tissue between the disk and the margin; common from the southwest through central U.S. to New England ....*Caloplaca sideritis* (Tuck.) Zahlbr.

54. Thallus dark olive gray or brownish gray, thick, consisting of lobed squamules; lacking a dark ring between the apothecial disk and margin; restricted to the arid southwest ....***Caloplaca pellodella* (Nyl.) Hasse**

55.(40) Soredia or isidia present 56

55. Soredia and isidia absent 64

56. Thallus isidiate, orange; on bark, Florida ....*Caloplaca epiphora* (Tayl.) Dodge

56. Thallus sorediate 57

57. On soil, especial in arid interior and grasslands ....*Caloplaca tominii* (Savicz) Ahlner

57. On bark, wood, or rock 58

58. Thallus leprose, lacking apothecia; soredia pale yellow or gray-yellow; in humid habitats on rocks or bark .... *Caloplaca chrysodeta* (Vainio *ex* Räsänen) Dombr. (syn. *Leproplaca chrysodeta* (Vainio *ex* Räsänen) J. Laundon)

58. Thallus not leprose, producing soredia in soralia; apothecia present or absent 59

59. Thallus continuous, with discrete round to irregular soralia on thallus surface; on bark 60

59. Thallus areolate to subsquamulose; on rocks or wood, rarely on bark 61

60. Soredia fine, with excavate to hemispherical, round soralia; apothecia dark orange, often with a yellow outer margin; on poplar bark in the boreal region .... *Caloplaca chrysophthalma* Degel. (syn. *Solitaria chrysophthalma* (Degel.) Arup. et al.)

60. Soredia coarse, forming irregular masses; apothecia rusty red-orange with a thin biatorine margin; mostly on deciduous trees, especially maple, and on cedar; mostly east coast to Great Lakes region; scattered western montane and arctic .... *Caloplaca xanthostigmoidea* (Räsänen) Zahlbr. (syn. *Gyalolechia xanthostigmoidea* (Räsänen) Søchting et al.; *Caloplaca discolor*)

61. Thallus areoles usually on a conspicuous orange prothallus; apothecia common; on rocks in strictly marine habitats (upper intertidal and salt spray zones); areoles often constricted at the base, some dissolving into granular soredia ....***Caloplaca flavogranulosa* Arup** (syn. *Polycauliona flavogranulosa* (Arup) Arup et al.)

61. Prothallus absent; apothecia present or absent; on rock or wood, rarely bark, marine or not 62

62. Thallus and soredia dark orange, areolate to subsquamulose; soredia mostly remaining along the areole margins or in patches on the surface; common on exposed wood or sometimes bark, never rock; from Colorado to the northeast ....*Caloplaca microphyllina* (Tuck.) Hasse

62. Thallus and soredia yellow to orange-yellow, areolate or not 63

63. Thallus areolate, with soredia developing first at the edges of the areoles, but frequently taking over the thallus; on rock, central to eastern regions .... *Caloplaca flavocitrina* (Nyl.) H. Olivier (syn. *Flavoplaca flavoctrina* (Nyl.) Arup et al.)

63. Thallus continuous or becoming areolate, soredia covering the thallus, not beginning at, or confined to, the margins of the areoles; on rocks or wood in maritime or inland localities .... ***Caloplaca citrina* (Hoffm.) Th. Fr.** (syn. *Flavoplaca citrina* (Hoffm.) Arup et al.)

[Note: Distinguishing the above species is not easy since there are intermediates. The North American material of *C. citrina* needs to be revisited.]

64.(55) On bark or wood 65

64. On rock or rock lichens, rarely on wood 67

65. Thallus continuous and rather smooth; apothecial margin paler than the disk; on bark or sometimes wood; widespread in central and eastern regions .... ***Caloplaca flavorubescens* (Hudson) J. R. Laundon** (syn. *Gyalolechia flavorubescens* (Hudson) Søchting et al.)

65. Thallus sparse, consisting of scattered small areoles 66

66. On wood (or rocks); maritime, along the west coast; apothecial margin the same color as the disk, without an outer margin .... ***Caloplaca inconspecta* Arup** (syn. *Polycauliona inconspecta* (Arup) Arup et al.)

66. On tree bark, especially poplars; not maritime, very widespread; apothecial margin often with a double margin: an inner, disk-colored ring surrounded by a thick, thallus-colored (gray to yellow) outer margin .... *Caloplaca pyracea* (Ach.) Th. Fr. (syn. *Athallia pyracea* (Ach.) Arup et al.)

67.(64) Thallus continuous to rimose, rimose-areolate or verruculose, not dispersed areolate or squamulose 68

67. Thallus discontinuous, verruculose to squamulose or dispersed areolate, the squamules sometimes becoming lobate 70

68. Growing on maritime rocks of all types in salt spray zone, rarely elsewhere; thallus orange with an orange prothallus; California to British Columbia .... *Caloplaca rosei* Hasse (syn. *Polycauliona rosei* (Hasse) Arup et al.)
[Note: Some forms of *C. marina* can resemble *C. rosei*. See Arup (1992).]

68. On calcareous or sometimes non-calcareous, non-maritime rock; lacking a prothallus 69

69. Thallus yellow, widespread east of Arizona, Nevada, and Idaho ....
***Caloplaca flavovirescens* (Wulfen) Dalla Torre & Sarnth.** (syn. *Gyalolechia flavovirescens* (Wulfen) Søchting et al.)

69. Thallus orange; throughout North America (see couplet 73) .... *Caloplaca subsoluta* (Nyl.) Zahlbr. (syn. *Squamulea subsoluta* (Nyl.) Arup et al.)

70.(67) Parasitic on rock lichens; thallus of tiny orange areoles scattered on host; apothecia also dark orange with orange margins; arctic-alpine .... *Caloplaca castellana* (Räsänen) Poelt (syn. *Pachypeltis castellana* (Räsänen) Søchting et al.)

70. Growing directly on rock 71

71. Thallus consisting of round or squamulose, convex areoles frequently constricted at the base, up to 2 mm in diameter, usually shiny or "waxy"; apothecia (0.6-)1-2 mm in diameter; on rocks, coastal California .... *Caloplaca bolacina* (Tuck.) Herre (syn. *Polycauliona bolacina* (Tuck.) Arup et al.)

71. Thallus composed of small areoles or squamules (mostly less than 1.0 mm in diameter) or granules; areoles can be contiguous or imbricate, but not constricted at the base, not waxy in appearance; apothecia mostly under 1.0 mm in diameter; on rocks, bark, or wood 72

72. Not maritime; thallus orange, areoles frequently lobed; layer below hypothecium (the exciple) composed of small rounded cells (pseudoparenchyma) 73

72. Mostly maritime, in salt spray zone; thallus orange to yellowish; areoles lobed or not; the exciple layer composed of elongate cells (prosoplectenchyma) 74

73. Thallus squamulose, often imbricate or lifting from the surface; areoles and squamules large, often up to 1 mm across; apothecia 0.4-1.0 mm in diameter, containing abundant algae in the margins; only known from the southwest, Texas to California (see couplet 31) ....*Caloplaca squamosa* (B. de Lesd.) Zahlbr. (syn. *Squamulea squamosa* (B. de Lesd.) Arup et al.)

73. Thallus rimose-areolate to dispersed areolate; areoles small, rarely larger than 0.5 mm across; apothecia mostly under 0.6 mm in diameter, often with few algae in margin; widespread in North America .... *Caloplaca subsoluta* (Nyl.) Zahlbr. (syn. *Squamulea subsoluta* (Nyl.) Arup et al.)

[Note: The name *C. velana* (A. Massal.) Du Rietz has often been used for this lichen in North America, but they may not be synonymous.]

74. Eastern; prothallus absent; not on bird rocks ....*Caloplaca microthallina* (Wedd.) Zahlbr. (syn. *Flavoplaca microthallina* (Wedd.) Arup et al.)

74. Western; prothallus present or absent 75

75. Thallus well developed, usually with an orange prothallus, often on bird rocks (see couplet 61) ....***Caloplaca flavogranulosa* Arup** (syn. *Polycauliona flavogranulosa* (Arup) Arup et al.)

75. Thallus well developed or very thin; prothallus absent or spotty 76

76. Apothecia biatorine; apothecial disks dark orange (var. *luteominia*) or red (var. *bolanderi*); hymenium containing abundant oil drops .... ***Caloplaca luteominia* (Tuck.) Zahlbr.** (syn. *Polycauliona luteominia* (Tuck.) Arup et al.)

76. Apothecia lecanorine; apothecial disks yellow-orange; hymenium clear, without oil drops 77

77. Thallus very poorly developed, consisting of scattered areoles less than 0.2 mm in diameter (see couplets 8, 16 and 66) .... ***Caloplaca inconspecta*** **Arup** (syn. *Polycauliona inconspecta* (Arup) Arup et al.)

77. Thallus usually verruculose and rough, sometimes poorly developed; areoles 0.2-0.7 mm in diameter; apothecia concave to flat with well-developed, often flexuose thalline margins ....*Caloplaca marina* (Wedd.) Zahlbr. var. *americana* Arup (syn. *Flavoplaca marina* (Wedd.) Arup et al.)

[Note: In *C. rosei,* which can be very similar, the apothecia are always flat and the apothecial margins are thin and even; in specimens of *C. flavogranulosa* lacking a good prothallus, the apothecia are flat to convex and the apothecial margins are thin to bumpy (crenulate).]

CANDELARIA Based on Westberg et al. (2011).

1. Soredia absent; apothecia abundant; East Temperate .... ***Candelaria fibrosa*** **(Fr.) Müll. Arg.**

1. Soredia present; apothecia rare 2

2. Lower cortex present; soredia first produced at lobe edges, later reducing lobes to a mass of soredia; spores numerous in ascus; lobes typically appressed; widespread .... ***Candelaria concolor*** **(Dickson) Stein**

2. Lower cortex absent; soredia produced on lower side of lobe tips; spores 8 per ascus; lobes typically ascending; western, mostly California to southeastern Alaska and Alberta ....*Candelaria pacifica* M. Westberg & Arup

CANDELARIELLA See Westberg (2007a, b, c), Westberg et al. (2011).

1. Thallus composed of tiny dispersed areoles breaking down into clumps of soredia, sometimes coalescing into a leprose crust 2

1. Thallus without soredia, but can be granular to areolate 3

2. Spores 12-32 per ascus; East Temperate to boreal; common .... ***Candelariella efflorescens*** **R. C. Harris & W. R. Buck**

2. Spores 8 per ascus; southeastern and Ozark and scattered in west; rare ....*Candelariella xanthostigmoides (*Müll. Arg.) R.W. Rogers (= "*C. reflexa*" of N. Am. authors)

[Note: In the absence of apothecia, these two species cannot be distinguished.]

3. Growing on soil or over mosses and plant debris 4

3. Growing on bark, wood, or rock 7

4. Thallus composed of very small granules, 0.1-0.2 mm in diameter, which tend to break down into even smaller "blastidia"; apothecia not common, spores 16-20 per ascus; on soil, infrequent, arctic-alpine .... *Candelariella granuliformis* M. Westb.

4. Thallus granular to areolate or subsquamulose, granules or areoles larger than 0.2 mm in diameter, not breaking into blastidia; apothecia common or rare 5

5. Asci with many spores; thallus areolate to subsquamulose, with a rough surface; on soil, arctic-alpine .... *Candelariella placodizans* (Nyl.) H. Magn.

5. Asci with 8 spores 6

6. Spores ellipsoid to teardrop-shaped, with at least one pointed tip, (9-)10-14(-16) x 4-7 µm; thallus areolate to subsquamulose; apothecia with thick, persistent margins; on tundra soil over rocks; arctic-alpine ....***Candelariella citrina* B de Lesd.** (syn. *C. terrigena;* Plate 179 in LNA)

6. Spores elongate ellipsoid, tips rounded, (11-)13-20(-24) x 4-6.5 µm; thallus granular, but often obscured by crowded apothecia; apothecia with thin, sometimes excluded margins; on plant debris and mosses, arctic-alpine and prairies ....*Candelariella aggregata* M. Westb.

7.(3) Sterile (without apothecia); thallus granular to areolate or verrucose 8

7. With apothecia 10

8. Thallus composed of convex areoles to lobed squamules; widespread in west; on siliceous rocks ....***Candelariella rosulans* (Müll. Arg.) Zahlbr.**

8. Thallus granular or with very small areoles 9

9. Granules rounded, ± uniform in size and scattered (rarely piled together), 0.03-0.15 mm in diameter; on bark or wood ....*Candelariella xanthostigma* (Ach.) Lettau

9. Granules or areoles round or irregular in shape, usually contiguous to overlapping, often becoming more or less lobed, larger than 0.15 mm in diameter; on non-calcareous rock or wood, rarely on bark ....***Candelariella vitellina* (Hoffm.) Müll. Arg.**

10.(7) Spores 16-32 per ascus 11

10. Spores 8 per ascus 13

11. Thallus areolate to irregularly granular; apothecia 0.5-1.5 mm diameter, with thick, prominent, often verrucose margins; on non-calcareous rock or wood .... ***Candelariella vitellina* (Hoffm.) Müll. Arg.**

11. Thallus minutely granular; apothecia less than 0.7 mm diameter, with relatively thin margins; on bark or wood 12

12. Apothecia 0.15-0.4 mm diameter, commonly present and abundant; thallus granules round or irregular, flat to ± convex, irregularly distributed, forming small round patches up to 10 mm in diameter; widespread but under-reported, western and northeast .... *Candelariella lutella* (Vainio) Räsänen

12. Apothecia 0.3-0.7 mm diameter, rarely seen and not abundant; thallus granules round and convex, evenly distributed, scattered to crowded, covering large irregular areas; widespread except in southeast, very common .... *Candelariella xanthostigma* (Ach.) Lettau

13.(10) Spores elongate and tapering, 25-46 x 3-5 µm, often appearing 2-celled; rare, montane .... *Candelariella spraguei* (Tuck.) Zahlbr.

13. Spores ellipsoid, 12-21(-25) x 4.5-7.5 µm 14

14. On bark and wood in arid interior; thallus lacking or gray; common .... *Candelariella antennaria* Räsänen

14. On rock or rarely on wood; thallus scanty or well developed, yellow 15

15. On non-calcareous rock; thallus usually well developed, granular to dispersed areolate or subsquamulose; spores elongate, length to width ratio usually more than 3:1, (12-)14-21(-25) x 4-6(-7) µm; southwestern and central prairies .... ***Candelariella rosulans* (Müll. Arg.) Zahlbr.**

15 . On calcareous rock, occasionally wood; thallus usually within substrate, absent from view, or poorly developed, granular; spores narrowly ellipsoid, length to width ratio 3:1 or less, 11-16.5 x 4-5(-6) µm; widespread .... ***Candelariella aurella* (Hoffm.) Zahlbr.**

CANDELINA

1. Medulla white .... ***Candelina submexicana* (B. de Lesd.) Poelt**

1. Medulla yellow .... *Candelina mexicana* (B. de Lesd.) Poelt

CATILLARIA (including *Catinaria*)

1. Growing on rock 2

1. Growing on bark or moss 3

2. On siliceous rocks; apothecia black; spores 6-15 x 2.5-5; widespread temperate to boreal .... *Catillaria chalybeia* (Borrer) A. Massal.

2. On calcareous rocks; apothecia dark brown; exciple almost colorless within, black at outer edge; widespread in the east, especially the Ozarks .... *Catillaria lenticularis* (Ach.) Th. Fr.

3. Spores 13-30 x 5-15 µm .... *Megalaria*

3. Spores 6-12 x 2-7 4

4. Spores 10-12.5(-15) x 5-7 µm; apothecia dark reddish brown to black, 0.3-0.6 mm in diameter; tips of paraphyses slightly expanded and pigmented brown; hypothecium colorless; on bark of different kinds, especially poplar and cedar; temperate to boreal in east and California .... *Catinaria atropurpurea* (Schaerer) Vězda & Poelt

4. Spores 8-12 x 2-5 µm; apothecia mostly black 5

5. Tips of paraphyses much expanded and with a dark brown pigment on the upper half of the tip cell; hymenium colorless; spores not, or very slightly, tapered; ascus with K/I+ dark blue tholus; on various trees, especially poplars and white cedar, prairies eastward, southern California .... *Catillaria nigroclavata* (Nyl.) Schuler

5. Tips of paraphyses not at all expanded; epihymenium and part of hymenium greenish or greenish brown; spores tapered; ascus K/I– or pale blue often with a dark-staining "net" around the axial mass; only on poplars; widespread in aspen woodlands .... *Arthonia patellulata* Nyl.

CETRARIA (including *Arctocetraria, Cetrariella,* and *Masonhalea*)

1. On trees and shrubs, occasionally on the ground, subalpine; branches rather flat; medulla PD– (protolichesterinic and lichesterinic acids) 2

1. On the ground, rarely on twigs, sometimes mixed in moss mats or low heath; medulla PD– or PD+ red (fumarprotocetraric acid) 3

2. Branches divided several times and dichotomous; pseudocyphellae marginal, not continuous and usually inconspicuous; western montane, especially coastal mountains .... ***Tuckermanopsis subalpina* (Imshaug) Kärnefelt**

2. Branches undivided or divided once or twice; long, continuous pseudocyphellae conspicuous along lower surface of margins; northwestern Alaska to northern Yukon .... *Masonhalea inermis* (Krog) Nelson et al. (syn. *Cetraria inermis, Tuckermanopsis inermis*)

3. Branches round in cross section; thallus entirely fruticose; with round to oval, depressed pseudocyphellae; arctic-alpine to boreal western North America, western grasslands, and along the east coast 4

3. Branches flat; thallus erect foliose; pseudocyphellae either irregular in shape or linear, depressed or not 5

4. Pseudocyphellae abundant, predominantly deep, up to 0.5 mm long and 0.3 mm wide; branches about 1 mm wide, often conspicuously ridged and grooved; surface typically dull .... ***Cetraria aculeata*** **(Schreber) Fr.**

4. Pseudocyphellae scattered and sparse, predominantly shallow, rarely longer than 0.3 mm or wider than 0.2 mm; branches about 0.5 mm wide, usually without conspicuous ridges and grooves; surface typically shiny .... *Cetraria muricata* (Ach.) Ach.

[Note: Intermediates between *C. aculeata* and *C. muricata* are not uncommon.]

5. Lower surface pruinose, with an irregular pattern of white and dark areas; medulla UV+ blue-white (alectoronic acid) .... ***Masonhalea richardsonii*** **(Hooker) Kärnefelt**

5. Lower surface not pruinose; medulla UV– (alectoronic acid absent) 6

6. Medulla KC+ red, C+ red or pink (gyrophoric acid); lobe margins more or less even, not toothed or ciliate; pseudocyphellae laminal 7

6. Medulla KC– (gyrophoric acid absent); lobe margins even, or toothed, or ciliate 8

7. Branches flat to concave, not rolled into tubes; pseudocyphaellae white, abundant and conspicuous; arctic to northern boreal and northern Appalachians ... ***Cetrariella delisei*** **(Bory *ex* Schaerer) Kärnefelt & Thell**

7. Branches strongly concave, rarely forming a tube; pseudocyphaellae sparse; western arctic .... *Cetrariella fastigiata* (Delise *ex* Nyl.) Kärnefelt & Thell

8. Medulla PD+ red 9

8. Medulla PD– 10

9. Pseudocyphellae marginal, forming an almost unbroken line along the lobe margins .... ***Cetraria laevigata*** **Rass.**

9. Pseudocyphellae laminal and irregular in shape, but often also along the margins here and there .... ***Cetraria islandica*** **(L.) Ach.**

10. Mainly temperate, Appalachian-Great Lakes to southern Canadian prairies; thallus gray-olive to olive-brown when dry; branching mainly dichotomous; pseudocyphaellae marginal, rarely laminal .... *Cetraria arenaria* Kärnefelt

10. Mainly arctic-alpine or boreal; thallus yellowish brown to reddish brown or olivaceous to almost black when dry 11

11. Pseudocyphellae inconspicuous or absent 12

11. Pseudocyphellae conspicuous 13

12. Thallus yellowish brown to olive-brown, never reddish; pseudocyphellae absent; margins with sparse cilia, mostly unbranched; branches flat to slightly concave, not forming tubes; containing rangiformic and norrangiformic acids; arctic tundra .... *Arctocetraria nigricascens* (Nyl.) Kärnefelt & Thell

12. Thallus glossy reddish brown to very dark brown or black; pseudocyphellae dark, intermittent along the margins; margins with abundant, long cilia, often dichotomously branched; branches strongly concave and often forming tubes; containing protolichesterinic and lichesterinic acids; common, arctic tundra .... *Cetraria nigricans* Nyl.

13. Pseudocyphellae mostly laminal, broad and irregular in shape 14

13. Pseudocyphellae mostly marginal 15

14. Pseudocyphellae large and white, usually conspicuous; containing protolichesterinic acid and (or) lichesterinic acid; arctic-boreal to montane .... rare PD– strain of ***Cetraria islandica* (L.) Ach.**

14. Pseudocyphellae small and dark, usually inconspicuous; containing rangiformic and norrangiformic acids; arctic to northern boreal .... *Arctocetraria andrejevii* (Oxner) Kärnefelt & Thell

15. Thallus very dark reddish brown to almost black, usually less than 2 cm high; branches with long, sometimes branched cilia; common, arctic tundra .... *Cetraria nigricans* Nyl.

15. Thallus light to dark brown, usually more than 2 cm high; branches with marginal tooth-like outgrowths but not with branched cilia 16

16. Branches strongly channeled; pseudocyphellae conspicuous, forming an almost continuous line at the lobe margins; common and widely distributed in boreal and montane areas .... ***Cetraria ericetorum* Opiz**

16. Branches more or less flat or slightly concave; pseudocyphellae inconspicuous, intermittent along the lobe margins; western mountains, subalpine .... ***Tuckermanopsis subalpina* (Imshaug) Kärnefelt**

CETRELIA

1. Soredia absent; on tundra in arctic Alaska ....*Cetrelia alaskana* (C. F. Culb. & W. L. Culb.) W. L. Culb. & C. F. Culb.
1. Soredia present along lobe margins; on trees or rocks 2

2. Medulla C+ red or pink, UV– (olivetoric acid); pseudocyphellae sparse and small (up to 0.3 mm in diameter) ....***Cetrelia olivetorum* (Nyl.) W. L. Culb. & C. F. Culb.**
2. Medulla C– 3

3. Medulla KC+ pink to red, UV+ blue-white (alectoronic acid); pseudocyphellae abundant and often large (0.15-0.6 mm in diameter) .... ***Cetrelia chicitae* (W. L. Culb.) W. L. Culb. & C. F. Culb.**
3. Medulla KC–, rarely KC+ faint pink (perlatolic or imbricaric acid); pseudocyphellae small and inconspicuous 4

4. Containing perlatolic acid; Appalachian-Great Lakes and Pacific Northwest ....*Cetrelia cetrarioides* (Duby) W. L. Culb. & C. F. Culb.
4. Containing imbricaric acid; Appalachian..................... *C. monachorum* (Zahlbr.) W. L. Culb. & C. F. Culb.

CHRYSOTHRIX Modified from Harris & Ladd (2008) based on specimens in CANL and NY.

1. Thallus fertile, esorediate, granular or leprose; asci arthonioid (fig. 14n), with 4-celled spores 2
1. Thallus without apothecia, leprose 3

2. Apothecia and thallus bright yellow (rhizocarpic and diffractaic acids); thallus not sorediate; Alaskan coast .... *Chrysothrix chrysophthalma* (P. James) P. James & J. R. Laundon
2. Ascomata blue-gray pruinose; thallus pale yellowish (usnic acid), granular to leprose in small patches; spore cells equal in size, 15-20 x 4-6 µm; on a variety of trees in the northeast .... *Chrysothrix caesia* (Flotow) Ertz & Tehler (syn. *Arthonia caesia*)

3. Thallus containing lecanoric acid (C+ pink) as well as rhizocarpic acid; growing on moss, peat, and sometimes rocks; thallus pale yellow green; widespread in east on vertical rock faces .... *Chrysothrix susquehannensis* Lendemer & Elix
3. Thallus lacking lecanoric acid, C–; growing on rock, bark, or wood 4

4. Photobiont *Stichococcus,* the cells subrectangular to elongate; thallus containing vulpinic acid

....sterile ***Chaenotheca furfuracea* (L.) Tibell**

4. Photobiont chlorococcoid, the cells ± isodiametric; thallus chemistry various 5

5. On bark or wood 6

5. On rock 11

6. Thallus thick, forming fluffy cushions with a thin or thick "medulla" of white to pale brownish hyphae and bright yellow granules on top; granules 30-45 µm in diameter; southwestern U.S., southern California, with scattered localities along west coast to Alaska; also Alberta; containing calycin and diffractaic acid ....*Chrysothrix granulosa* Thor

6. Thallus thin or moderately thick, never with a fluffy medulla, the yellow granules lying directly on the substrate 7

7. Granules coarse, 35-80 µm across; calycin and (or) pinastric acid; western 8

7. Granules small, 15-45 µm across; calycin and (or) leprapinic? acid or pinastric acid or rhizocarpic acid as the major substance; eastern 9

8. Contains calycin +/- pinastric acid; southwestern U.S., with scattered localities along west coast to B.C .... ***Chrysothrix candelaris* (L.) J. R. Laundon** *s. str.*

[Note: In Tasmania, it is mostly saxicolous: Elix and Kantvilas (2007).]

8. Contains calycin and diffractaic acid; Pacific Northwest .... unnamed chemotype of ***Chrysothrix candelaris* (L.) J. R. Laundon**

9. Thallus dull pale yellow to pale yellowish green (whitish yellow with age in herbarium), UV+ dull to bright orange (rhizocarpic acid); granules small, 15-25(-30) µm across, "loose," without binding hyphae; exclusively on conifers .... *Chrysothrix chamaecyparicola* Lendemer (= "*C. flavovirens*" of N. Am. authors)

9. Thallus usually bright yellow, UV– (calycin and leprapinic? acid or pinastric acid), rather thin except in center, often discontinuous; granules 25-45 µm; on hardwoods or conifers 10

10. Pinastric acid major; widespread in east and California ....*Chrysothrix xanthina* (Vainio) Kalb

10. Calycin and leprapinic? acid major; specimens on *Quercus* & palm, southern Coastal Plain or on *Abies*, Maine and Michigan .... *Chrysothrix insulizans* R. C. Harris & Ladd *s. lat.* (see also couplet 14)

11.(5) On calcareous rock; thallus K+ instantly magenta-purple (parietin)

..........................*Caloplaca chrysodeta* (Vainio *ex* Räsänen) Dombr. or ***Caloplaca citrina* (Hoffm.) Th. Fr.**

[Note: See *Caloplaca* key, couplets 58 and 63.]

11. On acidic siliceous rock; thallus K– to K+ slowly reddish 12

12. Thallus thick, attached to rock by rhizohyphae (these sometimes not evident); usually easily separated from rock 13

12. Thallus thin, lacking rhizohyphae, of scattered to contiguous granules or leprose; rarely easily separable from rock 14

13. Thallus UV± dull orangish (leprapinic? acid), of loosely aggregated granules with numerous projecting hyphae, yellow above, whitish to brown hypothallus below (but difficult to see); Ozarks, Pennsylvania, and southeastern U.S. .... *Chrysothrix onokoensis* (Wolle) R. C. Harris & Ladd

13. Thallus UV– (calycin, vulpinic acid, ± zeorin), usually of more tightly compacted granules, mostly without obvious projecting hyphae, ± uniform yellow in section, without a hypothallus; B.C., N.W.T., Ontario, Vermont .... *Chrysothrix chlorina* (Ach.) J. R. Laundon

14. Thallus K+ slowly reddish (calycin + leprapinic? acid), forming small ± round patches on rock, sometimes forming large continuous, rimose or rimose-areolate patches with rounded soralium-like outliers .... *Chrysothrix insulizans* R. C. Harris & Ladd

14. Thallus K– (pinastric or rhizocarpic acid), forming a thin, ± continuous, granular crust 15

15. Thallus bright yellow to bright greenish yellow, UV– (pinastric acid); normally on bark, rarely on rock .... *Chrysothrix xanthina* (Vainio) Kalb

15. Thallus pale yellow to pale greenish yellow, UV+ dull to bright orange (rhizocarpic acid) normally on rock, often in heavily shaded crevices of fieldstone walls, rarely on bark or wood .... ***Psilolechia lucida* (Ach.) M. Choisy**

CLADONIA (including *Cladina*) Follows, for the most part, Ahti and Stenroos (2013).

1. Podetia much branched 2

1. Podetia unbranched, or branched only once or twice, OR podetia absent 30

[Note: Some species with richly proliferating cups such as *C. crispata* or *C. verticillata* can appear to be "much branched," but they will key out from couplet 30.]

2. Podetia without a cortex; surface dull and webby .... 3 [reindeer lichens]

2. Podetia with a cortex; surface usually somewhat shiny 16

3. Podetia silver gray to pale greenish gray; thallus KC–, K– or K+ yellow 4
3. Podetia pale yellow-green or greenish yellow (usnic-yellow); thallus KC+ yellow, K– or K+ yellow 8

4. Thalli forming tight, rounded tufts without clearly defined main stems; southeastern coastal plain .... ***Cladonia evansii* Ahbayes** (syn. *Cladina evansii*)
4. Thalli not forming tight, rounded tufts; main stems usually obvious; mainly northern or western 5

5. Branching at tips divergent, not bent in one direction; thallus PD–, UV+ bright blue-white on lower parts (perlatolic acid) 6
5. Branching at tips at least partly bent in one direction giving the thallus a combed appearance; thallus PD+ red, K+ pale yellow, UV– (atranorin and fumarprotocetraric acid) 7

6. West coast, from California to Alaska; thallus K– (lacking atranorin) .... ***Cladonia portentosa* (Dufour) Coem.** (syn. *Cladina portentosa*)
6. Northeastern, New Jersey to Newfoundland; thallus K+ yellow (atranorin) .... *Cladonia terrae-novae* Ahti

7. Basal part of podetia and stereome blackened; pycnidial jelly red .... ***Cladonia stygia* (Fr.) Ruoss** (syn. *Cladina stygia*)
7. Basal part of podetia and stereome pale, more or less the same as upper portions; pycnidial jelly colorless .... ***Cladonia rangiferina* (L.) F. H. Wigg.** (syn. *Cladina rangiferina*)

8. Podetia broad, mostly 3-11 mm in diameter, appearing inflated .... ***Cladonia boryi* Tuck.**
8. Podetia slender, 0.5-2 mm in diameter, not appearing inflated 9

9.(3) Podetia branching mostly in twos (dichotomies) 10
9. Podetia branching mostly in threes and fours (trichotomies or tetrachotomies) 12

10. Podetia thick, 0.7-2 mm wide; podetial axils often open; stereome composed of distinct strands; thallus PD– or PD+ yellow (with or without psoromic acid; fumarprotocetraric acid absent) .... ***Cladonia pachycladodes* Vainio**
10. Podetia slender, usually under 0.7 mm wide; stereome continuous, not broken into strands 11

11. Thallus PD+ red, K–, UV– (fumarprotocetraric acid); podetial axils mostly closed, podetial surface smooth; eastern .... ***Cladonia subtenuis*** **(Abbayes) Mattick** (syn. *Cladina subtenuis*)

11. Thallus PD–, K+ yellow, UV+ white on lower parts (atranorin, perlatolic acid); podetial axils open or closed; podetial surface bumpy (flocculent); New Jersey to Newfoundland .... *Cladonia terrae-novae* Ahti (*syn. Cladina terrae-novae*)

12. Thallus forming tight, rounded tufts without obvious main stems; podetial tips with radiating branches around an open hole .... ***Cladonia stellaris*** **(Opiz) Pouzar & Vězda** (syn. *Cladina stellaris*)

12. Thallus forming loosely organized cushions with clearly recognizable main stems; tips without radiating branches around an open hole 13

13. Surface flocculent (with small fluffy clumps) except close to the branch tips; thallus UV+ bright blue-white on lower parts (perlatolic acid) ....go back to couplet 6

13. Surface compact, smooth or bumpy, not flocculent; thallus UV– (lacking perlatolic acid) 14

14. Thallus PD+ red (fumarprotocetraric acid, lacking fatty acids) .... ***Cladonia arbuscula*** **(Wallr.) Flotow** (syn. *Cladina arbuscula*)

14. Thallus PD– (containing fatty acids) 15

15. Branches very robust, 0.7-2 mm wide, often sprawling and strongly wrinkled; axils broadly open; branching usually in fours; containing pseudonorrangiformic acid; east coast, Maryland to Massachusetts .... ***Cladonia submitis*** **A. Evans** (syn. *Cladina submitis*)

15. Branches usually slender, 0.5-0.8 mm wide, always erect, generally smooth; axils often closed or only slightly open; branching usually in threes, usually containing rangiformic acid; widespread boreal to north temperate .... ***Cladonia mitis*** **Sandst.** (syn. *C. arbuscula* subsp. *mitis, Cladina mitis*)

16.(2) Thallus gray to gray-green, often browned in sunny habitats, KC–, K– or K+ yellow (usnic acid absent) 17

16. Thallus yellowish green or greenish yellow, KC+ yellow (usnic acid present) 21

17. Primary squamules large and persistent; podetia up to 25 mm tall; thallus K+ deep yellow (thamnolic acid) .... ***Cladonia floridana*** **Vainio**

17. Primary squamules usually disappearing (although there may be few to many on the podetia); podetia 20-60 (-120) mm tall; thallus K– 18

18. Podetia with granules and microsquamules especially at the tips .... ***Cladonia scabriuscula* (Delise) Nyl.**

18. Podetia without soredia or granules 19

19. Podetia lacking squamules; cortex and (or) medulla PD–, UV+ white; podetial surface typically pale, browned at tips, with flat areoles over a translucent stereome; thallus finely branched, cortex and (or) medulla KC+ pinkish violet, rapidly disappearing (merochlorophaeic acid), uncommon, boreal, mainly western .... *Cladonia wainioi* Savicz (syn. *C. pseudorangiformis*)

[Note: *Cladonia subfurcata* is very similar but the podetial surface is dark brown throughout, less commonly pale brown with dark tips, usually erect with side branches; stereome often blackened, especially at base; medulla KC–, UV+ white (squamatic acid); boreal to arctic (see couplet 139).]

19. Podetia usually having at least a few (often many) squamules, with a continuous cortex; thallus stiff due to a hard stereome, PD+ red, KC–, UV– (fumarprotocetraric acid) 20

20. Podetia with unequal branches that are frequently split lengthwise ....uncupped morphotype of ***Cladonia multiformis* G. Merr.**

20. Podetia with more or less equal, dichotomous branches, split or intact .... ***Cladonia furcata* (Hudson) Schrader**

21.(16) Apothecia red, almost spherical .... ***Cladonia leporina* Fr.**

21. Apothecia brown or absent 22

22. Stereome rudimentary or absent, not forming a well-defined, cartilaginous cylinder or strands; southeastern coastal plain 23

22. Stereome forming an intact, cartilaginous cylinder or network of cartilaginous cords or strands; southeastern or widespread 24

23. Podetial wall perforated with oval holes; podetial axils open; surface very shiny; very rare, Florida .... ***Cladonia perforata* A. Evans**

23. Podetial wall not perforated or longitudinally split; podetial axils closed; surface smooth but usually dull; common, southeastern .... ***Cladonia leporina* Fr.**

24. Medulla PD+ yellow, K–, UV+ white (baeomycesic and squamatic acids); branches slender, under 0.6 mm wide, tangled (resembling a reindeer lichen); southeastern, mostly coastal plain .... *Cladonia subsetacea* Robbins *ex* A. Evans

24. Medulla PD–, UV + or UV–; podetia 0.7-4 mm wide, thorny in appearance; not in southeastern coastal plain 25

25. Stereome cylindrical, smooth, not broken into strands; tiny needles never developing at branch tips 26

25. Stereome broken into broad or narrow strands; needle crystals sometimes forming at the branch tips of old specimens (triterpenes) 27

26. Podetia tall and slender, pointed at the tips or forming narrow but distinct cups; axils often closed or only partially open; barbatic acid present; boreal to arctic .... ***Cladonia amaurocraea*** **(Flörke) Schaerer**

26. Podetia forming densely branched cushions, entirely without cups; axils wide open; barbatic acid absent; with or without squamatic acid; temperate to arctic .... ***Cladonia uncialis*** **(L.) F. H. Wigg.**

27. Coastal B.C. to Alaska; stereome with broad strands .... *Cladonia kanewskii* Oksner

27. Eastern, temperate 28

28. Podetia slender, 0.5-1.5 mm in diameter, not appearing inflated, usually forming rather flattened mats; stereome cords broad and flat; East Temperate .... ***Cladonia dimorphoclada*** **Robbins**

28. Podetia broad, mostly 3-11 mm in diameter, appearing inflated, usually erect and cushion forming; stereome with broad or narrow cords 29

29. On rock, less frequently on soil; stereome barely broken into broad bands; podetial wall intact, not perforated; Appalachian .... ***Cladonia caroliniana*** **Tuck.**

29. On sandy soil or bogs and heaths; stereome broken into a meshwork of narrow cords; podetial wall perforated and fissured; northeastern coastal plain, rarely upland .... ***Cladonia boryi*** **Tuck.**

30.(1) Main thallus consisting of primary squamules; podetia absent or extremely small (under 4 mm high) 31

[Note: Many young or poorly developed specimens of *Cladonia* may key out here.]

30. Main thallus consisting of many podetia, usually over 4 mm tall; with or without a basal thallus of primary squamules 42

31. Upper surface of thallus yellow-green to olive, containing usnic acid (KC+ gold) 32

31. Upper surface of thallus grayish green, brownish, or olive; lower surface of squamules white; thallus KC– (usnic and barbatic acids absent) or KC+ green [see couplet 37] 34

32. Lower surface of squamules white; containing usnic acid alone; southeastern U.S. north to Connecticut .... *Cladonia piedmontensis* G. Merr.

32. Lower surface of squamules pale yellow 33

33. Lower surface of squamules smooth to fibrous, very pale yellow to yellowish white; squamules reflexed, with rounded digitate lobes resembling toes; containing barbatic as well as usnic acid; eastern to central regions and southern Alberta .... ***Cladonia robbinsii* A. Evans**

33. Lower surface of squamules cottony or powdery, pale to sulphur yellow; squamules rounded, often concave or contorted, not digitate; several chemical races; western .... *Cladonia luteoalba* Wheldon & A. Wilson

34. Primary thallus forming a lobed, almost foliose, closely attached, brown rosette; medulla K– (atranorin usually absent) .... ***Cladonia pocillum* (Ach.) Grognot**

34.(31) Primary thallus composed of discrete squamules, or if forming a rosette, then loosely attached and ascending; medulla K– or K+ pale yellow or red (atranorin, sometimes with norstictic acid) 35

35. Primary squamules very large, 10-20 mm long, 2-8 mm wide 36

35. Primary squamules mostly less than 8 mm long and 4 mm wide 37

36. Squamules usually forming a radiating rosette, curled up at the margins when dry; southeastern coastal plain .... ***Cladonia prostrata* A. Evans**

36. Squamules separate, not forming radiating rosettes; arctic and western montane to New Mexico .... ***Cladonia macrophyllodes* Nyl.**

37. Thallus C+ green, KC+ green (strepsilin); thallus olive-green, sometimes forming almost spherical, vagrant colonies; East Temperate, rarely boreal .... ***Cladonia strepsilis* (Ach.) Grognot**

37. Thallus C–, KC–; thallus gray-green, never forming spherical colonies 38

38. Thallus PD– or PD+ yellow or orange (usually with atranorin, often with psoromic or norstictic and stictic acids) 39

38. Thallus PD+ red (fumarprotocetraric acid) [but see couplet 40]; East Temperate 40

39. Thallus mostly gray, lower surface of squamules sometimes with a violet tinge; rarely producing podetia; on calcareous soils; mostly North Temperate to arctic .... ***Cladonia symphycarpa* (Flörke) Fr.**

39. Thallus greenish above; lower surface of squamules always white; podetia common; East Temperate

.... *Cladonia subcariosa* Nyl. *s. lat.* (including *C. polycarpia, C. polycarpoides;* see couplet 156.)

40. Growing directly on rock; lower side of squamules UV+ blue-white (sphaerophorin) .... *Cladonia petrophila* R. C. Harris

[Note: A chemotype of *C. petrophila* lacks fumarprotocetraric acid and is PD–.]

40. Growing on soil on ground or over rock; squamules UV– 41

41. Squamules finely divided; apothecia raised on short, non-corticate stalks 1-2 mm high; medulla K– .... ***Cladonia caespiticia* (Pers.) Flörke**

41. Squamules strap-shaped, lobed but not finely divided; apothecia produced directly on basal squamules; medulla K+ yellow (atranorin) .... ***Cladonia apodocarpa* Robbins**

42(30). Most or all podetia with cups or cup-like expansions 43

42. Most or all podetia lacking cups 97

43. Podetia sorediate, or with spherical, corticate granules 44

43. Podetia without soredia or granules 68

44. Thallus distinctly yellow-green or greenish yellow, KC+ yellow (containing usnic acid); most species have red apothecia 45

44. Thallus gray, gray-green, pale green, or olive, often becoming brown, KC– or KC+ pinkish orange (usnic acid absent); most species have brown apothecia 52

45. Podetia mostly over 25 mm tall, with relatively narrow cups; soredia farinose; apothecia red 46

45. Podetia 6-25(-40) mm tall, with narrow or broad cups; soredia farinose or granular; apothecia red or pale brown 47

46. Cups malformed, with often irregularly dentate margins; podetia typically fissured; medulla (not surface) UV+ blue-white (squamatic acid) .... ***Cladonia sulphurina* (Michaux) Fr.**

46. Cups usually well formed, with even, often dentate or proliferating margins; podetial wall intact or fissured; medulla UV– (zeorin) .... ***Cladonia deformis* (L.) Hoffm.**

47. Thallus PD+ orange, K+ yellow, UV– (thamnolic acid); cups very narrow; apothecia red 48

47. Thallus PD–, K–, UV– or UV+; cups broad or narrow; apothecia red or pale brown 49

48. Primary squamules usually sorediate on the lower surface near the margins; podetia 6-27 mm tall, usually covered with farinose soredia; northwest .... ***Cladonia umbricola*** **Tønsberg & Ahti**

48. Primary squamules always without soredia; podetia 15-40 mm tall, with granular soredia typically confined to upper half or tip; west coast .... *Cladonia transcendens* (Vainio) Vainio

49. Podetia slender, up to 25 mm tall, covered with farinose soredia, with narrow, flaring cups and marginal proliferations sometimes bearing pale brown pycnidia or apothecia; basal squamules sorediate on lower side; on wood (containing barbatic acid, lacking squamatic acid or zeorin); boreal to arctic .... *Cladonia bacilliformis* (Nyl.) Sarnth [see also couplet 121]

49. Podetia with narrow or broad cups; squamules sorediate or not; lacking barbatic acid 50

50. Thallus UV+ blue-white (squamatic acid; lacking didymic acid); primary squamules usually sorediate on the lower surface near the margins; podetia 6-27 mm tall, usually covered with farinose soredia; apothecia red; northwest .... ***Cladonia umbricola*** **Tønsberg & Ahti**

[Note: See also couplet 70, *C. straminea*, which contains didymic acid as well as squamatic acid and which has microsquamules and areoles that are sometimes granular.]

50. Thallus UV–, lacking squamatic acid, containing zeorin; primary squamules never sorediate; apothecia red or brown 51

51. Apothecia and pycnidia bright red; soredia granular on upper half of cups; cups relatively broad, goblet-shaped, edge of cups more or less even; East Temperate to boreal, infrequent in arctic .... ***Cladonia pleurota*** **(Flörke) Schaerer**

51. Apothecia and pycnidia pale brown (frequently absent); soredia farinose, more or less covering podetia; cups broad or narrow, edge of cups usually toothed; mostly boreal .... ***Cladonia carneola*** **(Fr.) Fr.**

52.(44) Cups opening by a wide hole like a funnel; cup margins bent inward ..... ***Cladonia cenotea*** **(Ach.) Schaerer**

52. Cups closed 53

53. Apothecia or pycnidia red; thallus PD+ orange, K+ yellow (thamnolic acid), or PD–, K–, UV+ white (squamatic acid) in some specimens of *C. umbricola* .... 54 (also see couplet 48)

53. Apothecia or pycnidia brown or absent; thallus PD+ red (fumarprotocetraric acid), or PD– in some specimens of *C. rei*, *C. chlorophaea,* and *C. albonigra* (but lacking thamnolic or squamatic acids) 55

54. Primary squamules large, rounded, lower surface uniformly sorediate; cups trumpet- to goblet-shaped, often toothed at margin; widespread, mostly inland .... ***Cladonia digitata*** **(L.) Hoffm.**

54. Primary squamules finely divided, sorediate mostly near margins of lower surface; podetia covered with soredia; cups narrow, not toothed; mostly northwest coast .... ***Cladonia umbricola*** **Tønsberg & Ahti**

55. Cups proliferating from the center as well as from the margins; podetia blackened at the base with a spotty appearance; soredia coarsely granular or with corticate granules; western 56

55. Cups proliferating from the margins (if at all); podetia not blackened at the base 57

56. Podetia 20-70 mm tall, surface roughened with granules and microsquamules; cups narrow, 1-4 mm across, some podetia without cups; thallus lacking grayanic or 4-*O*-methylcryptochlorophaeic acids .... ***Cladonia verruculosa*** **(Vainio) Ahti**

56. Podetia 10-40 mm tall, surface rather smooth; cups broad, 2-8 mm across, all podetia cupped; usually containing either grayanic or 4-*O*-methylcryptochlorophaeic acids .... *Cladonia albonigra* Brodo & Ahti

57. Cups broad, goblet-shaped 58

57. Cups narrow, trumpet-shaped, or irregular 61

58. Soredia granular; thallus gray-green or brownish, never gray; lacking fatty acids; very common and widespread 59

58. Soredia farinose or, in part, granular; thallus gray to gray-green or very pale geen; containing fatty acids 60

59. Thallus containing grayanic acid, ± fumarprotocetraric acid (UV+ white, PD+ red or PD–) .... ***Cladonia grayi*** **G. Merr. *ex* Sandst.** [Plate 35 in LNA]

59 Thallus lacking grayanic acid but always with fumarprotocetraric acid (UV–, PD+ red) .... ***Cladonia chlorophaea*** **(Flörke *ex* Sommerf.) Sprengel**

[Note: Two rarer species are almost identical morphologically: *C. merochlorophaea* Asahina with merochlorophaeic acid ± fumarprotocetraric acid; *C. cryptochlorophaea* Asahina with cryptochlorophaeic ± fumarprotocetraric acid, ± atranorin.]

60. Containing rangiformic or protolichesterinic acids; west coast .... ***Cladonia asahinae*** **J. W. Thomson** [see also couplet 62 below]

60. Containing bourgeanic acid; soredia confined to outside of cups on upper part of podetia; mostly East Temperate, rare on west coast .... *Cladonia conista* (Nyl.) Robbins

61.(57) Cups well-formed, trumpet-shaped 62
61. Cups flat, shallow, or irregular and aborted 63

62. Podetia with granular and farinose soredia; often with a bluish tint when wet; containing rangiformic or protolichesterinic acid; west coast .... ***Cladonia asahinae* J. W. Thomson**
62. Podetia with only farinose soredia; never with a bluish tint when wet; soredia covering podetia, base not squamulose; lacking fatty acids; widespread .... ***Cladonia fimbriata* (L.) Fr.**

63. Cups very narrow and rather flat, with star-like proliferations; podetia brownish to olive, granular sorediate; thallus PD– or PD+ red (homosekikaic acid ± fumarprotocetraric acid) .... ***Cladonia rei* Schaerer**
63. Cups poorly formed and irregular, without regular proliferations on the cup margins; podetia gray-green to pale green or brownish; soredia farinose or granular; lacking homosekikaic acid 64

64. Thallus PD– (barbatic acid, sometimes with zeorin) ....*Cladonia cyanipes* (Sommerf.) Nyl. (see couplet 108)
64. Thallus PD+ red (fumarprotocetraric acid) 65

65. Southeastern coastal plain; podetia covered with farinose soredia on upper half and granules or microsquamules on the lower half .... ***Cladonia subradiata* (Vainio) Sandst.** (see couplet 115)
65. Northern taxa, not on southeastern coastal plain; soredia not farinose above and granular below 66

66. Soredia covering podetia, entirely farinose; podetia often branched; thallus ashy gray to greenish gray; boreal-montane .... ***Cladonia subulata* (L.) F. H. Wigg.**
66. Soredia produced in irregular patches on the upper 1/2 to 3/4 of the podetia, often becoming continuous at the tip; soredia granular or farinose; podetia unbranched 67

67. Thallus greenish; cups infrequent, and if present, not inflated; widespread temperate to boreal .... ***Cladonia ochrochlora* (Flörke)**
67. Thallus becoming brownish in part; cups always present, with an inflated appearance; oceanic areas on both coasts and arctic ....*Cladonia cornuta* subsp. *groenlandica* (E. Dahl) Ahti

68.(43) Thallus yellowish green or greenish yellow (containing usnic acid); apothecia red 69
68. Thallus not yellowish (usnic acid absent); apothecia brown 72

69. Cups narrow or lacking; podetia and cup margins squamulose; thallus UV+ blue-white, PD–, K–, or UV–, PD+ orange, K+ deep yellow (barbatic acid absent; containing squamatic or thamnolic acid) .... ***Cladonia bellidiflora*** **(Ach.) Schaerer**

69. Cups broad; podetia and cup interior covered with round to irregular areoles, sometimes granular, not squamulose; cup margins smooth or proliferating; thallus PD–, K– 70

70. Medulla UV+ blue-white (squamatic and didymic acids, sometimes with barbatic acid); widespread, especially boreal to arctic .... *Cladonia straminea* (Sommerf.) Flörke (syn. *C. metacorallifera*)

70. Medulla UV– (barbatic acid or zeorin present; squamatic and thamnolic acids absent) 71

71. Contains barbatic acid; common, boreal to montane and arctic .... ***Cladonia borealis*** **S. Stenroos**

71. Contains zeorin; frequent, widespread but mostly arctic-alpine .... *Cladonia coccifera* (L.) Willd.

72.(68) Cups opening by a gaping hole; thallus K–, PD–, UV+ blue-white (squamatic acid); K+ yellow, PD+ orange, UV– (thamnolic acid); or K–, PD+ yellow, UV– (baeomycesic acid) 73

72. Cups closed or partially perforate; thallus K–, PD+ yellow or red (psoromic or fumarprotocetraric acid) 76

73. Contains baeomycesic acid; podetia perforate; Coastal Plain .... *Cladonia atlantica* A. Evans

73. Contains thamnolic and (or) squamatic acid 74

74. Podetial cortex absent; surface covered with large or extremely small squamules; containing squamatic acid .... ***Cladonia squamosa*** **(Nyl. *ex* Leighton) Vainio**

74. Podetial cortex present, continuous or in patches (areoles); surface with or without squamules 75

75. Podetia without longitudinal splits; cortex smooth and continuous; containing squamatic or, rarely, thamnolic acid; widespread boreal and coastal .... ***Cladonia crispata*** **Flotow** (syn. "*C. carassensis*" *of N.Am. authors*)

75. Podetia split with lacerations; cortex discontinuous, broken into areoles; containing thamnolic acid usually with barbatic acid; west coast .... *Cladonia artuata* S. Hammer

76.(72) Podetia blackened at the base 77

76. Podetial base more or less the same color as upper portions 81

77. Cups proliferating from the margins; widespread .... ***Cladonia phyllophora*** **Hoffm.**

77. Cups proliferating from the center 78

[Note: But see *C. stricta*, couplet 80.]

78. Eastern and southern coastal plain; thallus K–, PD+ red or yellow (fumarprotocetraric or psoromic acid; atranorin absent); primary squamules gray-green; podetia more or less uniform in color, or becoming brown at the tips .... ***Cladonia rappii*** **A. Evans** (= N. Am. records of *C. calycantha*)

78. Arctic regions; thallus K+ pale yellow (atranorin); PD+ red (fumarprotocetraric acid) 79

79. Podetia usually covered with squamules, with pale greenish gray areoles over a black stereome giving the upper part a mottled appearance; cups narrow, usually deformed .... ***Cladonia trassii*** **Ahti**

79. Podetia mostly lacking squamules 80

80. Podetia slender, dark brown, cups very infrequent, well formed but narrow, proliferating marginally or from the center; arctic heath .... *Cladonia stricta* (Nyl.) Nyl.

80. Podetia robust, gray, with well-developed cups proliferating from the center; subalpine to alpine stream banks .... *Cladonia uliginosa* (Ahti) Ahti

81.(76) Cups proliferating from the center 82

81. Cups not proliferating, or proliferating from the margins 85

82. Thallus K+ pale yellow (atranorin) .... ***Cladonia macrophyllodes*** **Nyl.**

82. Thallus K– (atranorin absent) 83

83. Cups poorly formed and asymmetrical, having both marginal and central proliferations; East Temperate .... ***Cladonia mateocyatha*** **Robbins**

83. Cups round and symmetrical, proliferating almost exclusively from the center 1-4 times 84

84. Cups flaring abruptly, very flat or deformed; eastern coastal plain ...... ***Cladonia rappii*** **A. Evans** [couplet 78]

84. Cups tapered, flaring gradually, remaining concave at least at the edges; widespread, temperate to arctic .... ***Cladonia verticillata*** **(Hoffm.) Schaerer** (syn. *Cladonia cervicornis* subsp. *verticillata*)

[Note : *Cladonia concinna* Ahti & Goward, with smooth podetia and narrow cups, 5-7 tiers, is found from California to B.C. along the coast. See Ahti 2007.]

85.(81) Primary thallus persistent 86

85. Primary thallus evanescent 92

86. Cups poorly formed and asymmetrical; thallus K– (atranorin absent) 87

86. Cups broad and goblet-shaped, or narrow and trumpet-shaped; thallus K+ yellow or K– 88

87. Podetia 10-60 mm tall; cups with only marginal proliferations, often squamulose at the margins ....***Cladonia phyllophora*** **Hoffm.**

87. Podetia under 20 mm tall; cups irregular, also having some central proliferations ....***Cladonia mateocyatha*** **Robbins**

88. Thallus K+ yellow (atranorin), usually gray or bluish gray; cups broad or very narrow 89

88. Thallus K– (mostly lacking atranorin), typically brownish; cups broad, more or less goblet-shaped 90

89. Podetia reaching 3-7 cm tall; cortex more or less smooth, lacking flattened areoles; cups lacking, very narrow, or occasionally broad; arctic-montane .... ***Cladonia ecmocyna*** **Leighton**

89. Podetia up to 3 cm tall; cortex broken into round, flattened areoles that cover the podetia and line the cups; cups goblet-shaped, always present; central to northeastern, on calcareous soil ....*Cladonia magyarica* Vainio

90. Primary thallus consisting of thick, radiating, closely appressed, almost foliose lobes; on calcareous soil ....***Cladonia pocillum*** **(Ach.) Grognot**

90. Primary thallus consisting of distinct, ascending squamules; on acidic soils 91

91. Cortex broken into round areoles that cover the podetia and line the cups; podetia under 30 mm tall ....***Cladonia pyxidata*** **(L.) Hoffm.**

91. Cortex continuous, smooth; podetia 20-50 mm tall ....***Cladonia gracilis*** **subsp.** ***turbinata*** **(Ach.) Ahti**

92.(85) Cups and (or) podetia perforated or fissured, usually irregular in form 93

92. Cups and podetia closed (imperforate) 94

93. Cups with numerous perforations, like a sieve, cortex compact, rather smooth and shiny ....***Cladonia multiformis*** **G. Merr.**

93. Cups without sieve-like perforations, but podetia irregularly perforate or fissured; surface dull, almost velvety in texture, especially at tips between areoles ....***Cladonia phyllophora*** **Hoffm.**

94. Thallus K+ pale yellow (atranorin) .... ***Cladonia ecmocyna*** **Leighton**

94. Thallus K– (usually lacking atranorin) 95

95. Podetia up to 150 mm tall and 3 mm in diameter, never squamulose ...... ***Cladonia maxima*** **(Asahina) Ahti**

95. Podetia 30-80 mm tall and 0.5-1.5 mm in diameter, often squamulose in part 96

96. Cups typically squamulose at the margins; surface dull, almost velvety in texture, especially at tips between areoles .... ***Cladonia phyllophora*** **Hoffm.**

96. Cups without marginal squamules; podetia smooth and sometimes shiny throughout .... ***Cladonia gracilis*** **subsp.** ***turbinata*** **(Ach.) Ahti**

97.(42) Podetia sorediate, or with spherical, corticate granules 98

97. Podetia without soredia or granules 123

98. Apothecia and (or) pycnidia red 99

98. Apothecia and (or) pycnidia brown, or absent 102

99. Primary squamules with soredia on lower surface; thallus yellow-green or greenish yellow or gray-green (± usnic acid), K+ yellow (thamnolic acid) or K–, UV+ white (squamatic acid); humid northwestern forests .... ***Cladonia umbricola*** **Tønsberg & Ahti**

99. Primary squamules with marginal soredia; thallus gray, gray-green, or brown (usnic acid lacking) 100

100. Soredia powdery, covering the podetia from top to bottom; apothecia present or very frequently absent; K+ yellow, PD+ orange (thamnolic acid) or K–, PD–, KC+ orange (barbatic acid); widespread and common .... ***Cladonia macilenta*** **Hoffm.** (syn. *C. bacillaris*)

100. Soredia coarsely granular, sometimes sparse; apothecia almost always present; eastern 101

101. Primary squamules rounded, only slightly lobed; podetia corticate at least on the lower half; thallus PD–, K– (barbatic acid) .... ***Cladonia floerkeana*** **(Fr.) Flörke**

101. Primary squamules finely lobed; podetia entirely without a cortex, revealing the cartilaginous translucent stereome; thallus PD+ orange, K+ yellow (thamnolic acid), or, less commonly, PD–, K– (barbatic acid) .... ***Cladonia didyma*** **(Fée) Vainio**

102.(98) Soredia coarsely granular, confined to tip of podetia; podetia slender, tall, branched once or twice .... ***Cladonia scabriuscula* (Delise) Nyl.**

102. Soredia or granules covering at least the upper half of podetia 103

103. Podetia with a continuous, more or less smooth cortex on the lower 1/3 to 2/3; soredia usually in discrete patches 104

103. Podetial cortex more or less broken up with areoles, granules, or soredia, not smooth and continuous except sometimes very close to the base; soredia, granules, or squamules diffuse 106

104. Podetia branched once or twice with mostly unequal branches, commonly split or fissured; uncommon; northeastern and northwestern .... *Cladonia farinacea* (Vainio) A. Evans

104. Podetia usually unbranched, rarely fissured; common and widespread 105

105. Thallus with gray-green tones predominating; stereome relatively thick and tough .... ***Cladonia ochrochlora* Flörke**

105. Thallus mostly brownish, especially on upper half; stereome usually relatively thin .... ***Cladonia cornuta* (L.) Hoffm.**

106.(103) Primary thallus disappearing; podetia 30-100 mm tall, often branched once or twice near the tip, covered with powdery soredia; boreal to arctic 107

106. Primary thallus persistent; podetia usually under 35 mm tall 109

107. Thallus PD+ red, UV– (fumarprotocetraric acid); common .... ***Cladonia subulata* (L.) F. H. Wigg.**

107. Thallus PD– 108

108. Thallus UV+ blue-white, KC– (squamatic acid; lacking zeorin); base of podetia remaining pale; rare .... *Cladonia glauca* Flörke

108. Thallus UV–, KC+ orange (barbatic acid and zeorin); base of podetia often becoming blackened with a purplish tint; frequent .... *Cladonia cyanipes* (Sommerf.) Nyl.

109.(106) Podetia more or less covered with corticate granules or granular soredia, often mixed with microsquamules 110

109. Podetia covered with farinose soredia at least on the upper half 114

110. Podetia covered with corticate granules and tiny round areoles, largely ecorticate elsewhere; usually on soil 111

110. Podetia with granular soredia 112

111. Thallus UV+ white, PD–, K– (perlatolic acid); podetial squamules often decumbent, like shingles .... *Cladonia decorticata* (Flörke) Sprengel

111. Thallus UV–, PD+ yellow, K+ red (atranorin and norstictic acid); PD+ yellow, K– (psoromic acid); or PD– to + pale yellow, K+ yellow (atranorin alone or with low concentrations of norstictic acid); podetial squamules not decumbent .... *Cladonia acuminata* (Ach.) Norrlin

112. Granular soredia not mixed with squamules, mainly on upper half of podetia, which has a continuous cortex on the lower half; primary squamules large, lobed but not finely divided; thallus PD+ red (fumarprotocetraric acid) .... ***Cladonia ochrochlora* Flörke**

112. Granular soredia mixed with finely divided podetial squamules over entire podetial surface, which is largely without a cortex; primary squamules finely divided, becoming granular sorediate (with blastidia) at the margins and often reduced to a granular mass 113

113. Thallus PD+ orange, K+ yellow (thamnolic acid); podetia short, up to 30 mm tall; growing on wood .... ***Cladonia parasitica* (Hoffm.) Hoffm.**

113. Thallus PD+ red, K– or brownish (fumarprotocetraric acid); podetia 5-50 mm tall, usually bent or distorted; on wood or soil .... *Cladonia ramulosa* (With.) J. R. Laundon

114.(109) Podetia with fine soredia above, but mixed with granules and microsquamules on lower half 115

114. Podetia uniformly farinose sorediate 116

115. Thallus UV+ white (grayanic acid); East Temperate .... *Cladonia cylindrica* (A. Evans) A. Evans

115. Thallus UV– (lacking grayanic acid); southeastern coastal plain .... ***Cladonia subradiata* (Vainio) Sandst.**

116. Podetia tall and slender, up to 100 mm, sometimes branched, pale gray to whitish above and brownish below; on soil 117

116. Podetia less than 50 mm, rarely branched; yellowish or greenish to greenish gray; on wood, rock, and soil 118

117. Thallus PD–, K–, UV+ white (squamatic acid); widespread but infrequent .... *Cladonia glauca* Flörke

117. Thallus PD+ red, K– or brownish, UV–; common and widespread except for southeast .... ***Cladonia subulata*** **(L.) F. H. Wigg.**

118. Thallus PD+ red, K– (fumarprotocetraric acid); common and widespread 119

118. Thallus PD+ orange, K+ yellow, **or** PD–, K– 120

119. Podetia entirely uniformly sorediate except at the very base, often strongly tapered .... ***Cladonia coniocraea*** **(Flörke) Sprengel**

119. Podetia corticate on the lower ¼ - ½, tapered or cylindrical .... ***Cladonia ochrochlora*** **Flörke**

120. Thallus PD+ orange, K+ yellow (thamnolic acid); widespread .... ***Cladonia macilenta*** **Hoffm.** *s. str.*

120. Thallus PD–, K–, KC+ yellow or orange (barbatic acid) 121

121. Thallus yellowish (usnic acid) .... *Cladonia bacilliformis* (Nyl.) Sarnth. [see couplet 49]

121. Thallus greenish to brownish, without yellow tint (lacking usnic acid) 122

122. Podetia slender, tapering to a point; apothecia pale brown; red spots on squamules where damaged by mites; humid western forests, rare in northeast .... *Cladonia norvegica* Tønsberg & Holein

122. Podetia stubby or slender, usually blunt; sterile; red spots absent; widespread .... sterile material of ***Cladonia macilenta*** **Hoffm.** (syn. *C. bacillaris*)

123.(97) Thallus yellow-green or greenish yellow (usnic acid) 124

[Note: In *Cladonia alaskana* (below), the yellow tint is often hidden by brown pigments.]

123. Thallus not yellowish (usnic acid absent) 131

124. Primary thallus disappearing; podetia slightly branched, surface dull, PD+ red (fumarprotocetraric acid); base black, creating a spotted appearance when covered with pale areoles; axils commonly open, rarely flaring; western arctic .... *Cladonia alaskana* A. W. Evans

124. Primary thallus persistent; podetia unbranched, PD– or PD+ orange, base not blackened 125

125. Apothecia pale brown; thallus PD– 126

125. Apothecia red 127

126. Growing on soil or rock; primary squamules dense, crenulate, or deeply lobed

.... ***Cladonia robbinsii* A. Evans** [see also couplet 33]

126. Growing on wood; primary squamules scattered, not lobed .... ***Cladonia botrytes* (K. G. Hagen) Willd.**

127. Primary squamules sorediate; medulla UV+ white, PD– (squamatic acid) 128

127. Primary squamules without soredia; medulla UV–, PD– or PD+ orange 129

128. Soredia on lower surface of squamules; containing didymic acid; common throughout the coastal plain .... ***Cladonia incrassata* Flörke**

128. Soredia on margins of squamules; containing grayanic acid; rare, North Carolina to Florida .... *Cladonia anitae* W. L. Culb. & C. F. Culb.

129. Western and arctic regions; podetial surface areolate, abundantly squamulose .... ***Cladonia bellidiflora* (Ach.) Schaerer**

129. Eastern; podetia with a smooth to areolate, cortex squamulose or without any squamules 130

130. Thallus K–, PD– (barbatic acid); very common, East Temperate extending westward in boreal regions .... ***Cladonia cristatella* Tuck.**

130. Thallus K+ deep yellow, PD+ orange (thamnolic acid); Florida and Georgia .... *Cladonia abbreviatula* G. Merr.

131.(123) Primary thallus disappearing 132

131. Primary thallus persistent 142

132. Podetial axils closed 133

132. Podetial axils open 137

133. Thallus K+ pale yellow .... ***Cladonia ecmocyna* Leighton**

133. Thallus K– 134

134. Podetia 30-80 mm tall and 0.5-1.5 mm in diameter; base not blackened; eastern boreal, uncommon .... ***Cladonia gracilis* (L.) Willd. subsp. *gracilis***

134. Podetia up to 150 mm tall and 3 mm in diameter 135

135. Base of podetia distinctly blackened; western arctic to northwestern B.C., rare in northeastern mountains

.... *Cladonia gracilis* (L.) Willd. subsp. *elongata* (Wulfen) Vainio

135. Base of podetia not at all blackened 136

136. Podetia up to 3 mm wide, rarely perforate or lacerate; common on the northeastern coast .... ***Cladonia maxima* (Asahina) Ahti**

136. Podetia rarely over 2 mm wide; commonly perforate and lacerate; common on oceanic west coast .... *Cladonia gracilis* (L.) Willd. subsp. *vulnerata* Ahti

137.(132) Podetial axils wide open, forming funnels; thallus PD–, or PD+ yellow or orange (squamatic, baeomycesic or thamnolic acid; fumarprotocetraric acid absent) 138

137. Podetial axils partially open or wide open, not forming funnels; thallus PD+ red (fumarprotocetraric acid) 140

138. Containing baeomycesic acid (K–, PD+ yellow); Coastal Plain .... *Cladonia atlantica* A. Evans

138. Containing squamatic and (or) thamnolic acid 139

139. Podetia with pointed tips, often dark brown throughout; stereome often blackened, especially at base; containing squamatic acid; northern boreal to arctic .... *Cladonia subfurcata* (Nyl.) Arnold

139. Podetial tips pointed or funneled; stereome not blackened even at base .... See couplet 75

140.(137) Thallus K+ pale yellow (atranorin); podetia stout, 1-5 mm in diameter ....***Cladonia turgida* Hoffm.**

140. Thallus K– (atranorin absent); podetia mostly slender, under 2 mm in diameter 141

141. Branching unequal; cortex mostly continuous .... ***Cladonia multiformis* G. Merr.**

141. Branching in more or less equal dichotomies; cortex usually breaking up into small irregular patches especially on the upper half of the podetia ..... ***Cladonia furcata* (Hudson) Schrader** (western morphotype)

142.(131) Apothecia red 143

142. Apothecia brown or absent 144

143. Primary squamules small, white below, often dissolving into soredia; common on trees, southeastern coastal plain ....*Cladonia ravenelii* Tuck.

143. Primary squamules large, with a broad yellow band on lower surface, sometimes sorediate on the margins; Florida .... *Cladonia hypoxantha* Tuck.

144. Podetia mostly 1-3 mm tall, entirely without a cortex, developing from squamules that are much divided and lacey; thallus PD+ red, K– (fumarprotocetraric acid) .... ***Cladonia caespiticia* (Pers.) Flörke**

144. Podetia mostly 3-40 mm tall, with or without a cortex 145

145. Growing on wood or bark 146

145. Growing on soil, moss, or rock 148

146. Primary squamules granular at the margins often reducing the primary thallus to a granular crust; thallus PD+ orange, K+ deep yellow (thamnolic acid) .... ***Cladonia parasitica* (Hoffm.) Hoffm.**

146. Primary squamules finely divided but not granular 147

147. Thallus PD+ yellow, K– (baeomycesic acid); coastal plain .... ***Cladonia beaumontii* (Tuck.) Vainio**

147. Thallus usually PD–, K–, UV+ white (squamatic acid) or sometimes PD+ orange, K+ deep yellow (thamnolic acid); widespread .... ***Cladonia squamosa* Hoffm.**

148.(145) Podetia without a cortex 149

148. Podetia at least partly corticate, sometimes broken into areoles or flat squamules 151

149. Podetia abundantly lacerate, with scattered areoles and microsquamules; thallus K+ yellow, PD+ yellow or red or PD–, UV– (atranorin with or without norstictic or psoromic acid) .... ***Cladonia symphycarpa* (Flörke) Fr.** ("*C. dahliana* Kristinsson" can be used for psoromic acid chemotype.)

149. Podetia not or moderately lacerate; thallus UV+ white 150

150. Podetia covered with abundant large to tiny, mostly shingled squamules; thallus PD–, K– or PD+ orange, K+ yellow (squamatic or thamnolic acid); podetial tips blunt, often with a gaping hole; very common and widespread .... ***Cladonia squamosa* Hoffm.**

150. Podetial squamules small and scattered; podetia pointed at tips, sometimes fissured on the sides; PD–, K–, (perlatolic acid); boreal to arctic .... *Cladonia decorticata* (Flörke) Sprengel

151(148). Medulla C+ green, KC+ green (strepsilin); thallus distinctly olive in color; podetia more or less inflated; eastern .... ***Cladonia strepsilis* (Ach.) Grognot**

151. Medulla C–, KC– 152

152. Thallus PD–, K–, UV+ blue-white (squamatic acid); podetial squamules shingle-like; west coast, Alaska to Washington .... *Cladonia singularis* S. Hammer

152. Thallus PD+ or K+, UV– 153

153. Thallus K+ yellow, orange, or red 154

153. Thallus K– (or brownish); PD+ red (fumarprotocetraric acid) or PD+ yellow 159

154. Thallus K+ red, PD+ yellow (norstictic acid) 155

154. Thallus K+ yellow, PD– or PD+ yellow or red 157

155. Podetia rare, stocky, 2-5 mm in diameter, 10-20 mm tall, walls abundantly lacerate and fissured; containing atranorin .... ***Cladonia symphycarpa* (Flörke) Fr.**

155. Podetia usually abundant, slender, 0.6-2.5 mm in diameter, 8-20 (-30) mm tall, walls intact or sometimes split; with or without atranorin or stictic acid 156

156. Containing atranorin .... *Cladonia subcariosa* Nyl. *s. str.*

156. Lacking atranorin .... ***Cladonia polycarpoides* Nyl.** (Plate 258 in LNA)

[Note: Specimens that also contain stictic acid can be called *C. polycarpia* G. Merr.]

157. Podetia slender, branched at tips, not split or lacerate, cortex continuous; thallus PD+ orange, K+ yellow (thamnolic acid); California to Alaska .... *Cladonia poroscypha* S. Hammer

157. Podetia robust, lacerate, mostly unbranched; cortex in patches; lacking thamnolic acid 158

158. Primary squamules large, 2-8 x 1-6 mm; podetia rare, 10-15 mm tall; thallus PD+ yellow (psoromic acid) or PD– (atranorin alone) .... ***Cladonia symphycarpa* (Flörke) Fr.**

158. Primary squamules small, thick, 1-3 x 0.5-2 mm; podetia usually abundant, 10-30 mm tall; thallus PD+ red (fumarprotocetraric acid) or PD– (atranorin alone) .... ***Cladonia cariosa* (Ach.) Sprengel**

159.(153) Podetia covered with peltate areoles or squamules (attached in the center); thallus K–, PD+ bright yellow (psoromic acid); primary squamules 3-8 mm in diameter; arctic-alpine .... *Cladonia macrophylla* (Schaerer) Stenh.

159. Podetial squamules and areoles not peltate; thallus PD+ red (fumarprotocetraric acid) or PD+ yellow (psoromic acid) 160

160. Podetia blackened at base; arctic regions; apothecia rarely present; thallus PD+ red 161

160. Podetial base more or less the same color as upper portions; apothecia always present; East Temperate 162

161. Podetia usually covered with squamules sticking out at right angles from podetia, with pale, flat areoles over a black stereome giving the podetium a mottled appearance .... ***Cladonia trassii* Ahti**

161. Podetia mostly lacking squamules, dark brown, with convex areoles, slender .... *Cladonia stricta* Nyl.

162. Podetia minutely warty, slender, often twisted; apothecia much broader than podetia; thallus PD+ red .... ***Cladonia peziziformis* (With.) J. R. Laundon**

162. Podetia smooth, stout, not twisted; apothecia slightly or not broader than podetia 163

163. Thallus PD+ red; East Temperate .... ***Cladonia subcariosa* Nyl.** (including *C. sobolescens*), [Plate 265 in LNA]

163. Thallus PD+ yellow; mostly northeastern seaboard .... *Cladonia brevis* (Sandst.) Sandst.

[Note: Although sometimes included within *C. subcariosa s. lat.*, there is some genetic evidence that it is distinct.]

COENOGONIUM (syn. *Dimerella*)

1. Thallus crustose, spores 7-14 x 2-4 µm; widespread 2

1. Thallus filamentous, forming cottony tufts or shelves, spores 6-10 x 2-3 µm; subtropical 3

2. Apothecia 0.1-0.3(-0.4) mm in diameter, typically concave; apothecia pale orange to beige or almost white .... *Coenogonium pineti* (Schrad. *ex* Ach.) Lücking & Lumbsch (syn. *Dimerella pineti*)

2. Apothecia (0.3)0.8--1.5 mm in diameter, typically flat, pale orange to orange-brown .... ***Coenogonium luteum* (Dickson) Kalb & Lücking** (syn. *Dimerella lutea*)

3. Spores 1-celled, in two rows within the asci .... *Coenogonium interpositum* Nyl.

3. Spores 2-celled 4

4. Thalli forming horizontal shelves with the apothecia mostly on the lower side of the shelves .... *Coenogonium linkii* Ehrenb.

4. Thalli tufted, not forming horizontal shelves; apothecia on outer surface of tufts 5

5. Thallus filaments loose, forming a fluffy mat; apothecia raised on short stalks .... *Coenogonium interplexum* Nyl.

5. Thallus filaments dense, forming a compact mat; apothecia not raised on short stalks .... ***Coenogonium implexum*** **Nyl.**

COLLEMA (including *Blennothallia, Enchylium, Lathagrium*: see Otálora et al. 2013)

1. Growing on bark 2
1. Growing on rock, soil, or moss 10

2. Lobes small, under 3 mm wide, crowded into small cushions; spores fusiform, 2-celled, (13-)15-24(-26) x 3-4.5(-6) µm .... *Enchylium conglomeratum* (Hoffm.) Otálora et al. (syn. *Collema conglomeratum*)
2. Lobes over 3 mm wide, not cushion-forming; spores fusiform to needle-shaped, 4- to 16-celled 3

3. Isidia absent 4
3. Isidia present 8

4. Spores 4-celled, 20-40 x 3-4.5 µm; interior northwestern montane .... *Collema curtisporum* Degel.
4. Spores 6- to 15-celled, 50-100 µm long 5

5. Apothecia pruinose; eastern 6
5. Apothecia without pruina 7

6. Apothecial cortex absent .... ***Collema pulcellum*** **Ach. var.** ***subnigrescens*** **(Müll. Arg.) Degel.**
6. Apothecial cortex present .... *Collema pulcellum* var. *leucopeplum* (Tuck.) Degel.

7. Tissue below the hypothecium in mature apothecia composed of round to angular cells; eastern .... *Collema pulcellum* var. *pulcellum* Ach.
7. Tissue below the hypothecium in mature apothecia composed of long, cylindrical cells; East Temperate and Pacific coast .... ***Collema nigrescens*** **(Hudson) DC.**

8.(3) Isidia cylindrical in mature thalli, but granular and globose on young thalli, unbranched or branched .... ***Collema furfuraceum*** **(Arnold) Du Rietz**
8. Isidia spherical or globular in mature thalli 9

9. Thallus relatively flat, although sometimes with folds, always isidiate ....***Collema subflaccidum*** **Degel.**
9. Thallus with round to elongate pustules or blisters, rarely isidiate .... ***Collema nigrescens*** **(Hudson) DC.**

10.(1) Lobes more or less uniform in thickness, not thicker at the margins; lobes mostly (1-)2-6(-10) mm wide, with or without erect marginal lobules 11

10. Lobe with distinctly thickened margins, often producing erect, swollen lobules; small species with lobes mostly 0.5-3 mm wide 18

11. Lobes erect and branched, finely divided to lobulate at the margins, under 3 mm wide; spores 4-celled to muriform when mature, 18-32(-40) x 8-13 µm .... ***Lathagrium cristatum* (L.) Otálora et al.** (syn. *Collema cristatum*)

11. Lobes prostrate or on edge and overlapping, branched or rounded, often over 3 mm wide; spores muriform or not muriform 12

12. Isidia absent; thallus pustulate; on siliceous rocks; mainly East Temperate .... *Collema ryssoleum* (Tuck.) A. Schneider

12. Isidia or lobules present; thallus pustulate or not; on various kinds of rock 13

13. Isidia flattened like squamules, overlapping 14

13. Isidia cylindrical or globular, not flattened 15

14. Lobes broad, often over 6 mm across; spores fusiform, 4- to 6-celled, 6-8.5 µm wide; widely distributed .... *Collema flaccidum* (Ach.) Ach.

14. Lobes less than 3 mm across; spores ellipsoid, 4-celled to rarely submuriform, 26-34 x 13-16 µm; mainly southwestern .... *Blennothallia crispa* (Hudson) Otálora et al. (syn. *Collema crispum*)

15. Isidia cylindrical on mature specimens, unbranched or branched; lobes strongly ridged and pustulate, 5-10 mm wide .... ***Collema furfuraceum* (Arnold) Du Rietz**

15. Isidia spherical or globular, rarely cylindrical; lobes more or less smooth or folded, not pustulate, (1-)2-6 mm wide 16

16. On siliceous rock; lobes mostly flat, with small, gobular isidia (rarely becoming cylindrical); apothecia rare; spores 6- to 8-celled; mainly East Temperate and southwest .... ***Collema subflaccidum* Degel.**

16. On calcareous rock; lobes crowded, often concave, with large globular isidia; apothecia relatively abundant; widespread 17

17. Spores fusiform, 4-celled ....
Lathagrium undulatum **(Flotow) Otálora et al. var.** ***granulosum*** **(Degel.),** ***ined.*** (syn. *Collema undulatum* var. *granulosum*)

17. Spores ellipsoid, muriform ....*Lathagrium fuscovirens* (With.) Otálora et al. (syn. *Collema fuscovirens*)

18.(10) On calcareous rock 19

18. On soil or bryophytes 21

19. Spores 2-celled, (8.5-)12-22(-24) x 4-7(-8.5) µm; lobes narrow and finely divided
....“*Collema*” *texanum* Tuck. **[see also couplet 23]**

19. Spores (2-)4-celled or submuriform 20

20. Lobes flat, crowded, usually on one edge, without lobules or isidia; apothecia abundant; spores (2-)4-celled, narrow, 6.5-8.5 µm wide ...... ***Enchylium polycarpon*** **(Hoffm.) Otálora et al.** (syn. *Collema polycarpon*)

20. Lobes with cylindrical, erect branches; globular isidia sometimes present; spores muriform to 4-celled, ellipsoid, 8-13 µm wide .... ***Lathagrium cristatum*** **(L.) Otálora et al.** (syn. *Collema cristatum*)

21.(18) Lobes prostrate; lobules, when present, rounded, strap-shaped, or finely divided.......................................22

21. Lobes erect, often with cylindrical lobules 24

22. Lobes 2-4(-6) mm broad, adnate; spores submuriform, (20-)26-32 x 8.5-15 µm (Degelius 1974)
....*Enchylium bachmannianum* (Fink) Otálora et al. (syn. *Collema bachmannianum*)

22. Lobes narrower than 2.5 mm; spores 2- to 4-celled, rarely submuriform 23

23. Spores 2-celled, (8.5-)12-22(-24) x 4-7(-8.5) µm; isidia sparse to numerous, globular to cylindrical or squamiform; montane southwest and Alabama .... “*Collema*” *texanum* Tuck.

23. Spores mostly 4-celled and fusiform,sometimes submuriform, 15-25 x 6.5-10 µm; globular isidia sometimes present on lobe surface; widespread .... ***Enchylium tenax*** **(Ach.) Otálora et al.** (syn. *Collema tenax*)

24. Spores 2-celled; lobules cylindrical, appearing globular from above, but true isidia absent
....*Enchylium coccophorum* (Tuck.) Otálora et al. (syn. *Collema coccophorum*)

24. Spores 4-celled to submuriform; lobules erect, sometimes branched; globular isidia sometimes present
.... ***Lathagrium cristatum*** **(L.) Otálora et al.** (syn. *Collema cristatum*)

CRYPTOTHECIA (including *Herpothallon*)

1. Prothallus bright red, PD+ deep reddish purple, K+ dark purple-red; thallus C– (chiodectonic and confluentic acids); sterile
   .... ***Herpothallon rubrocinctum*** **(Ehrenb: Fr.) Aptroot et al.** (syn. *Cryptothecia rubrocincta*)
1. Prothallus white, PD–, K–, C+ pink (gyrophoric acid); usually fertile, but without clearly defined ascomata; spores muriform, 55-70 x 23-30 µm, 1(-2) per ascus .... ***Cryptothecia striata*** **G. Thor**

CYPHELIUM

1. Thallus bright greenish yellow (pulvinic acid pigments) 2
1. Thallus pale or yellowish gray to dark gray to greenish or brownish 5

2. Apothecia largely immersed in the thallus with hardly any margin showing; yellow pigment is rhizocarpic acid 3
2. Apothecia prominent, with margins clearly visible 4

3. Spores always 2-celled, broadly ellipsoid; widely distributed ....*Cyphelium tigillare* (Ach.) Ach.
3. Spores soon becoming submuriform, almost spherical, 17-24 x 13-18 µm; very common in the northcentral prairies and infrequent in southern California
   .... *Cyphelium notarisii* (Tul.) Blomb. & Forssell (syn. *C. brachysporum*)

4. Rim of apothecium yellow pruinose; southwest, southern Rockies, and Great Lakes region to Maritimes; yellow pigment is vulpinic acid ....***Cyphelium lucidum*** **(Th. Fr.) Th. Fr.**
4. Apothecium entirely black, without pruina; western montane at high elevations; yellow pigment is rhizocarpic acid ....*Cyphelium pinicola* Tibell

5. Apothecia buried in thallus verrucae, which are constricted at the base and contain an orange pigment that turns red with K; spores 20-23 x 10-14 µm; no lichen substances; California and northern British Columbia .... *Cyphelium trachylioides* (Nyl. *ex* Branth & Rostrup) Erichsen
5. Apothecia prominent with margins showing; lacking orange pigment in the base of the verrucae; spores smaller 6

6. Yellow pruina present on apothecial surface and margin; spores 7-10 x 5-6 µm; southern California, rare .... *Cyphelium chloroconium* (Tuck.) Zahlbr.
6. Lacking yellow pruina on apothecia; spores 13-22 x 8-11 µm 7

7. Thallus gray to yellowish gray; apothecia 0.7-1.5 mm in diameter; containing usnic acid, atranorin, and unidentified pigments; California to Pacific Northwest, common ...... ***Cyphelium inquinans*** **(Sm.) Trevisan**

7. Thallus gray to dark greenish brown; apothecia 0.4-0.8 mm in diameter; lacking lichen substances; northern Rocky Mountains .... *Cyphelium karelicum* (Vainio) Räsänen

DACTYLINA (including *Allocetraria*)

1. Stalks mostly unbranched, entirely hollow, 20-70 mm tall, 2-6(-14) mm in diameter 2

1. Stalks branched; medulla webby to dense, mostly under 20 mm tall and 2 mm in diameter 3

2. Medulla PD–, KC–, cortex C–; medulla C+ pink (gyrophoric acid) ..... ***Dactylina arctica*** **(Richardson) Nyl.**

2. Medulla PD+ red, KC+ pink (physodalic and physodic acids); cortex (not medulla) C+ pink (gyrophoric acid) .... *Dactylina beringica* C. D. Bird & J. W. Thomson

3. Stalks yellowish to pinkish violet, pruinose on young tips, with irregular, divergent branching, almost hollow or partially filled with cobwebby hyphae; medulla KC+ pink, usually PD+ red (physodic acid, usually with physodalic acid) .... ***Dactylina ramulosa*** **(Hooker) Tuck.**

3. Stalks greenish yellow, not pruinose, branching in regular dichotomies; medulla densely filled with white hyphae; medulla KC–, PD– (fatty acids) .... ***Allocetraria madreporiformis*** **(Ach.) Kärnefelt & Thell**

DENDROGRAPHA (including *Lecanographa, Schismatomma,* and *Sigridea*) Based on Ertz & Tehler (2011).

1. Thallus crustose 2

1. Thallus fruticose 7

2. Spores halonate, 7- to 8-celled, 19-30 x 4-6 µm; apothecia round to irregular, marginless or appearing lecanorine, but with a dark brown to black exciple; thallus cortex and medulla C+ red, KC+ red (lecanoric and gyrophoric acids); on rocks, coastal southern California
.... ***Lecanographa hypothallina*** **(Zahlr.) Egea & Torrente**

2. Spores not halonate; apothecia truly lecanorine or appear to be lecanorine; thallus cortex and medulla C+ or C–, KC+ or KC–; on rocks or bark 3

3. Hypothecium not well developed, appearing yellowish; apothecia largely marginless, white pruinose; thallus brownish, very thin, C–; spores 4-celled, 18-30 x 5.5-8.5 µm (Fink 1935); on bark; southeastern, north to New Jersey .... *Schismatomma glaucescens* (Nyl. *ex* Willey) R. C. Harris

3. Hypothecium well developed, brown to black; some sort of apothecial margin usually evident 4

4. Hypothecium lens-shaped, not extending to the substrate; thallus cortex or medulla C+ red, KC+ red (erythrin, lecanoric acid); California or Florida .... *Dirina*

4. Hypothecium extending like a peg to the substrate (seen in a section through the center of the apothecium); thallus cortex of North American species C–, KC–; California 5

5. Thallus PD–; ascomatal disks irregular in shape, lobed or slightly elongate; on bark or rock .... *Dendrographa franciscana* (Zahlbr. *ex* Herre) Ertz & Tehler (*syn. Roccellina franciscana*)

5. Thallus PD+ yellow or red-orange; ascomatal disks round; on bark or wood 6

6. Thallus creamy white when fresh, PD+ yellow (psoromic acid); on trees along coast and into chaparral, San Francisco to Los Angeles .... *Sigridea californica* (Tuck.) Tehler (syn. *Schismatomma californicum*)

6. Thallus tan to greenish brown, PD+ red-orange (unknown substance); on bark or wood, Santa Catalina Island .... ***Dendrographa conformis* (Tehler) Ertz & Tehler** (syn. *Roccellina conformis*)

7. Branches all round in cross section (terete); medulla cartilaginous making the thallus quite stiff 8

7. Branches flattened in part, terete only on young parts; medulla cottony making the thallus rather flaccid 9

8. Fertile, with ascomata ....*Dendrographa alectoroides* Sundin & Tehler f. *alectorioides*

8. Sterile, ascomata absent .... *Dendrographa alectoroides* f. *parva* Sundin & Tehler

9. Fertile, with black apothecia-like ascomata; thallus dark ....*Dendrographa leucophaea* (Tuck.) Darbish. f. *leucophaea*

9. Sterile, often with fungus-induced galls that resemble soralia; thallus pale ....***Dendrographa leucophaea* f. *minor* (Darbish.) Sundin & Tehler**

DERMATOCARPON

1. Lower surface covered with rhizines or tomentum 2

1. Lower surface lacking rhizines or tomentum 3

2. Lower surface with short, thick, dark brown to black rhizines; western montane, rare in the Great Lakes region and east ....*Dermatocarpon moulinsii* (Mont.) Zahlbr.

2. Lower surface with a fine tomentum consisting of bead-like (moniliform) brown hyphae; rare, on calcareous rock, mostly in shade, Missouri and Texas .... *Dermatocarpon tomentulosum* Amtoft

3. Thallus with multiple, almost squamulose lobes having multiple holdfasts 4

3. Thallus composed of single lobes with single holdfast (secondary lobes, if present, without holdfasts) 8

4. Center of thallus with contorted, convex, "intestine-like" lobes; on dry calcareous rocks; mainly western .... *Dermatocarpon intestiniforme* (Körber) Hasse

4. Center of thallus more or less flat, not contorted; on acidic rocks, often in aquatic or semiaquatic habitats, or in seepage areas 5

5. Spores mostly shorter than 15 μm; shade intolerant, but frequent in poorly drained or seepage areas; thallus thin, pale brown to black; southeastern .... *Dermatocarpon arenosaxi* Amtoft

5. Spores mostly longer than 15 μm; submerged in water at least periodically 6

6. Lower surface strongly wrinkled; western .... *Dermatocarpon rivulorum* (Arnold) Dalla Torre & Sarnth.

6. Lower surface smooth or, less frequently, wrinkled 7

7. Thallus bright green when wet; common; thallus thick; widespread, especially in northeast .... **Dermatocarpon luridum (With.) J. R. Laundon**

7. Thallus remaining brown or greenish brown when wet; west coast and western Great Lakes region .... *Dermatocarpon meiophyllizum* Vainio

8. Lower surface roughly papillate like coarse sandpaper, usually black (fig. 30) .... ***Dermatocarpon reticulatum* H. Magn.**

8. Lower surface smooth, without papillae, pale brown to black 9

9. Perthecia large, pear-shaped, (0.25-)0.35-0.6(-0.8) mm in diameter appearing as brown bumps on the upper thallus surface; lower surface also with round bumps corresponding to the embedded perithecia; spores (9-)12-15(-19) x (4-)5-7(-9); Eastern Temperate, common; calcareous or non-calcareous rocks .... *Dermatocarpon muhlenbergii* (Ach.) Müll. Arg.

9. Perithecia small (0.15-0.35 mm in diameter), globular or slightly tear-shaped, with black ostioles that are level with the thallus surface or are somewhat immersed 10

10. Lobes typically deeply dissected, although umbilicate and attached at a single point; in Ozarks and elsewhere in east; thallus dark brown but with a pruina-like bloom; lower surface typically verrucose or veined, less frequently smooth .... *Dermatocarpon dolomiticum* Amtoft

10. Lobes more or less entire, not deeply dissected; upper surface usually strongly white pruinose; lower surface smooth, verrucose, or reticulate veined 11

11. Southwestern; perithecia 0.2-0.35 mm in diameter with brown to black ostioles that are level with the thallus surface or are somewhat immersed; spores 10-14 x 5-7 µm .... *Dermatocarpon americanum* Vainio (Plate 305 in LNA?)

11. North Temperate to boreal; perithecia 0.15-0.35; spores (9.0-)12.2-16.5(-18.7) x 5.0-6.2(-8.5) µm, narrowly ellipsoid .... ***Dermatocarpon miniatum* (L.) W. Mann** *s. str.*

[Note: This description of *D. miniatum* is based mainly on Canadian specimens in CANL. Heiðmarsson & Breuss (2004) regard Plate 305 in LNA to be *D. americanum* Vainio, based on the red reaction of the medulla with IKI, with *D. miniatum* apparently reacting IKI–. Amtoft et al. (2008) recognize a widespread eastern species in this group as *D. muhlenbergii* with large perithecia, (0.3-)0.4-0.6(-0.8) mm diameter) that appear as brown bumps on the upper thallus surface. Southwestern specimens of *D. americanum* have smaller perithecia (0.2-0.33 mm in diameter) with black ostioles that are level with the thallus surface or are somewhat immersed. Amtoft et al. say that the IKI reaction of the medulla is unreliable, so it is unclear where the true *D. miniatum s. str.* fits in. Complicating the issue is *D. taminium* Heiðmarsson, common in Arizona, which has long spores, 14-18(-22) x (5-)6-8.5(-10) µm, but otherwise resembles *D. miniatum*.]

DIMELAENA

1. Thallus greenish yellow (usnic acid); widespread .... ***Dimelaena oreina* (Ach.) Norman**

1. Thallus white to brown; California to Washington 2

2. Thallus creamy white to brownish gray; medulla PD–, K– .... ***Dimelaena radiata* (Tuck.) Müll. Arg.**

2. Thallus dark brown 3

3. Hypothecium colorless; medulla PD–, K–; common .... *Dimelaena thysanota* (Tuck.) Hale & W. L. Culb.

3. Hypothecium dark brown; medulla PD+ orange, K+ yellow (stictic acid); rare .... *Dimelaena californica* (H. Magn.) Sheard

DIPLOSCHISTES

1. Apothecia opening by a narrow pore, i.e., perithecium-like 2

1. Apothecia at first crater-like, but later broadening, not pore-like 3

2. Thallus gray; ostiole opening with radial cracks; widely distributed, especially in southern and central U.S. .... *Diploschistes actinostomus* (Ach.) Zahlbr.

2. Thallus dark red-brown; South Carolina, Texas, and California .... *Diploschistes aeneus* (Müll. Arg.) Lumbsch

3. Thallus heavily pruinose, thick and areolate with areoles up to 3 mm across; spores 4-8 per ascus; mainly on bare soil in the arid southwest .... ***Diploschistes diacapsis* (Ach.) Lumbsch**

3. Thallus without, or with little pruina, thin or thick, with areoles less than 1.5 mm across; spores 4- or 4-8 per ascus; widely distributed 4

4. Spores constantly 4 per ascus; at first growing over and parasitic on lichens, later independent on soil or mosses .... ***Diploschistes muscorum* (Scop.) R. Sant.**

4. Spores 4-8 per ascus; growing directly on rock .... ***Diploschistes scruposus* (Schreber) Norman**

DIRINA See Tehler, Ertz, & Irestedt (2013)

1. Growing on bark (or on rock in non-North American material), thallus lacking soredia; cortex C+ red (erythrin and lecanoric acid), medulla C–; Florida .... *Dirina paradoxa* (Fée) Tehler (syn. *Dirina approximata* subsp. *hioramii*)

1. Growing on rock; thallus usually sorediate; not in Florida 2

2. Coastal California; cortex and medulla C+ red (erythrin and lecanoric acid); spores 23-29 x 5-6 µm .... ***Dirina catalinariae* Hasse f. *sorediata* Tehler**

2. Ozarks and southern Appalachians; cortex C+ red, medulla C–; spores 20-24 x 4-6 µm .... *Dirina massiliensis* Durieu & Mont.

DIRINARIA

1. Thallus non-sorediate 2

1. Thallus sorediate 4

2. Apothecial disks reddish purple caused by a purple pruina; lobes 0.2-0.7 mm wide .... ***Dirinaria purpurascens* (Vainio) B. Moore**

2. Apothecial disks black, without pruina 3

3. Lobes 1-2(-4) mm wide; spores 5-8 µm wide; medulla UV– (sekikaic acid compounds); common, on bark, southeastern coastal plain to Baja California ....***Dirinaria confusa* D. D. Awasthi**

3. Lobes typically 0.5-1(-2) mm wide; spores 7-10 µm wide; medulla UV+ white (divaricatic acid); on bark or rock, Florida to Baja California, rare in the east .... *Dirinaria confluens* (Fr.) D. D. Awasthi

4. Soredia granular, originating from the breakdown of pustules or hollow warts, covering the thallus surface .... ***Dirinaria aegialita* (Afz. *ex* Ach.) B. Moore**

4. Soredia farinose, discrete, in hemispherical mounds, rarely confluent 5

5. On rock, southeastern, rare in the northeast, largely avoiding the coastal plain; lobes less than 0.5 mm wide ....*Dirinaria frostii* (Tuck.) Hale & W. L. Culb.

5. On bark, southeastern coastal plain and southern Arizona; lobes mostly 0.5-1 mm wide 6

6. Lobes usually remaining discrete at least at the periphery, flat, not folded .... ***Dirinaria picta* (Sw.) Clem. & Shear**

6. Lobes growing together and appearing lobate-crustose at the periphery, often longitudinally folded and somewhat convex .... *Dirinaria applanata* (Fée) D. D. Awasthi

ENDOCARPON Based on Breuss (2002a) and Lendemer (2007).

1. Thallus dwarf fruticose, consisting of flattened and erect branches 2-6 mm high, sometimes rounded and recurved and appearing inflated; perithecia buried in tips often with a prominent ostiole; spores dark brown when mature, 44-60 x 18-24 µm; hymenial algae cylindrical; on rocks in dry or seepage areas; Rocky Mountains and western Nevada .... ***Endocarpon pulvinatum* Th. Fr.** (syn. *E. tortuosum*)

1. Thallus squamulose or crustose; hymenial algae globose or cylindrical; spores colorless to very pale brownish 2

2. Thallus crustose, rimose-areolate, pale brown to grayish brown; hymenial algae cylindrical; spores (4-)8 per ascus, 15-28 x 9-12 µm; East Temperate .... ***Willeya diffractella* (Nyl.) Müll. Arg.** (syn. *Endocarpon diffractellum, Staurothele diffractella*) [Plate 808 in LNA]

2. Thallus squamulose; algae more or less globose; spores brown when mature, 2 per ascus 3

3. Rhizines present, black; spores 35-65 x 13-26 µm; southwestern, on calcareous or non-calcareous soil .... *Endocarpon pusillum* Hedl. [*not* Plate 321 in LNA; see below]

3. Lacking rhizines (although tufts of hyphae can be present); spores 25-35 x 10-16 µm; widespread in east and southwest, on calcareous rock, or rarely, tree bases or soil 4

4. Squamules lobed, usually crowded and overlapping, typically lifting at the edges, 0.5-1.5 mm in diameter; lower surface black .... ***Endocarpon pallidulum* (Nyl.) Nyl.** [Plate 321 in LNA, as "*E. pusillum*"]

4. Squamules scattered to adjacent but not overlapping; adnate over entire lower surface, 0.5-1 mm in diameter; lower surface pale .... *Endocarpon petrolepidium* (Nyl.) Nyl.

ERIODERMA (including *Leioderma*)

1. Thallus without soredia, usually fertile with tiny hemispherical brown apothecia on the upper surface of the lobes; medulla PD+ orange (eriodermin); very rare, Newfoundland to coastal Maine .... *Erioderma pedicellatum* (Hue) P. M. Jørg.

1. Thallus with marginal soredia, apothecia absent 2

2. Tomentum on upper surface webby; medulla and cortex PD–, K+ pale orange (containing ursolic acid); extremely rare, Oregon to Vancouver Island .... ***Leioderma sorediatum* D. J. Galloway & P. M. Jørg.**

2. Tomentum on upper surface made up of erect hairs 3

3. Medulla PD+ orange (eriodermin); lower surface yellowish white; rare, coastal Oregon to Alaska .... ***Erioderma sorediatum* D. J. Galloway & P. M. Jørg.**

3. Medulla PD–; lower surface brownish; rare, Appalachians at high elevations .... *Erioderma mollissimum* (Samp.) Du Rietz

EVERNIA

1. Branches clearly flattened and dorsiventral throughout, with a pale, almost white lower surface; round soralia on the lobe margins and upper surface, containing coarse, white to blue-gray soredia .... ***Evernia prunastri* (L.) Ach.**

1. Branches irregular or angular in cross section, mostly not flattened or dorsiventral; sorediate or not 2

2. Thallus sorediate on ridges along the branches; bushy to pendent .... ***Evernia mesomorpha* Nyl.**

2. Thallus without soredia, pendent or prostrate 3

3. Branches stiff and brittle, with a tough, unbroken cortex; medulla dense; prostrate on calcareous soil in western arctic .... *Evernia perfragilis* Llano

3. Branches usually soft, not brittle; cortex thin and cracked; medulla loose (in pendent specimens on trees) or dense (in prostrate specimens on calcareous soil); Rocky Mountains .... ***Evernia divaricata* (L.) Ach.**

FLAVOCETRARIA

1. Lobes flat with a network of depressions and sharp ridges or at least wrinkled; base sometimes turning dark yellow .... ***Flavocetraria nivalis* (L.) Kärnefelt & Thell**

1. Lobes curled inward forming a channel, smooth; base often becoming blotched with red-violet 2

2. Thallus 20-60(-80) mm high; lobes 2-6(-8) mm wide, margins undulating and crisped, not finely toothed, inrolled and often forming tubes for part of the length; lacking pruina or hood-shaped, "hooked" tips; widespread arctic and western boreal ....***Flavocetraria cucullata* (Bellardi) Kärnefelt & Thell**

2. Thallus 20-30(-40) mm high; lobes (0.5-)1-2(-3) mm wide, margins becoming finely toothed, inrolled and forming tubes along almost the entire length; many lobes showing tiny hood-shaped, "hooked" tips that are sometimes pruinose; forests of central Alaska .... *Flavocetraria minuscula* (Elenk. & Savicz) Ahti et al.

FULGENSIA [Species of *Fulgensia*, together with some *Caloplaca* species, are classified in the genus *Gyalolechia* by Arup et al. (2013).]

1. Spores 2-celled .... ***Fulgensia desertorum* (Tomin) Poelt**

1. Spores 1-celled 2

2. Thallus areolate, not noticeably lobed at the margin; schizidia often present .... ***Fulgensia bracteata* (Hoffm.) Räsänen**

2. Thallus clearly lobed at the margin; schizidia present or absent 3

3. Lobes long and slightly ascending at the tips; schizidia absent; rare .... *Fulgensia fulgens* (Sw.) Elenkin

3. Lobes short, entirely adnate; schizidia abundant on thallus surface; common .... *Fulgensia subbracteata* (Nyl.) Poelt

FUSCIDEA Based in large part on A. Fryday (2008).

1. Growing on bark or wood; sorediate; soralia at first discrete, later coalescing in the center 2

1. Growing on rock; with or without soredia 3

2. Soredia PD+ red (fumarprotocetraric acid), UV– .... *Fuscidea arboricola* Coppins & Tønsberg

2. Soredia PD–, UV+ white (divaricatic acid) .... *Fuscidea pusilla* Tønsberg

3. Thallus sorediate 4

3. Thallus without soredia 6

4. Thallus thin, dispersed areolate; soralia discrete, with gray soredia; apothecia with a whitish inner ring between the disk and margin; spores broadly ellipsoid, 7-10 x 5-7 µm; rare; New England and Quebec .... *Fuscidea gothoburgensis* (H. Magn.) V. Wirth & Vězda

4. Thallus thick, continuous, or composed of patches of contiguous areoles; soralia at first discrete, later coalescing, with brown to dark gray soredia where soralia are not abraded 5

5. Spores consistently 1-celled, almost globose, 8-9 µm across; thallus pale gray; apothecia pale brown to black, sometimes pruinose; Appalachians .... *Fuscidea appalachensis* Fryday

5. Spores 1- to 2-celled, slightly curved, 9-12 x 4-5 µm; thallus pale gray to dark brown; apothecia dark brown to black, never pruinose. Eastern Temperate .... *Fuscidea recensa* (Stirton) Hertel, V. Wirth & Vězda [not Plate 338 in LNA; see below]

6.(3) Medulla IKI+ blue 7

6. Medulla IKI– 8

7. Thallus thick, dark brown, with convex areoles; apothecia constricted at base; mountains of northeastern North America .... *Fuscidea lowensis* (H. Magn.) R. Anderson & Hertel

7. Thallus thin to thick, rimose-areolate, pale brown to gray-brown; apothecia immersed in thallus, the disks level with the thallus surface, with a whitish inner ring next to the margin; humid coastal Pacific Northwest .... *Fuscidea thomsonii* Brodo

8. Medulla C+ pink, PD+ yellow, K– (alectorialic acid); spores broadly ellipsoid, 8-10 x 5-6 µm; mountains of New England, rarely coastal .... *Fuscidea scrupulosa* (Eckf.) Fryday

8. Medulla C–, PD–, K–, UV+ white (divaricatic acid) 9

9. Spores 1-celled, straight, broadly ellipsoid, 8-10 x 5-6 µm; thallus gray, relatively thin, areolate on a broad black prothallus; humid coastal Pacific Northwest .... *Fuscidea mollis* (Wahlenb.) V. Wirth & Vězda

9. Spores 1- to 2-celled, often curved or bean-shaped, 9-11(-12) x 3.5-4.5(-6) µm; thallus gray or brownish gray to dark brown, rimose-areolate, with a thin black prothallus; often forming patchy mosaics on the rock; common in Appalachians north to Newfoundland
.... ***Fuscidea recensa* var. *arcuatula* (Arnold) Fryday** [Note: Plate 338 in LNA, as "*F. recensa*"]

GRAPHIS (including *Acanthothecis, Dyplolabia, Fissurina,* and "*Graphina*") Based largely on Harris (1995).

1. Spores muriform 2
1. Spores transverse septate 8

2. Walls of the lirellae black, thick, sometimes coated with a thin layer of thalline tissue 3
2. Walls of lirellae pale to colorless 4

3. Lirellae slender, partly immersed, with a narrow disk, branched; spores 1 per ascus, 65-90 x 20-30 µm;
.... *Graphis xylophaga* (R. C. Harris) Lendemer (syn. *Graphina xylophaga*)
3. Lirellae round to elliptical, cup-like, with a broad disk; spores 6-8 per ascus, 33-48 x 13-16 µm
.... *Glyphis scyphulifera* (Ach.) Staiger (syn. *Gyrostomum scyphuliferum* (Ach.) Nyl.)

4. Spores 1 or rarely 2 per ascus; lirellae crowded or scattered, up to 3 mm long, sinuous; spores 85-105 x 20-28 µm .... *Fissurina cypressi* (Müll. Arg.) Lendemer (syn. *Graphina cypressi*)
4. Spores 2 to 8 per ascus 5

5. Spores small, mostly under 35 µm long 6
5. Spores large, over 35µm long, mostly 2-4 spores per ascus; thallus pale gray, K+ or K–, PD+ yellow to red; lirellae white 7

6. Spores 15-20(-35) x 6-9(-11) µm, 8 spores per ascus, IKI± weakly violet; lirellae appearing like fissures in the thallus, disk not exposed; thallus dark olive to yellow-brown, rather shiny, K–, PD–
.... *Fissurina incrustans* Fée (syn. *Graphina incrustans*)
6. Spores 20-30(-40) x (10-)13-19 µm; 6-8 per ascus, IKI+ dark violet; lirellae prominent, slender, sinuous or branching; thallus PD+ yellow (psoromic acid)
.... *Fissurina columbina* (Tuck.) Staiger (syn. *Graphina columbina*)

7. Spores ellipsoid, 35-78 x 18-36 µm; walls of lirellae usually multi-layered and disintegrating; thallus PD+ orange, K+ yellow to red (constictic acid or salazinic acid); coastal plain

....................***Acanthothecis peplophora* (M. Wirth & Hale) E. Tripp & Lendemer** (syn. *Graphina peplophora*)

7. Spores fusiform, 110-140 x 15-17 µm; walls not layered or disintegrating; thallus PD+ red-orange, K– (protocetraric acid); common in Florida
....*Acanthothecis leucopepla* (Tuck.) E. Tripp & Lendemer (syn. *Graphina abaphoides*)

8.(1) Lirellae white or thallus-colored; spores 4-celled 9
8. Lirellae black, C–; spores 6- to 12-celled 11

9. Lirellae prominent, thickly white pruinose, C+ red (lecanoric acid); walls of lirellae black; very common in southeastern coastal plain ....***Dyplolabia afzelii* (Ach.) A. Massal.** (syn. *Graphis afzelii*)
9. Lirellae not prominent, seen as raised fissures in the thallus, C–; walls of lirellae colorless or pale 10

10. Spores 17-24(-27) x 8-12 µm; common on trees in southeast and in Pacific Northwest coastal forests
....*Fissurina insidiosa* C. Knight & Mitten (syn. *Graphis insidiosa*)
10. Spores 15-17 x 6-7 µm; southeastern coastal plain ....*Fissurina illiterata* (R. C. Harris) Lendemer

11.(8) Walls of lirellae with long ridges, in layers like a French pastry; Florida 12
11. Walls of lirellae uniform, not ridged or layered 15

12. Spores 6- to 8-celled, 20-30 x 7-9 µm; thallus very thin, UV+ yellow (lichexanthone); common
....*Graphis lucifica* R. C. Harris
12. Spores (6-)9- to 12-celled; thallus UV– (lacking lichexanthone); rare 13

13. Thallus superficial and easily seen, K+ red (norstictic acid); spores 10- to 12-celled, 32-55 x 6-12 µm; east and west coasts .... *Graphis elegans* (Borrer *ex* Sm.) Ach.
13. Thallus superficial or within the bark and almost invisible, K–; southeastern 14

14. Inner tissue of the lirellae orange; spores (6-)9- to 10-celled, 25-30 x 6-8(-10) µm
....***Graphis endoxantha* Nyl.** (syn. *Graphis subelegans*)
14. Inner tissues of lirellae colorless; spores 10- to 12-celled, 35-45 µm long
....*Graphis striatula* (Ach.) Sprengel

15.(11) Epihymenium orange, K+ purple; spores 8-celled, 23-30 x 6-7 µm; North Carolina to Florida
....*Graphis inversa* R. C. Harris

15. Epihymenium colorless, K– 16

16. Spores 6- to 8-celled; thallus K+ red, PD+ yellow (norstictic acid); common only in Florida and nearby coastal plain .... *Graphis librata* C. Knight
16. Spores 8- to 11-celled 17

17 Lirellae prominent or partly immersed, black, sometimes appearing pruinose; thallus K–, PD– (no lichen substances); extremely common and widespread .... ***Graphis scripta* (L.) Ach.**
17. Lirellae mostly immersed, gray, appearing pruinose; thallus K+ yellow to red, PD+ orange (stictic acid and [or] norstictic acid); common in Florida and Louisiana .... *Graphis caesiella* Vainio

GYALECTA (including *Gyalidea, Petractis, Pachyphiale, Ramonia*)

1. On bark 2
1. On soil, peat, moss, or rock 6

2. Spores 1-celled, many per ascus, 5-7 x 2 µm; frequent in southeast .... *Ramonia microspora* Vězda
2. Spores 2- or more celled, 8 or many per ascus 3

3. Spores transversely septate 4
3. Spores muriform, 8 per ascus 5

4. Spores 4- to 8-celled, 24-28(-35) x 3.0-3.8(-4.7) µm, 16-32(-48) per ascus; apothecia reddish brown, 0.2-0.3 mm in diameter, with prominent disk-colored margins; infrequent, Great Lakes to Maine .... *Pachyphiale fagicola* (Hepp) Zwackh
4. Spores (2-)4-celled, 12.5-15 x 4-5 µm, 8 per ascus; apothecia pale pink or orangish, up to 0.4 mm in diameter with ragged margins; Florida to Texas .... *Ramonia rappii* Vězda

5. Spores 14-21(-30) x (5-)7-9 µm; apothecia (0.1-)0.2-0.3(-0.4) mm in diameter .... *Gyalecta truncigena* (Ach.) Hepp
5. Spores 20-30(-40) x 10-14(-18) µm; apothecia 0.5-0.8 mm in diameter; California ...*Gyalecta herrei* Vězda

6.(1) On soil, peat, or moss; apothecia orange, usually over 0.5 mm in diameter 7
6. On rock 8

7. Apothecia flat to slightly concave, superficial, margins thin; thallus dark gray-green; spores transverse septate, fusiform, 4-celled, (10-)14-22 x 2.5-4.5 µm; rare, western .... *Gyalecta friesii* Flotow *ex* Körber

7. Apothecia deeply concave, sunken into substrate; thallus grayish, thin; spores 4-celled or occasionally submuriform, elongate ellipsoid, (12-)15-23 x 5-7.5 µm; frequent, widespread, especially arctic-alpine .... *Gyalecta foveolaris* (Ach.) Schaerer

8. Spores transverse septate, 4-celled, (12-)15-22(-24) x 5.0-7.2 µm; apothecia flat, with prominent pale to yellow-brown margins; photobiont *Cystococcus* (lacking orange oil globules, cell walls not shining in polarized light); on non-calcareous rocks, rare, west coast .... *Gyalidea hyalinescens* (Nyl.) Vězda

8. Spores muriform; apothecia deeply concave, pit-like; photobiont *Trentepohlia*; on calcareous rock 9

9. Apothecia less than 0.4 mm in diameter, pale; margins with conspicuous radial cracks; thallus gray; spores 18-24(-27) x 9-12(-14) µm; infrequent, in Ozark region .... *Gyalecta farlowii* Tuck. *ex* Nyl. (syn. *Petractis farlowii*) [Note: Classification follows Harris & Ladd 2005.]

9. Apothecia 0.5-1.0 mm in diameter, pinkish orange; margins smooth or cracked; thallus pink; spores 11-17(-25) x 8-10 µm; common and widespread .... ***Gyalecta jenensis* (Batsch) Zahlbr.**

GYALIDEOPSIS (including *Echinoplaca, Gomphillus,* and *Jamesiella*) See Lücking et al. (2007).

[Note: The following key uses only non-ascomatal characters because apothecia are often rare in this group. An exception is *Gyalideopsis vainioi* Kalb & Vězda, which is identified by its distinctive purplish-black apothecia usually with pruinose disks, but typically without mature spores; hyphophores are black, tiny, adnate, and rarely seen.]

1. Hyphophores colorless or white, sometimes slightly brownish toward the summit 2

1. Hyphophores mostly dark brown or black 4

2. Hyphophores long and tapered, up to 1.2 mm long, developing a capitulum-like tip that expands into a star-shaped disk; diahyphae ("conidia" of a kind) thread-like; on moss, southeastern ....***Gomphillus americanus* Essl.**

2. Hyphophores tiny, less than 0.4 mm tall; diahyphae not thread-like; on bark or twigs 3

3. Hyphophores white, at first subspherical, becoming cylindrical, under 0.4 mm tall, lacking an expansion at the tip; diahyphae segmented into units about 8-10 µm long; Appalachians at high elevations, and Washington .... *Jamesiella anastomosans* (P. James & Vězda) Lücking et al. (syn. *Gyalideopsis anastomosans*)

3. Hyphophores translucent to white, sometimes brownish close to the summit; forming a yellowish, globular, gelatinous expansion at the tip; diahyphae branched and segmented into units about 5-8 µm long, constricted at the septae; Florida .... *Echinoplaca areolata* Lücking & W. R. Buck

4. Hyphophores brown, with fan-shaped expansion at summit; under 1 mm tall 5
4. Hyphophores black, slender and hair-like, up to 2 mm tall 6

5. Diahyphae branched, constricted at septae (moniliform); on bark or wood, Florida .... *Gyalideopsis americana* Lücking & W. R. Buck
5. Diahyphae thread-like; on soil or humus, New Brunswick to Carolinas and Ozarks .... *Gyalideopsis moodyae* Lendemer & Lücking

6. Diahyphae branched, constricted at septae (moniliform); on bark and twigs, widespread in southeast .... *Gyalideopsis buckii* Lücking et al.
6. Diahyphae thread-like 7

7. Diahyphal mass subterminal, lying against the stalk (seen under the microscope, in water); on bark and twigs, Pacific Northwest and Florida .... *Gyalideopsis epicorticis* (A. Funk) Tønsberg & Vězda
7. Diahyphal mass forming a colorless, candleflame-like expansion at the stalk tip (seen under the microscope, in water); on bark or wood, rarely rock; widespread in southeast .... *Gyalideopsis submonospora* Lücking & W. R. Buck

HAEMATOMMA

1. Growing on rock 2
1. Growing on bark 3

2. Thallus sorediate, becoming leprose, yellowish (usnic acid, zeorin); rare, Pacific Northwest and southern Rockies .... *Haematomma ochroleucum* (Necker) J. R. Laundon
2. Thallus rarely sorediate, but if so, then in discrete soralia, never leprose, gray (lacking usnic acid); Texas to southern Arizona .... ***Haematomma fenzlianum* A. Massal.**

3. Thallus sorediate 4
3. Thallus not sorediate 5

4. Soralia irregular in shape, with greenish soredia; containing sphaerophorin in the thallus and russulone in the rare apothecia .... *Haematomma americanum* Kalb & Staiger

4. Soralia round and hemispherical, clearly delimited, with white soredia; containing placodiolic acid derivatives in the thallus and haematommone in the very rare apothecia .... *Haematomma guyanense* Kalb & Staiger

5. Epihymenium K+ persistent red (russulone) 6
5. Epihymenium K+ violet or purple that quickly disappears (haematommone); spores under 55 µm 7

6. Spores mostly under 50 µm long; apothecia 0.3-0.8(-1.0) mm in diameter, mostly sunken into thallus .... ***Haematomma persoonii* (Fée) A. Massal.**
6. Spores over 50 µm long; apothecia (0.4-)0.7-1.3 mm in diameter, superficial .... *Haematomma rufidulum* (Fée) A. Massal.

7. Contains placodialic acid .... ***Haematomma accolens* (Stirton) Hillm.**
7. Contains isoplacodialic and isopseudoplacodialic acids .... *Haematomma flexuosum* Hillm.

HYPERPHYSCIA

1. Sorelia absent, but with abundant apothecia; East and Central Temperate .... ***Hyperphyscia syncolla* (Tuck. *ex* Nyl.) Kalb**
1. Soredia present, apothecia rare 2

2. Thallus surrounded by a black prothallus; lower surface black; rare, Florida and Georgia to Louisiana .... *Hyperphyscia minor* (Fée) D. D. Awasthi
2. Thallus lacking a prothallus; lower surface pale 3

3. Soredia all laminal (on lobe surface), often excavate, irregular in shape; widespread Temperate .... *Hyperphyscia adglutinata* (Flörke) Mayerh. & Poelt [*not* Plate 370 in LNA; see below]
[Note: *Phaeophyscia insignis is very similar but has a few rhizines, lobes under 0.3 mm wide and circular soralia; see Phaeophyscia* key, couplet 12.]
3. Soredia mainly along the lobe margins but also on the lobe surface in places, soredia piled high, rarely excavate; central U.S. from North Dakota to Texas, rare on California coast .... ***Hyperphyscia confusa* Essl. et al.** [Plate 370 in LNA as *H. adglutinata*.]

HYPOCENOMYCE (including *Carbonicola, Fulgidea, Pycnora, Toensbergia,* and *Xylopsora*) See Bendiksby & Timdal (2013).

1. Squamules sorediate on the edges of areoles or undersurface of squamules; apothecia infrequent 2
1. Squamules (or areoles) not sorediate; black apothecia usually abundant 6

2. Thallus olive; cortex and medulla C+ pink, KC+ pink or red, PD– (lecanoric acid) .... ***Hypocenomyce scalaris* (Ach. *ex* Lilj.) M. Choisy**
2. Thallus brown to greenish brown or gray-brown, or creamy white; cortex and medulla C–, PD– or PD+ red, or C+ pink and PD+ yellow..............3

3. Thallus creamy white to brownish, PD+ yellow, C+ pink (alectorialic acid); rare 4
3. Thallus brownish, squamulose, PD+ red or PD–, C–; farinose sorediate on lower surface of upturned edges of squamules 5

4. Thallus subsquamulose with marginal, labriform soralia; Washington, Yukon .... *Toensbergia leucococca* (R. Sant.) Bendiksby & Timdal (syn. *Pycnora leucococca*)
4. Thallus crustose, areolate; granular sorediate at edges of areoles; southwest and Great Lakes .... *Pycnora sorophora* (Vainio) Hafellner

5. Thallus PD + red (fumarprotocetraric acid); Appalachian-Great Lakes and southwest ....*Carbonicola anthracophila* (Nyl.) Bendiksby & Timdal (syn. *Hypocenomyce anthracophila*)
5. Thallus PD–; western (1 record from New Jersey) .... *Carbonicola myrmecina* (Ach.) Bendiksby & Timdal (syn. *Hypocenomyce castaneocinerea*)

6. Thallus brownish gray, squamulose 7
6. Thallus yellowish gray, crustose with convex areoles, PD+ yellow, C+ pink, UV– (alectorialic acid); western and Appalachians ,.............. *Pycnora praestabilis* (Nyl.) Hafellner (syn. *Hypocenomyce praestabilis*) [Note: Most old North Amercan records of *Pycnora xanthococca* (Sommerf.) Hafellner (syn.*Hypocenomyce xanthococca*), which is rare in the west, refer to *P. praestabilis* (Timdal 1984).]

7. Thallus PD–, K–, C–, sometimes UV+ white (unknown substance); apothecia frequently with sterile tissue forming bumps and ridges on disk; widespread temperate ....*Xylopsora friesii* (Ach.) Bendiksby & Timdal (syn. *Hypocenomyce friesii*)

7. Thallus PD+ yellow, K+ yellow (rarely K–), C+ pink, UV– (alectorialic acid, usually accompanied by thamnolic acid); Arizona, South Dakota, and Alaska ....*Fulgidea oligospora* (Timdal) Bendiksby & Timdal (syn. *Hypocenomyce oligospora*)

HYPOGYMNIA (including *Menegazzia and "Cavernularia"*)

1. Lower surface of lobes perforated by numerous tiny holes, resembling a honeycomb; thallus small, lobes less than 1 mm wide, solid 2
1. Lower surface of lobes with few or no perforations; lobes over 1 mm wide, hollow 3

2. Soredia present on upper surface of lobe tips; apothecia rare ....***Hypogymnia hultenii (*Degel.) Krog** (syn. *Cavernularia hultenii)*
2. Soredia absent; apothecia abundant, very broad compared to the width of the lobes ....*Hypogymnia lophyrea* (Ach.) Krog (syn. *Cavernularia lophyrea*)

3. Lobes perforated on the upper surface with round holes; medulla PD+ orange, K+ yellow (stictic acid) 4
3. Lobes perforated at tips, or not perforated at all; medulla PD– or PD+ red, K– (stictic acid absent) 5

4. Mature soralia round to oval, flat to hemispherical, usually without a hole at the center, almost entirely laminal, not splitting into irregular segments at the margins of the soralium; soredia finely granular to farinose ....*Menegazzia terebrata* (Hoffm.) A. Massal. [*not* Plate 510 in LNA; see below]
4. Mature soralia very irregular, collar-like with a hole in the center, the margins of the soralia breaking into segments and producing coarsely granular fragments and (or) granular soredia; soredia mostly at the lobe tips, but also laminal in part .....***Menegazzia subsimilis* (H. Magn.) R. Sant.** [Plate 510 in LNA as *M. terebrata*]

5. Soredia present 6
5. Soredia absent 15

6. Soredia on expanded, turned-back, hood-shaped lobe tips (lip-shaped soralia) 7
6. Soredia on the thallus surface, sometimes close to the lobe tips (but not in lip-shaped soralia) 9

7. Medullary ceiling dark brown to grayish-black; lobules present on the lobe margins; medulla PD– ....***Hypogymnia vittata* (Ach.) Parrique**
7. Medullary ceiling white; lobules absent; medulla PD+ red (physodalic and protocetraric acids) 8

8. Non-sorediate lobe tips often perforate; black lower surface usually extending up sides of lobes and visible from above; medulla K– or K+ slowly yellow (lacking 3-hydroxyphysodic acid); infrequent except in Maritime Provinces, boreal .... *Hypogymnia incurvoides* Rass.

8. Non-sorediate lobe tips not perforate; black lower surface usually confined to lower side of lobes and not visible from above; medulla K+ brownish red (containing 3-hydroxyphysodic acid); very common and widespread, north temperate to boreal .... ***Hypogymnia physodes* (L.) Nyl.**

9.(6) Thallus ascending, small, rarely more than 3-4 cm across; lobes branching in short dichotomies, rarely perforated; soredia farinose, on upper surface of lobe tips; medulla and soralia PD– .... ***Hypogymnia tubulosa* (Schaerer) Hav.**

9. Thallus closely appressed or somewhat ascending, typically 3-15 cm across; lobes rounded, square, or truncated, often perforated at tips; soredia granular, on thallus surface or tips of lobes 10

10. Medulla PD+ red (physodalic acid); infrequent, in lowland coastal forests of Pacific Northwest .... *Hypogymnia oceanica* Goward

10. Medulla PD–; arctic and boreal-montane habitats in west 11

11. Soredia/isidia mostly on the upper surface of the lobe tips .... ***Hypogymnia bitteri* (Lynge) Ahti**

11. Soredia or isidia mostly on the older parts of the thallus surface, rarely on the tips 12

[Note: The distinguishing features of the following segrates of *H. austerodes* (see Goward et al. 2012) are very subtle, and all of the taxa are morphologically variable. The group requires genetic study to confirm the status of these segregates.]

12. Thallus with ascending peripheral lobes, the secondary lobes growing out of the main thallus generally reaching the edge of the thallus .... *Hypogymnia protea* Goward et al.

12. Thallus appressed to the substrate throughout, with secondary lobes rarely reaching the periphery 13

13. Thallus typically shiny at the periphery, often chestnut brown except in shady habitats; soredia fine, arising from the disintegration of the thallus cortex, not from pustules .... ***Hypogymnia dichroma* Goward** [Plate 375 in LNA, as "*H. austerodes*"]

13. Thallus brown to greenish or gray, shiny or dull, soredia coarse, at least some arising from pustules 14

14. Thallus producing at least some pustular "isidia"; surface more or less smooth, not verrucose .... *Hypogymnia austerodes* (Nyl.) Räsänen [*not* Plate 375 in LNA; see above]

14. Thallus lacking isidia; thallus surface typically verrucose .... *Hypogymnia verruculosa* Goward

15.(5) Eastern U.S. and Canada (Appalachians) .... ***Hypogymnia krogiae*** **Ohlsson**

15. Western or northern North America 16

16. Medullary ceiling white 17

16. Medullary ceiling light to dark brown to grayish-black 21

17. On soil or rock (rarely wood); lobules present along the lobe margins; pycnidia sparse or very inconspicuous; in alpine or arctic tundra .... ***Hypogymnia subobscura*** **(Vainio) Poelt**

[Note: Plate 387 in LNA illustrating *H. subobscura* may well represent a depauperate form of *H. austerodes* (Goward et al. 2012)]

17. On conifer bark or wood 18

18. Medulla PD+ red (diffractaic, physodalic, and protocetraric acids present); thallus loosely attached, ascending or pendent, sometimes forming round cushions of overlapping lobes; lobes convex, perforated at tips 19

18. Medulla PD– (diffractaic, physodalic, and protocetraric acids absent); thallus closely appressed or ascending; lobes flat or concave, rarely perforated 20

19. Lobes forked in more or less regular dichotomies, more or less even, not constricted at intervals; thallus rather stiff and shrubby; apothecia abundant .... ***Hypogymnia imshaugii*** **Krog**

19. Lobes branching irregularly, constricted at irregular intervals; thallus usually somewhat pendent, with upturned tips; apothecia rare .... ***Hypogymnia duplicata*** **(Ach.) Rass.**

20. Lobes somewhat ascending or, at least, peripheral lobes upturned, 1.5-3(-8) mm wide; thallus greenish gray, containing vittatolic acid (a depsidone), lacking apinnatic acid (a fatty acid) .... *Hypogymnia recurva* Goward et al.

20. Lobes flat, appressed, the largest lobes 1-2(-2.5) mm wide; thallus pale gray; lacking vittatolic acid, containing apinnatic acid .... ***Hypogymnia wilfiana*** **Goward et al.** [Plate 383 in LNA as "*H. metaphysodes*"; see couplet 25]

21.(16) Lobules present along the lobe margins; medulla PD+ red or PD– 22

21. Lobules absent; medulla PD– 24

22. Medulla PD– (physodalic and protocetraric acids absent); lobes mostly less than 3 mm wide .... ***Hypogymnia occidentalis*** **L. Pike**

22. Medulla PD+ red (physodalic and protocetraric acids); lobes usually more than 3 mm wide at the tips 23

23. Lobules strap-shaped; diffractaic acid absent; rare, in coastal forests ...... ***Hypogymnia heterophylla*** **L. Pike**

23. Lobules rounded; diffractaic acid present; very common in Pacific Northwest .... ***Hypogymnia enteromorpha*** **L. Pike**

24.(21) Lobes rarely perforated 25

24. Lobes perforated at tips 26

25. Older lobes becoming very wrinkled, often sinuous, convex throughout; apothecia abundant; medulla PD– (hypoprotocetraric acid present) .... ***Hypogymnia rugosa*** **(G. Merr.) L. Pike**

25. Lobes relatively smooth; lobe tips concave; apothecia occasional; medulla PD– (hypoprotocetraric acid absent, 3-hydroxyphysodic acid present) .... *Hypogymnia canadensis* Goward & McCune

[Note: This species, together with *H. wilfiana* and *H. recurva*, represent most North American records of *H. metaphysodes*.]

26. Lobes 1.5-2 mm wide, more or less even, not constricted at intervals; medulla KC+ red (diffractaic and physodic acids) .... ***Hypogymnia inactiva*** **(Krog) Ohlsson**

26. Lobes 3-4 mm wide, constricted at irregular intervals; medulla KC– (diffractaic and physodic acids absent) .... ***Hypogymnia apinnata*** **Goward & McCune**

HYPOTRACHYNA (including only species with forked rhizines)

[Note: Species of *Hypotrachyna* with unbranched rhizines formerly classified in the genus *Parmelinopsis* (*H. horrescens, H. minarum,* and *H. spumosa*) are keyed out in the *Parmelia* key. (See Divakar et al. 2013)]

1. Thallus forming subfruticose tufts 2-4 cm across; lobes flattened to strongly convex, with marginal cilia, producing dark green soredia on upper surface of lobe tips; soralia C+ faint pink (gyrophoric acid); Appalachian .... ***Hypotrachyna catawbiense*** **(Degel.) Divakar et al.** (syn. *Everniastrum catawbiense*)

1. Thallus foliose throughout, with flat lobes; soredia present or absent 2

2. Soredia, or soredia-like fragments (schizidia), originating from the breakdown of pustules or hollow warts, present on lobe surface 3

2. Soredia or soredia-like fragments absent 12

3. Rhizines mostly unbranched, a few dichotomous, slender and short; lobes ascending and frequently curled inward (revolute); medulla C+ red 4

3. Rhizines mostly dichotomously branched, sometimes more than once, robust; lobes adnate to ascending, flat (not revolute); medulla C+ red or C–; mainly Appalachian, also California 5

4. Soredia powdery on the upper surface of the lobe tips .... ***Hypotrachyna revoluta* (Flörke) Hale**

4. Soredia originating from pustules (schizidia) on the upper surface of much of the thallus .... *Hypotrachyna afrorevoluta* (Krog & Swinscow) Krog & Swinscow

5. Soredia powdery to granular, not originating from pustules 6

5. Soredia or soredia-like fragments orginating from pustules (schizidia); medulla C+ red or C–; on bark 8

6. Thallus yellow-green (usnic acid present); medulla K+ yellow becoming red (salazinic acid); powdery soredia on upper surface of older lobes; Pacific Northwest .... ***Hypotrachyna sinuosa* (Sm.) Hale**

6. Thallus gray to greenish gray (usnic acid absent); medulla K– 7

7. Medulla C+ red (lecanoric and evernic acids); rare, Appalachian, mostly on rocks .... *Hypotrachyna rockii* (Zahlbr.) Hale

7. Medulla C– or C+ pale pink-orange (barbatic acid complex); Southwestern, on bark .... *Hypotrachyna laevigata* (Sm.) Hale

8.(5) Cortex K–, UV+ yellow (lichexanthone); medulla K+ brown-red, PD– (lividic acid) .... ***Hypotrachyna osseoalba* (Vainio) Park & Hale**

8. Cortex K+ yellow, UV– (atranorin); unpigmented medulla K–, PD+ or PD– 9

9. Medulla white with orange patches, especially under the pustules, PD+ red-orange, K– where white and K+ purple where pigmented (protocetraric acid and anthraquinone) .... ***Hypotrachyna croceopustulata* (Kurok.) Hale**

9. Medulla white throughout, PD– 10

10. Medulla K+ brownish red, C–, KC+ deep brown-red (lividic acid complex); common, southcentral U.S. .... *Hypotrachyna pustulifera* (Hale) Skorepa

10. Medulla K–, C+ pink to red 11

11. Medulla C+ red (lecanoric and evernic acids); pustules breaking up into soredia-like schizidia; fairly common, Appalachians and Ohio Valley .... *Hypotrachyna taylorensis* (M. E. Mitch.) Hale

11. Medulla C+ pink (gyrophoric and hiascic acids); pustules not fragmenting and producing true schizidia except after abrasion or crushing; lobe tips with maculae; infrequent; northeast and Appalachians .... *Hypotrachyna showmanii* Hale

12.(2) Isidia or isidia-like pustules present (sometimes sparse), apothecia infrequent; Appalachian, on bark 13

12. Isidia absent; apothecia abundant and conspicuous 15

13. "Isidia" hollow, formed from erect pustules .... *Hypotrachyna showmanii* Hale [See couplet 11]

13. Isidia solid, not pustular 14

14. Isidia cylindrical, laminal; medulla C– or C+ pale orange (barbatic acid complex) ....*Hypotrachyna imbricatula* (Zahlbr.) Hale

14. Isidia flattened like lobules, on the lobe margins and sometimes on the surface; medulla C+ red (anziaic acid) .... *Hypotrachyna prolongata* (Kurok.) Hale

15.(12) Thallus spotted with maculae; medulla K–, KC+ red, C+ red (evernic and lecanoric acids); southwestern U.S. .... ***Hypotrachyna pulvinata* (Fée) Hale**

15. Thallus lacking maculae; medulla K+ pinkish brown, KC+ purple-brown, C– (lividic acid); eastern U.S. .... ***Hypotrachyna livida* (Taylor) Hale**

IMSHAUGIA

1. Isidia absent; black pycnidia sometimes abundant and raised, scattered on lobe surface or marginal; apothecia abundant, up to 10 mm broad ....***Imshaugia placorodia* (Ach.) S. F. Meyer**

1. Isidia abundant on thallus surface; pycnidia sparse and inconspicuous; apothecia rare ....***Imshaugia aleurites* (Ach.) S. F. Meyer**

IONASPIS (including *Hymenelia* and *Eiglera*) Based on Lutzoni and Brodo (1995) and Fletcher et al. (2009a).

1. Apothecial disks black; epihymenium shades of green 2

1. Apothecia pink to orange-brown or gray, rarely darkening to black; epihymenium shades of gray, yellow, or brown, negative or changing to orange-yellow in nitric acid 6

2. Epihymenial pigment unchanged or dissolving in nitric acid; thallus epilithic, smooth to rimose; spores almost globose, 8-10 x 7 µm; on siliceous rocks .... *Ionaspis suaveolens* (Fr.) Th. Fr. *ex* Stein.

2. Epihymenial pigment changing to wine-red to purple in nitric acid 3

3. Ascus tip with an K/I+ blue tholus; thallus pale orange or yellowish when fresh; photobiont chlorococcoid, 7-12 µm in diameter; spores 12-18 x 7-9 µm; northern arctic ....*Eiglera flavida* (Hepp) Hafellner

3. Ascus tip K/I–; thallus whitish to pale yellow-white, gray, or pinkish, often endolithic; photobiont *Trentepohlia,* 15-30 µm in diameter; mostly arctic-alpine, rarely temperate 4

4. On siliceous rocks; thallus epilithic, smooth to rimose; spores 10-12(-14) x 5-8.5 µm; apothecia 0.3-0.5 mm in diameter .... *Hymenelia cyanocarpa* (Anzi) Lutzoni

4. On calcareous rocks; thallus endolithic or epilithic 5

5. Thallus usually epilithic, creamy white, membranous to rimose; apothecia sunken in thallus or almost sessile, with flat disks, sometimes with a thin margin, 0.2-0.5(-0.8) mm in diameter; spores broad, 11-15 x 7.5-9.0 µm; hymenium 60-100 µm high .... *Hymenelia heteromorpha* (Kremp.) Lutzoni

5. Thallus endolithic; apothecia immersed in pits in rock with pore-like disks, 0.2-0.5 mm in diameter; spores subspherical, 10-15(-19) x (8-)10-12.5 µm; hymenium 120-150 µm high .... *Hymenelia melanocarpa* (Kremp.) Arnold

6(1). Thallus (and apothecial disks) distinctly orange ranging to pale brown 7

6. Thallus pale pinkish (when fresh) to yellowish white or dark tan to olive; apothecial disks white, pink, or gray-brown 8

7. Common, in streams or on lake shores; thallus pale orange, thin, continuous; apothecia rarely with distinct margins; Appalachian-Great Lakes region and Pacific Northwest .... **Ionaspis lacustris (With.) Lutzoni**

7. Rare, on dry rocks; thallus dark rusty orange, rimose-areolate, often patchy; apothecia typically with distinct red-brown margins; perhaps eastern .... *Hymenelia ceracea* (Arnold) Poelt & Vězda

8. On dry rocks, often partly shaded siliceous boulders; temperate to southern boreal; apothecia almost white; spores 12-16(-17) x 5-8 µm; Appalachian-Great Lakes region, rare on west coast ......*Ionaspis alba* Lutzoni

8. On periodically splashed, moist, or submerged rocks; boreal to arctic-alpine; apothecia pink to dark brown, rarely white 9

9. Apothecia pale pinkish (when fresh); spores 11.5-15(-17) x 6-8(-10) µm; usually on calcareous (rarely siliceous) wet or dry rocks; widespread ....*Hymenelia epulotica* (Ach.) Lutzoni (syn. *Ionaspis epulotica*)

9. Apothecia smoky gray-brown to brown; spores 8-15 x 5-8 µm; on siliceous rocks in mountain streams; western mountains 10

10. Thallus pinkish to white, often indistinct, thallus and exciple K– ....***Ionaspis lavata* H. Magn.**

10. Thallus dark tan to olive, thallus and exciple K+ violet (sedifolia-gray) ....*Ionaspis odora* (Ach.) Th. Fr.

LECANACTIS (including *Cresponea*).

1. Thallus well developed, PD+ yellow, C– (psoromic acid); spores (20-)24-28 x 3-7 µm; conidia 5-6 x 1 µm; southern California coast ....*Lecanactis californica* Tuck.

1. Thallus usually thin, PD–, C– 2

2. On shaded rock; spores 5- to 7-celled, 20-27 x 5-6 µm; apothecia lightly white pruinose when young; Ozarks and southeast ....*Cresponea premnea* (Ach.) Egea & Torrente var. *saxicola* (Leighton) Egea & Torrente

2. On bark; spores 4- to 6-celled 3

3. Spores 28-44 x 3-7 µm; apothecia and pycnidia with a yellowish pruina that usually turns C+ red (C– in the Pacific Northwest); conidia 10-17 x 2-4 µm ....***Lecanactis abietina* (Ach.) Körber**

3. Spores 11-22 x 3-5.5 µm; apothecia and pycnidia with a whitish to yellow or greenish pruina or commonly without pruina, C– 4

4. Spores (11-)13-17(-22) x 3.0-5.0 µm, 4-celled; paraphyses tips with pigmentation within cells; conidia 5.0-7.5 x 1.5-1.8 µm; frequent in northeast, rare in southern California ....*Cresponea chloroconia* (Tuck.) Egea & Torrente

4. Spores 15-22 x 4.5-5.5 µm, 4- to 5(-6)-celled; paraphyses tips with pigmentation coating cells; frequent in southeast ....*Cresponea flava* (Vainio) Egea & Torrente

LECANIA (including *Halecania*) Based in part on Naesborg (2008).

1. On rock or soil 2

1. On bark 9

2. Spores (2-)4-celled, 12-16(-18) x 4-6 µm; apothecia dark brown to black, always pruinose; on limestone, mostly from Great Lakes westward ....*Lecania nylanderiana* A. Massal.

2. Spores 2-celled; apothecia pale to dark brown or black, sometimes pruinose 3

3. Thallus rimose-areolate, not at all squamulose or lobed 4

3. Thallus areolate to almost squamulose, or clearly lobed at the margins, brown or gray due to pruina, apothecia brown, not pruinose on calcareous or siliceous rocks; California 5

4. On siliceous rocks; greenish gray to almost black, granular to sorediate in places; thallus or amphithecial medulla PD+ red (argopsin); spores halonate; asci *Catillaria*-type; northeastern ....*Halecania pepegospora* (H. Magn.) van den Boom

4. On calcareous rocks pale greenish to gray 8

5. Spores halonate; asci *Catillaria*-type; spores narrowly ellipsoid, (9-)10-14(-16) x 3-4 µm; thallus rimose-areolate, occasionally subsquamulose; brown to gray-brown; on siliceous rock, Arizona and California ....*Halecania australis* Lumbsch

5. Spores not halonate; asci *Bacidia*-type; spores ellipsoid, 10-16 x 4-7 µm 6

6. On soil or occasionally calcareous rock; spores broadly ellipsoid, 6-7 µm wide; thallus verrucose to almost squamulose, reddish brown to gray, sometimes pruinose in part, verrucae or areoles more or less separate; apothecial margins thin and soon disappearing; California .... *Lecania dudleyi* Herre

6. On siliceous rock; spores ellipsoid, mostly under 6 µm wide; thallus without pruina; apothecial margins well developed 7

7. Thallus rimose to areolate or subsquamulose, grayish brown to yellowish brown or gray, areoles not overlapping; spores (3-)4-5.5 µm wide; coastal and sometimes maritime ....*Lecania fructigena* Zahlbr.

7. Thallus areolate to squamulose, brown, often with overlapping scales; spores broad, (4-)5-6 µm wide; common in coastal and inland areas of southern California .... ***Lecania brunonis* (Tuck.) Herre**

8.(4) Thallus areolate with granular to sorediate (blastidiate) edges to the areoles; spores (8-)10-12(-14.5) x 3.0-4.0 µm, length/width ratio 2.5-3.0; mostly central to western ....*Lecania erysibe* (Ach.) Mudd

8. Thallus not at all sorediate; spores 10-15(-18) x 4-6 µm, length/width ratio 2.0-2.5:1; mostly in central regions .... "*Lecania perproxima*" of many authors

[Note: *Lecania inundata* (Hepp *ex* Körber) M. Mayerhofer is a similar species of central U.S. and southern California that has nodulose areoles and spores 11-18 x 4.5-6(-7.5) μm.]

9.(1) Thallus with bright green, punctiform, convex soralia on a thin, uniform thallus; common on hardwoods in the northeast; almost always sterile; spores 1(-2)-celled
.... *Lecania croatica* (Zahlbr.) Kotlov [see Key C, couplet 47; and *Biatora* key, couplet 7]

9. Thallus lacking soredia; spores 2- to 4-celled 10

10. Spores mostly 4-celled, 12-23 x 4-6 μm, straight or slightly bent; apothecia pale to dark brown to black, margin very thin and typically paler than disk 11

10 Spores mostly 2-celled 12

11. Apothecia lecanorine; amphithecium containing many algae, disks sometimes lightly pruinose; spores 8-16 per ascus; central parts of continent to California ....*Lecania fuscella* (Schaerer) Körber

11. Apothecia biatorine, algae, if present, only in base; disks without pruina; spores 8 per ascus; widespread
....*Lecania naegelii* (Hepp) Diederich & van den Boom

12. Spores bent or bean-shaped, (10.5-)12.5-14.5(-16) x 4.0-5.0(-6.0) μm; apothecia dark brown to black, scattered; North Temperate, especially Great Lakes region ....***Lecania dubitans* (Nyl.) A. L. Sm.**

12. Spores straight, 9-13(-16) x 3.0-4.5 μm; apothecia pale reddish brown, rarely darker, often crowded; on dry, exposed, often nutrient-rich bark; widespread ....*Lecania cyrtella* (Ach.) Th. Fr.

LECANORA (includes *Palicella*) Based in part on Printzen (2001), Ryan et al. (2004), Harris and Ladd (2005), Śliwa (2007), Pérez-Ortega et al. (2010).

1. Growing on bark, wood, mosses, or dead vegetation 2

1. Growing directly on rock, bones, or antlers 57

2. Thallus within substrate, absent from view, or indistinct 3

2. Thallus clearly visible 12

3. Spores 12-32 per ascus. On bark, especially poplars, mostly in Great Lakes region
....*Lecanora sambuci* (Pers.) Nyl.

3. Spores 8 per ascus 4

4. Amphithecial cortex either thin or indistinct, or relatively uniform in thickness laterally and at the base 5

4. Amphithecial cortex gelatinous and very distinct, much thicker at base than laterally 8

5. Apothecial disks lightly to heavily pruinose, brown, without a yellowish tint; lacking usnic or isousnic acid .... ***Lecanora hagenii* (Ach.) Ach.**

5. Apothecial disks or margins yellow or yellowish, pruinose or not; containing usnic and (or) isousnic acid 6

6. Apothecia yellow; margin disk-colored, smooth; algae very sparse in the apothecial margin, sometimes essentially absent (i.e., biatorine) .... ***Lecanora symmicta* (Ach.) Ach.**

[If apothecia darkening to black, see also couplet 14]

6. Apothecia beige to pale yellowish brown or reddish brown to black; lecanorine margin present at least in young apothecia 7

7. Spores (7-)9-11(-14) x (3-)4-5 µm; apothecial margin level with disk, rough; hymenium 40-55 µm high; apothecia crowded .... ***Lecanora albellula* Nyl.** (syn. *Lecanora piniperda*)

7. Spores (5.5-)6.5-11(-13) x (2.5-)3-4 µm; apothecial margin prominent in young apothecia, later level with disk; hymenium 30-45(-50) µm high; apothecia scattered .... *Lecanora subintricata* (Nyl.) Th. Fr.

8.(4) Apothecial disks red-brown, not at all pruinose .... ***Lecanora zosterae* Nyl. var. *zosterae***

8. Apothecial disks brown to black, or bluish because of light to heavy pruina 9

9. On bryophytes or decaying vegetation; arctic-alpine .... *Lecanora zosterae* var. *palanderi* (Vainio) Śliwa

9. On bark 10

10. Apothecial disks yellowish, never brown, with varying amounts of pruina or without pruina; some apothecial margins becoming cracked and crenulate; epihymenial granules superficial and between the paraphyses tips, insoluble in both K and HNO_3; cortex thick; west temperate .... *Lecanora juniperina* Śliwa

10. Apothecial disks orange to brown, sometimes black in *L. wetmorei,* pruinose, often heavily; apothecial margins smooth or rough, more or less even; epihymenial granules superficial, insoluble in K and soluble in HNO_3; cortex thin or thick 11

11. Amphithecial cortex up to 3-5X thicker at base than laterally; western temperate at high elevations

.... *Lecanora wetmorei* Śliwa

11. Amphithecial cortex thin or thick, uniform or up to 2.5X thicker at base than laterally; common and widespread at all elevations .... ***Lecanora hagenii*** (Ach.) Ach.

12.(2) Thallus with abundant soredia; apothecia rare or absent .... See Key C, couplets 20 and 53

12. Thallus producing apothecia; soredia present or absent 13

13. Most apothecia essentially biatorine when mature, without algae in margins (or only a few in apothecial base) 14

13. Apothecia all lecanorine, with algae relatively abundant in margins 19

14. Apothecial margin or thallus PD+ or PD–; apothecia always distinct and scattered; common on conifers in western mountains; usnic acid absent 15

14. Apothecial margin or thallus PD–, K– or K+ yellow; apothecia sometimes crowded and fusing; usnic acid present, sometimes only as traces 17

15. Spores narrowly ellipsoid, 8.5-10(-11) x 3.0-4.5(-5.0) µm; thallus and apothecial sections K+ yellow or red, PD+ yellow (stictic or norstictic acid); thallus usually well developed, minutely areolate to almost granular, very pale creamy brown to almost white .... *Lecanora cadubriae* (A. Massal.) Hedl.

[See also *Lecidea turgidula* in *Lecidea* key, couplets 5 and 15]

15. Spores broadly ellipsoid to spherical; apothecial margin (in section under the microscope) K–, PD+ red or PD– ; thallus thin, continuous to rimose-areolate 16

16. Spores broadly ellipsoid, 6-10(-13) x 4-6(-8) µm; apothecial margin PD+ red (fumarprotocetraric acid); apothecia pale to very dark brown to black .... *Lecanora fuscescens* (Sommerf.) Nyl.

16. Spores globose, (4.0)5.0-7.0 µm in diameter; apothecial margin PD–; apothecia dark red-brown to black .... *Lecanora boligera* (Norman *ex* Th. Fr.) Hedl.

[Note: Ryan et al. (2004) reports *L. boligera* as PD+ red, but all CANL specimens are PD–.]

17.(14) Apothecia yellowish green to vivid yellow (containing abundant usnic acid, lacking atranorin); spores narrowly ellipsoid, 7-16 x 3-6 µm; widespread on bark or wood of all kinds, boreal to temperate .... ***Lecanora symmicta* (Ach.) Ach.**

17. Apothecia varying from pale yellowish brown to very dark brown or black; containing atranorin 18

18. Spores ellipsoid, length to width ratio ca. 2:1, 11.5-14.5(-16.5) x (4.5-)5.0-6.5 µm; thallus pale yellowish gray; apothecial margins paler than, or same as, disk; usnic acid sometimes only present as a trace; Appalachians, on conifers
.... *Palicella filamentosa* (Stirton) Rodr. Flakus & Printzen (syn. *Lecanora filamentosa*, *L. ramulicola*)

18. Spores narrowly ellipsoid, length to width ratio ca. 2.2-2.5:1, (8.0-)10-13(-14) x 3-5 µm; thallus whitish, rimose to areolate (on bark) or largely within substrate (on wood); apothecial margins darker than, or same as, disk; usnic acid abundant; Pacific Northwest, on conifers and wood
.........*Palicella schizochromatica* (Pérez-Ortega et al.) Rodr. Flakus & Printzen (syn. *Lecanora schizochromatica*)

19.(13) Thallus and (or) apothecial margin yellowish, KC+ gold (usnic or isousnic acid); cortex K– (atranorin absent); spores broad or narrowly ellipsoid 20

19. Thallus and apothecial margin cortex KC–, K+ yellow (atranorin) 30

20. Apothecial disks with a light to heavy yellow pruina; apothecial margin prominent, bumpy, and uneven; southeastern coastal plain ....***Lecanora cupressi*** **Tuck.**

20. Apothecial disks not pruinose or lightly gray pruinose 21

21. Amphithecial cortex gelatinous and distinct, 20-100 µm thick at base; apothecia constricted at the base; apothecial margin prominent and flexuose at least in young apothecia; western 22

21. Amphithecial cortex gelatinous, under 20 µm thick and uniform in thickness, or indistinct and poorly developed; apothecia sessile, usually not constricted at the base; apothecial margin flush with disk or becoming thin and disappearing in maturity, not flexuose 25

22. Medulla PD+ yellow (psoromic acid) 23

22. Medulla PD– (lacking psoromic acid) 24

23. Apothecial margin remaining prominent, cortex up to 100 µm at base; spores ellipsoid, 10.5-12.5 x 5.0-5.5 µm .... *Lecanora varia* (Hoffm.) Ach.

23. Apothecial margin becoming even with disk, cortex up to 65 µm at base; spores very broad, (6.5-)8.0-9.5(-11) x (5.0-)5.5-6.5(-7.5) µm ....*Lecanora densa* (Śliwa & Wetmore) Printzen

24. Spores broadly ellipsoid to subglobular, averaging 9.5-10 µm long; apothecia constricted at base; tips of paraphyses brown, HNO_3– ....*Lecanora laxa* (Śliwa & Wetmore) Printzen

24. Spores ellipsoid, averaging over 10 μm long; apothecia not constricted at base; tips of paraphyses black, HNO_3+ violet; montane .... *Lecanora mughicola* Nyl.

25.(21) Apothecia yellow-brown to black, never yellowish green or greenish yellow, margin more or less smooth, not sorediate; amphithecium filled with algae; thallus very thin, pale yellowish gray 26

25. Apothecia usually yellow-green or greenish yellow, margin smooth, sorediate or granular; algae sparse or relatively abundant in amphithecium; thallus pale yellowish green 27

26. Spores narrowly ellipsoid, (7-)9-11(-14) x (3-)4-5 μm; disks varying from black to yellow-brown; apothecial margins ± even with disks, not prominent; amphithecial cortex poorly developed, not gelatinous .... **Lecanora albellula Nyl.** (syn. *L.piniperda*)

26. Spores ellipsoid to broadly ellipsoid, 7-13 x 4-7 μm; disks always brown to yellow-brown; apothecial margins usually prominent; amphithecial cortex gelatinous, 5-20 μm thick .... *Lecanora saligna* (Schaerer) Zahlbr.

27. Apothecial margin essentially biatorine laterally, containing very few algae at base, never sorediate or granular .... ***Lecanora symmicta* (Ach.) Ach.**

27. Apothecial margin with algae abundant at least where thalline tissues of the margin are well developed, but amphithecial cortex absent 28

28. Thallus and apothecia C+ orange (thiophaninic acid); apothecial margin typically smooth and even; western .... *Lecanora confusa* Almb.

28 Thallus and apothecia C–; apothecial margin verruculose, granular, or sorediate 29

29. Thallus PD+ red (fumarprotocetraric acid); thallus thick, granular to verruculose, often becoming granular sorediate; apothecia 0.5-1.5 mm in diameter, with a persistent lecanorine margin having a gelatinous cortex at the base; margin smooth to verruculose, occasionally sorediate; lacking crystals caused by zeorin; locally abundant from Nova Scotia to Maine, the Pacific Northwest, and southern California; rare elsewhere .... *Lecanora conizaeoides* Nyl. *ex* Crombie

29. Thallus PD–; thallus thin, areolate to granular; apothecia 0.3-0.7(-1.0) mm in diameter, wth a thin granular margin that becomes excluded in old apothecia; cortex absent; thallus and apothecial margins often fuzzy with zeorin crystals; East Temperate and California .... ***Lecanora strobilina* (Sprengel) Kieffer**

30.(19) Amphithecial cortex lacking with medullary hyphae growing out to the margin; apothecia heavily pruinose making them pinkish or violet 31

30. Amphithecial cortex present, indistinctly or distinctly delimited from the medulla; apothecia usually yellowish brown to reddish brown or black, with or without pruina 37

31. Spores 12-16 per ascus; amphithecial medulla PD+ deep yellow (psoromic acid); Lake Superior region and rare in western Canada ....*Lecanora cateilea* (Ach.) A. Massal.

31. Spores 8 per ascus, amphithecial medulla PD+ yellow or red (norstictic, protocetraric, or virensic acid) or PD– 32

32. Amphithecial medulla PD+ yellow or PD–, K+ red (norstictic acid, sometimes sparse), lacking virensic or protocetraric acids; mostly on lignum, sometimes bark; southern Rockies, Dakotas, Minnesota ....*Lecanora caesiorubella* subsp. *saximontana* Imshaug & Brodo

32. Amphithecial medulla PD+ red (virensic or protocetraric acid), norstictic acid present or absent, or PD– 33

33. Apothecia 0.3-1.5 mm in diameter, flat; amphithecium forming a cellular pseudocortex at edge; mainly on conifers; apothecial sections K+ red in base (norstictic acid) .... *Lecanora albella* (Pers.) Ach. var. *rubescens* (Imshaug & Brodo) Lumbsch

33. Apothecia 0.8-3 mm in diameter, amphithecial hyphae distinguishable at outer surface, usually not forming a cellular pseudocortex; mainly on hardwoods or lignum 34

34. Apothecial disks C+ yellow; apothecial sections K+ red (norstictic acid); coastal plain, Massachusetts to Texas ....***Lecanora subpallens* Zahlbr.** (syn. *L. caesiorubella* subsp. *prolifera*)

34. Apothecial disks C– 35

35. California; apothecial sections K+ red (norstictic acid) ....***Lecanora caesiorubella* subsp. *merrillii* Imshaug & Brodo**

35. Appalachian-Great Lakes region or southeast; apothecial sections K– or K+ yellow, lacking norstictic acid 36

36. Appalachian-Great Lakes region; apothecial sections PD+ red (virensic acid) ....*Lecanora caesiorubella* Ach. subsp. *caesiorubella*

36. Southeastern; apothecial sections PD– or PD+ red (protocetraric acid)

....*Lecanora caesiorubella* subsp. *glaucomodes* (Nyl.) Imshaug & Brodo

37.(30) Amphithecium containing numerous small crystals, lacking large crystals; amphithecial cortex thick and gelatinous 38

37. Amphithecium containing mainly large, irregular crystals (sometimes sparse; examine several sections); amphithecial cortex thick or thin 43

38. Epihymenium yellow-brown, coarsely granular; disk heavily pruinose, C+ orange; mostly western U.S. and western Great Lakes region ....*Lecanora carpinea* (L.) Vainio

38. Epihymenium red-brown, not at all granular; disk not pruinose, C– 39

39. Amphithecial cortex lacking crystals, but medulla is filled with crystals 40

39. Amphithecial cortex indistinctly delimited from medulla with crystals extending from one into the other 41

40. Growing on logs and fences near the sea on both coasts .... ***Lecanora xylophila* Hue** (syn. *L. grantii*)

40. On trees, rarely wood and then not maritime; coastal California .... *Lecanora horiza* (Ach.) Lindsay

41. Growing on soil and vegetation; arctic-alpine ....*Lecanora epibryon* (Ach.) Ach.

41. Growing on bark 42

42. Apothecia (0.5-)0.7-2.0(-3.0) mm in diameter, flat; spores (12-)13-18(-21) x 7-10(-11) µm; widespread, especially on poplars and ash ....***Lecanora allophana* Nyl.**
[Note: If sorediate, see Key C, couplet 53]

42. Apothecia 0.3-0.6(-1.2) mm in diameter, often becoming convex; spores 9.5-13(-14.5) x (5.5-)6.0-8.0 µm; mostly on maple and beech, rarely poplar; Appalachian-Great Lakes .... *Lecanora glabrata* (Ach.) Malme

43.(37) Epihymenium clear red-brown, lacking crystals or granules; mainly eastern, rare in west ....*Lecanora argentata* (Ach.) Malme (syn. *L. subrugosa*)
[Note: The synonymy is based on molecular evidence with European specimens (Malíček 2014).]

43. Epihymenium granular 44

44. Tiny brown granules, insoluble in HNO_3, distributed in upper third of hymenium; apothecia not pruinose 45

44. Coarse granules, soluble in HNO_3, limited to the surface of the epihymenium, not extending into the hymenium; apothecia pruinose or not 47

45. Spores ellipsoid, 10-13 x 6-7.5 µm; amphithecial cortex not much wider at the base than at the sides of the amphithecium; apothecial disks pale yellow-brown to reddish brown; apothecial sections PD–; eastern temperate, Pacific Northwest, California, and Arizona .... ***Lecanora hybocarpa* (Tuck.) Brodo**

45. Spores broadly ellipsoid, usually over 7.5 µm wide; amphithecial cortex much wider at the base than at the sides of the amphithecium; apothecial disks dark reddish brown to black; apothecial sections PD+ red or PD–; mostly boreal forests 46

46. Spores 13-17.5 x 8-11 µm; apothecial sections PD– .... ***Lecanora circumborealis* Brodo & Vitik.**

46. Spores (9.0-)11.0-15.0 x (6.5-)7.5-9.5(-11.0) µm; apothecial sections PD+ red (fumarprotocetraric acid), rarely PD– .... *Lecanora pulicaris* (Pers.) Ach.

47(44). Apothecial disks UV+ yellow (lichexanthone), heavily pruinose giving disks a yellowish pink to bluish color; containing zeorin; southeastern .... *Lecanora miculata* Ach.

47. Apothecial disks UV–, lightly pruinose or not pruinose, reddish brown to black 48

48. Apothecial disks C+ orange; southeast coastal plain .... *Lecanora louisianae* B. de Lesd.

48. Apothecial disks C– 49

49. Apothecia immersed in the thallus for a long time, finally superficial 50

49. Apothecia mostly superficial with well-developed margins 51

50. Spores ellipsoid, 9-13 x 5-7 µm; apothecia small and very crowded, 0.3-0.8 mm in diameter; apothecial margin smooth, continuous; epihymenium PD–; southeast coastal plain .... *Lecanora leprosa* Fée

50. Spores broadly ellipsoid, mostly 10-14.5 x 7-8.5 µm, apothecia not usually crowded together, 0.7-1.5 mm in diameter; apothecial margin “beaded”; epihymenium PD+ orange (pannarin) .... ***Lecanora cinereofusca* H. Magn.**

51. Apothecia margin cortex less than 15 µm thick; apothecial margin strongly verrucose or "beaded"; apothecia never pruinose; epihymenium PD+ orange (pannarin, with orange crystals slowly developing, as seen under the microscope) .... ***Lecanora cinereofusca* H. Magn.**

51. Apothecia margin cortex 20-50 µm thick; apothecial margin even or slightly bumpy; apothecia often pruinose; epihymenium PD– 52

52. Apothecia dark red-brown, 0.3-0.8 mm in diameter; spores small, 8.5-12 x 5.5-7.0(-8.0) µm; epihymenial granules often sparse; southeast .... *Lecanora pseudargentata* Lumbsch
52. Apothecia shades of brown, usually 0.5-1.5(-3) mm in diameter; spores mostly 11-17 x 6.0-10 µm; epihymenial granules usually abundant 53

53. Thallus very thin; apothecia thin and flat, varying from yellowish through olive-brown to black, sometimes on the same thallus; Pacific Northwest .... ***Lecanora pacifica* Tuck.**
53. Thallus thick, verrucose to granular; apothecia thin or thick, usually pale to very dark reddish to yellowish brown, more or less uniform on one thallus 54

54. Apothecia flat and closely attached to thallus, margin thin but prominent, yellowish, disks typically without pruina; amphithecial crystals often sparse or even absent in some sections; thallus verruculose, continuous; Appalachian-Great Lakes .... *Lecanora wisconsinensis* H. Magn.
54. Apothecia thick, typically constricted at base, margins thick, not yellowish, disks often lightly pruinose; amphithecial crystals abundant; thallus thin to thick; widespread 55

55 On hard weathered wood, never bark; apothecia dark brown to black .... ***Lecanora cenisia* Ach.**
55. On bark, rarely wood; apothecia orange- to yellowish brown 56

56 Apothecia (0.5-)0.7-1.3(-3.0) mm in diameter, spores (10-)12-17(-18) x 7.0-9.5(-11) µm, thick walled, thallus thick, verruculose to granulose .... *Lecanora rugosella* Zahlbr.
56. Apothecia 0.4-0.8 mm in diameter, spores (9.0-)11-13(-15) x 6.0-7.5 µm, thin-walled; thallus smooth to uneven and somewhat verruculose .... *Lecanora chlarotera* Nyl.

57.(1) Thallus endolithic, mostly within the rock or substrate and out of sight, with only the apothecia on the surface 58
57. Thallus clearly visible, continuous or areolate 66

58. On calcareous rock, or on bone or antlers; apothecial margins white, K–, KC– 59
58. On siliceous rock 64

59. Apothecial margins split by distinct fissures; disks pruinose or not pruinose 60

59. Apothecial margins more or less smooth to slightly bumpy (verruculose), not distinctly split; disks not pruinose, or slightly pruinose when young 62

60. Apothecial disks lightly to heavily pruinose; epihymenium granular. Great Lakes region and northeast as well as western temperate ..... *Lecanora crenulata* Hooker

60. Apothecial disks not, or only very lightly, pruinose; epihymenium not granular. Western temperate to western montane 61

61. Epihymenium deep red-brown; thallus entirely within substrate ..... *Lecanora flowersiana* H. Magn.

61. Epihymenium brown to olive or blue-green; thallus often partly visible ..... *Lecanora percrenata* H. Magn.

62. Epihymenium typically not granular; growing on bones and antlers and sometimes on limestone; arctic-alpine ..... *Lecanora zosterae* var. *beringii* (Nyl.) Śliwa

62 Epihymenium granular; on calcareous rock; widespread, temperate to arctic 63

63. Epihymenial granules soluble in K; apothecial disks C+ yellow or orange, UV+ yellow-orange (5-chloro-3-*O*-methylnorlichexanthone); apothecial sections PD– (pannarin absent) ..... ***Lecanora semipallida* H. Magn.** [Plate 420 in LNA, as *L. dispersa*; see also couplet 95.]

63. Epihymenial granules insoluble in K; apothecial disks C–, UV– or sometimes + pale yellow to greenish (2,7-dichlorlichexanthone); apothecial sections often PD+ orange (± pannarin) ..... *Lecanora dispersa* (Pers.) Sommerf. [*not* Plate 420 in LNA]

64.(58) Apothecial disks and margins yellowish, K–, KC+ gold (usnic acid) ..... ***Lecanora polytropa* (Hoff.) Rabenh.**

64. Apothecial disks yellowish brown to black, margins white, K–, KC– (lacking usnic acid) 65

65. Epihymenium deep red-brown, not granular; western temperate ..... *Lecanora flowersiana* H. Magn.

65. Epihymenium yellow to brown, granular; widespread ..... *Lecanora dispersa* (Pers.) Sommerf. [*not* Plate 420 in LNA]

66.(57). On shoreline rocks or near the sea 67

66. On non-maritime rocks 71

67. Thallus white, not lobed; cortex C–, K+ yellow or K– 68

67. Thallus yellowish, lobed or not; cortex C+ orange, K– (xanthones) 69

68. Thallus K+ yellow (atranorin); apothecia bright red-brown; epihymenium not granular; spores mostly 13-17 x 6-8.5 µm; common on rocks on the east coast, rare on rocks on the west coast .... ***Lecanora xylophila*** **Hue** (see also couplet 40)

[Note. If apothecia are black, see *L. gangaleoides*, couplet 106.]

68. Thallus K– (containing 2,7-dichlorlichexanthone); apothecia dull reddish brown; epihymenium granular; spores 9.0-12.4 x 4.8-6.2 µm; frequent in Pacific Northwest, rare in east .... *Lecanora schofieldii* Brodo

[Note: See Brodo (2010) for key to less common maritime *Lecanora* species.]

69. Thallus minutely fruticose, with round branches .... *Lecanora phryganitis* Tuck.

69. Thallus crustose 70

70. Thallus not lobed; California; apothecia yellowish, heavily pruinose .... ***Lecanora pinguis*** **Tuck.**

70. Thallus lobed; common on northern Pacific coast, rare in the northeast; apothecia red-brown, not or slightly pruinose .... *Lecanora straminea* Ach.

71.(66) Thallus distinctly lobed at periphery or with lobed areoles 72

71. Thallus not lobed 89

72. Ascus tips K/I± uniformly very pale blue (*Aspicilia*-type); thallus cortex often K+ blood red (norstictic acid) .... *Lobothallia*

72. Ascus tips K/I+ dark blue (*Lecanora*-type); thallus cortex K–, or K+ persistently yellow 73

73. Apothecia bright red-brown; epihymenium entirely without granules; thallus pale yellow to gray .... ***Lecanora argopholis*** **(Ach.) Ach.** [see also couplets 96 and 99]

73. Apothecia yellow, yellow-brown to dull reddish-brown, greenish, or black; epihymenium usually granular; thallus yellow-green, greenish yellow, or honey-brown 74

74. Thallus red-brown to honey-brown, often shiny; apothecia without pruina but rarely seen; California to Pacific Northwest 75

74. Thallus yellowish green or greenish yellow; apothecia with or without pruina, usually present and abundant 76

75. Lobes mostly flat to slightly convex, not inflated and strongly folded; woodlands of California foothills; typically containing rangiformic acid .... *Lecanora mellea* W. A. Weber

75. Lobes strongly convex, inflated and strongly folded; California to Washington and Idaho, mostly montane; containing protolichesterinic acid .... *Lecanora pseudomellea* B. D. Ryan

76. Thallus consisting of lobed squamules or areoles, not forming rosettes with lobed margins 77

76. Thallus forming lobed rosettes, almost foliose in appearance 80

77. Apothecia lightly pruinose, orange-pink; Appalachian-Great Lakes, rare in west .... ***Rhizoplaca subdiscrepans* (Nyl.) R. Sant.**

77. Apothecia without or with light pruina, yellowish brown or yellow-green 78

78. Apothecia waxy yellow-green, without any pruina; containing placodiolic acid; Great Lakes region to the N.W.T. and southwest .... *Lecanora opiniconensis* Brodo

78. Apothecia yellow to yellow brown or black, lightly pruinose 79

79. Lobes typically flat, blackened at edges; sometimes containing placodiolic acid, rarely with fatty acids; California to B.C. mountains .... *Lecanora semitensis* (Tuck.) Zahlbr.

79. Lobes flat to very convex, not blackened at edges; containing fatty acids; southwestern, rare in northeast .... *Lecanora weberi* B. D. Ryan

80.(76) Growing on calcareous rock; apothecia yellow to yellow-brown, without pruina 81

80. Growing on siliceous rock or sandstone 82

81. Thallus very pale green to yellowish green, heavily white pruinose; containing roccellic acid; southwestern, central, and northeastern regions .... *Lecanora valesiaca* (Müll. Arg.) Stizenb.

81. Thallus yellow-green or shades of gray-green or olive; without or with spotty pruina; containing zeorin; widespread .... ***Lecanora muralis* (Schreb.) Rabenh.**

82. Thallus lobes mostly flat to slightly convex 83

82. Thallus lobes usually somewhat to strongly convex; common throughout western U.S. and into southern Canada .... 88 (but see also couplet 87)

83. Apothecia heavily yellowish pruinose; medulla usually PD+ yellow (psoromic acid); arid southwest 84
83. Apothecia not pruinose; medulla PD– or PD+ 85

84. Lobe margins blackened with black granules (thallospores) .... *Lecanora nashii* B. D. Ryan
84. Lobe margins usually not blackened, lacking black thallospores .... *Lecanora bipruinosa* Fink

85. Cortex C+ pink, KC+ pink (gyrophoric acid and zeorin); lobes short, flat to slightly convex, divergent and spreading, without raised margins; Ozarks .... *Lecanora gyrophorica* Lendemer
85. Cortex C–, KC– 86

86. Lobes usually with thickened or slightly raised margins making the lobe tips somewhat concave; lobe margins usually whitish or pruinose; containing zeorin; widespread temperate, often on rocks frequented by birds .... ***Lecanora muralis* (Schreber) Rabenh.**
86. Lobes flat to slightly convex, not concave; margins not white; lacking zeorin 87

87. Thallus frequently pruinose, at least in older parts; medulla PD– (placodiolic acid); throughout the arid west .... *Lecanora phaedrophthalma* Poelt
87. Thallus rarely pruinose; medulla usually PD+ orange or yellow (rangiformic acid often with pannarin or psoromic acid); California mountains .... *Lecanora sierrae* B. D. Ryan & T. H. Nash

88.(82) Apothecia black to yellowish green, typically heavily yellowish pruinose but sometimes without pruina; lobes convex, but usually not strongly sinuous or folded, rarely pruinose; medulla PD+ yellow (psoromic acid) or C+ red (lecanoric acid); common in southwest .... ***Lecanora novomexicana* H. Magn.**
88. Apothecia not pruinose; thallus verrucose or rugose, or with sinuous folds, often pruinose; medulla PD–, C– (containing zeorin); throughout the arid west .... ***Lecanora garovaglii* (Körber) Zahlbr.**

89.(71) Thallus grayish brown to yellow- or red-brown, shiny .... *Protoparmelia*
89. Thallus shades of gray or pale green to yellow-green; if brownish, only at margins 90

90. On calcareous rock 91
90. On non-calcareous rock 97

91. Apothecial disks moderately or heavily pruinose; thallus thick, white 92
91. Apothecial disks not, or lightly, pruinose 93

92. Apothecia concave, immersed in areoles or later prominent; ascus tip K/I–; spores 14-24 x 10-16 µm; arctic-alpine; no lichen substances ....***Aspicilia candida* (Anzi) Hue**

92. Apothecia flat, convex, or becoming hemispherical; ascus tips K/I+ dark blue (*Lecanora*-type); spores 9-13(-15) x 4.5-7.5 µm; northeastern; containing 2,7-dichlorlichexanthone .... *Lecanora albescens* (Hoffm.) Branth & Rostrup

93. Apothecia under 1.3 mm in diameter; thallus thin 94

93. Apothecia 0.6-3 mm in diameter; thallus usually thick, even when in scattered areoles 96

94. Apothecial margin split by conspicuous fissures; epihyenium not granular ...*Lecanora percrenata* H. Magn.

94. Apothecial margin smooth, not split by fissures; epihymenium granular 95

95. Apothecia black or almost black; disks UV–, C–, containing 2,7-dichlorlichexanthone. Arctic-alpine .... *Lecanora torrida* Vainio

95. Apothecia yellowish brown to pale brown; disks UV+ yellow-orange, C+ yellow or orange (5-chloro-3-*O*-methylnorlichexanthone); northeastern to western temperate and arctic .... ***Lecanora semipallida* H. Magn.** [see also couplet 63]

96.(93) Apothecial disks black or almost black; epihymenium green ....***Lecanora marginata* (Schaerer) Hertel & Rambold**

96. Apothecial disks red-brown; epihymenium brown ...... ***Lecanora argopholis* (Ach.) Ach.** [see also couplet 99]

97.(90) Thallus leprose, yellow-green (usnic acid, zeorin), with a conspicuous white, fibrous prothallus; apothecia very rare ....***Lecanora thysanophora* R. C. Harris**

97. Thallus continuous or dispersed areolate, not leprose; prothallus present or absent; apothecia abundant 98

98. Apothecial disks shiny red-brown, never pruinose; epihymenium clear red-brown, not at all granular 99

98. Apothecia yellowish to brown or black, dull, pruinose or not; epihymenium granular on the surface or not granular 101

99. Thallus thick, areolate, usually with a yellowish tint; amphithecium filled with very small crystals; apothecia 1-3 mm in diameter, constricted at the base; containing xanthones and sometimes zeorin, gangaleoidin, or fatty acids; western .... ***Lecanora argopholis* (Ach.) Ach.**

99. Thallus thin, continuous to dispersed areolate, pale gray; amphithecium containing large crystals in clumps; apothecia less than 1 mm in diameter 100

100. Thallus continuous, smooth to rimose-areolate; apothecia more or less immersed in the thallus when young, then sessile; containing zeorin; East Temperate and southern California .....*Lecanora subimmergens* Vainio

100. Thallus dispersed areolate to areolate; apothecia sessile; containing "galactinulin" (a mixture of one to three 2'-*O*-methylated perlatolic acids); widely distributed .... *Lecanora pseudistera* Vainio

101.(98) Thallus white to pale gray or greenish gray (lacking usnic acid) 102

101. Thallus greenish yellow or yellowish green (containing usnic acid) 108

102. Apothecial disks C+ yellow, heavily white pruinose 103

102. Apothecial disks C–, pruinose or not 104

103. Apothecial with a distinct blue-black ring just inside the thalline margin; western montane .... *Lecanora bicincta* Ramond

103. Apothecia lacking a dark ring inside the thalline margin; widespread, especially in the west .... ***Lecanora rupicola* (L.) Zahlbr.**

104. Apothecia remaining immersed in thallus (cryptolecanorine), lacking margins; disks black, never pruinose .... ***Lecanora oreinoides* (Körber) Hertel & Rambold**

104. Apothecia superficial when mature with well-developed margins; disks brown to black, pruinose or not 105

105. Epihymenium lacking crystals or granules; apothecia dark smoky brown to black, without pruina; containing gangaleoidin 106

105. Epihymenium coarsely granular on the surface; apothecia pale to dark brown, rarely black, often pruinose; lacking gangaleoidin 107

106 Epihymenium brown to olive- or greenish brown; apothecia dark brown to pitch black; Great Lakes to Maritime provinces, rare in boreal forest; on shaded rock walls and protected cliff faces .... *Lecanora argentea* Oxner & Volkova (syn. *L. fuliginosa*)

106. Epihymenium blue-green; apothecia pitch black; common in California .... *Lecanora gangaleoides* Nyl.

107. Apothecia large, 0.9-2.5 mm in diameter, heavily pruinose, constricted at base; on coastal rocks and cliffs from California to Oregon; containing norgangaleoidin ....*Lecanora californica* Brodo

107. Apothecia 0.5-1.0(-2.0) mm in diameter, pruina, when present, thin or on young apothecia; widespread on exposed, sunny rocks; usually containing roccellic acid or other fatty acids ....***Lecanora cenisia* Ach.**

108.(101) Apothecia 2-7 mm in diameter, constricted at the base; disks shades of orange or pink, lightly pruinose; thallus well developed, dispersed areolate to almost squamulose ....***Rhizoplaca subdiscrepans* (Nyl.) R. Sant.**

108. Apothecia 0.3-1.5 mm in diameter, disks yellowish or darkening, not orange or pink, pruinose or not; thallus areolate, sometimes scanty 109

109. Apothecia 0.5-1.5 mm in diameter, often constricted at the base, typically pruinose; spores 8-12 x 4-5(-6) µm; thallus areoles dispersed or contiguous, often edged in black; western in coastal ranges and Rockies ....*Lecanora semitensis* (Tuck.) Zahlbr.

109. Apothecia 0.3-0.8(-1.3) mm in diameter, sessile to immersed, not pruinose; spores 8-15 x 5-7 µm; thallus areoles never edged in black; mostly northern and montane 110

110. Apothecia sessile; disks pale yellow to pale orange; thallus composed of scattered, very small, flat, adnate areoles; common and widespread except in central and southeastern regions ....***Lecanora polytropa* (Hoffm.) Rabenh.**

110. Apothecia mostly immersed in thallus; disks yellow to greenish brown or even black; thallus well developed, continuous, areolate; arctic, high montane and western Great Lakes region ....*Lecanora intricata* (Ach.) Ach.

LECIDEA (includes *Lecidea*-like crusts with a lecideine or black biatorine margin, colorless, non-septate spores, 8 per ascus: *Bryobilimbia, Calvitimela, Carbonea, Immersaria, Lecidella, Miriquidica, Rimularia, Schaereria,* and *Tremolecia*)

1. Parasitic on crustose lichens; thallus brown, areolate; epihymenium olive-brown; hypothecium dark brown-black; exciple black; asci *Rimularia*-type; spores 8-14 x 4-6(-7) µm; medulla C+ pink (gyrophoric acid) on *Lecanora rupicola s. lat.*; southwest ....*Rimularia insularis* (Nyl.) Rambold & Hertel

1. Not parasitic 2

2. Growing on bark, wood, soil, peat, etc., not rock 3

2. Growing directly on rock 17

3. On wood, soil, or peat 4
3. On bark 10

4. Thallus indistinct or not visible (within the substrate) 5
4. Thallus clearly visible, thin to thick 6

5. Hypothecium dark brown to black; tips of paraphyses capitate; spores 5-9.5 x 2.5-3.5 µm .... *Lecidea plebeja* Nyl.
5. Hypothecium colorless or very pale yellowish brown; tips of paraphyses not capitate; spores (8-)8.5-11(-14) x 2.5-4.5(-5) µm; mostly on hard wood, widespread ...."*Lecidea*" *turgidula* Fr. [see couplet 15]

6. Thallus dark: red-brown, dark brown, or dark gray to olive-black, verrucose, granular, or isidiate, PD–; apothecia brown to black, flat to convex, often with persistent margins at least in young apothecia 7
6. Thallus pale: whitish to creamy or pale gray, PD+ or PD–; apothecia black, becoming marginless and convex 8

7. Hypothecium dark brown, not distinct from dark brown exciple; spores 10-15 x 4-7 µm ....*Placynthiella*
7. Hypothecium colorless; exciple pale within; spores 7-15 x 3-7 µm ....*Trapeliopsis*

8. Epihymenium brown; hypothecium colorless; on hard, weathered wood, never soil; thallus smooth, continuous, PD+ red, KC– (fumarprotocetraric acid); spores 7-10 x 2.8-4.5 µm; boreal to montane .... *Ramboldia elabens* (Fr.) Kantvilias & Elix (syn. *Pyrrhospora elabens*)
8. Epihymenium blue-green; hypothecium red-brown; thallus PD–, sometimes KC+ orange or reddish (xanthones); on mosses, soil, or peat, arctic-alpine 9

9 Thallus coarsely granular to almost isidioid; spores narrow, (11-)15-19(-26) x 5-7(-9) µm .... *Frutidella caesioatra* (Schaerer) Kalb (see *Micarea* key)
9. Thallus continuous, sometimes verruculose, not granular; spores broad, 9-16 x (5-)6-8.5 µm ....*Lecidella wulfenii* (Hepp) Körber

10.(3) Hypothecium dark brown to black 11
10. Hypothecium colorless to yellowish or reddish brown 12

11. Spores narrowly ellipsoid, (8.8-)10-13(-14) x 4.0-6.0(-7.5) µm; hypothecium dark purplish brown to black, merging with exciple which is dark internally and almost colorless externally; paraphyses sticking together when mounted in water; epihymenium brown; boreal forest region and west coast .... "*Lecidea*" *albofuscescens* Nyl.

11. Spores globose, (5-)6-8(-10) µm in diameter; hypothecium brown to orange-brown; epihymenium green; California to southeastern B.C., Montana, and Idaho .... *Schaereria dolodes* (Nyl. *ex* Hasse) Schmull & T. Sprib.

12. Thallus leprose, blue-gray; spores globose, 5-7.5 µm in diameter; containing divaricatic acid (sometimes UV+ blue-white); widespread, especially on pine bark .... *Lecidea nylanderi* (Anzi) Th. Fr.

12. Thallus not leprose, although sometimes sorediate; spores ellipsoid 13

13. Epihymenium green to blue-green; paraphyses easily separating when mounted in water; spores broadly ellipsoid, mostly 11-16 x 6-10 µm; widespread arctic to temperate, from Lake Superior and James Bay westward .... *Lecidella euphorea* (Flörke) Hertel

[Note: *Lecidella elaeochroma* (Ach.) M. Choisy is regarded as distinct from *L. euphorea* by some lichenologists (e.g., Knopf & Leuckert 2004:313), but their characteristics overlap too much to be confidently separated here.]

13. Epihymenium brown to olive or greenish; paraphyses sticking together when mounted in water; spores narrowly ellipsoid to ellipsoid, 7-13.5 x 2.8-7.5 µm 14

14. Spores 2.5-4.5 µm wide; apothecia black 15

14. Spores 4-7.4 µm wide; apothecia black or very dark brown, dull, flat, with persistent, thin, raised margins; thallus thin or thick, greenish gray, dull, PD+ red or PD– 16

15. Apothecia shiny black, often convex with a disappearing margin; thallus thick, pale gray, often shiny, PD+ red (fumarprotocetraric acid); on wood, rarely on conifer bark .... *Ramboldia elabens* (Fr.) Kantvilas & Elix

15. Apothecia dull, usually with a bluish haze on the surface, especially when wet; thallus very thin or disappearing, PD–; on wood or conifer bark; widespread .... "*Lecidea*" *turgidula* Fr. [see couplet 5]

16. Paraphyses branched and anastomosing, with a series of bead-like end cells (moniliform); spores 7.5-13.5 x 4.5-7.5 µm, ellipsoid, not contricted; apothecia black, round to somewhat irregular; thallus PD–, sometimes sorediate, without a prothallus; Eastern Temperate .... *Rimularia caeca* (Lowe) Rambold & Printzen

16. Paraphyses unbranched, with capitate but not bead-like tips; spores 7-9 x 4-5 µm, often constricted at the center, rarely septate; apothecia very dark brown, almost black, with flexuous margins; thallus PD+ red

(fumarprotocetraric acid); always with greenish to brownish soralia, finally coalescing in the center, usually with a conspicuous brown prothallus; Appalachian-Great Lakes distribution
.... *Fuscidea arboricola* Coppins & Tønsberg

17.(2) Cephalodia present .... *Amygdalaria*
17. Cephalodia absent 18

18. Thallus distinctly rusty orange or bright orange, well developed 19
18. Thallus white, gray, brown, olive, or yellowish, not orange, or indistinct (within rock) 21

19. Apothecia concave, immersed in thallus, usually less than 0.5 mm in diameter
.... ***Tremolecia atrata* (Ach.) Hertel**
19. Apothecia flat when mature, sessile, usually more than 0.5 mm in diameter 20

20. Spores 14-24 x 6-11 µm, halonate (visible at least in ink preparation); epihymenium olive-brown; paraphyses abundantly branched; medulla IKI–
.... ***Porpidia flavocaerulescens* (Hornem.) Hertel & A. J. Schwab**
20. Spores 9-15 x 4.5-8 µm, without a halo; epihymenium green; paraphyses mostly unbranched except for tips; medulla IKI+ blue .... orange form of ***Lecidea lapicida* (Ach.) Ach.**

21.(18) Hypothecium pigmented dark brown to black 22
21. Hypothecium colorless or pale yellowish to orange, or pale brown 43

22. Epihymenium brown 23
22. Epihymenium greenish, olive, or blue, at least in part, or after treatment with K 26

23. Spores halonate (seen in an ink preparation); asci *Porpidia*-type (fig. 14g) .... *Porpidia* key
23. Spores not halonate; asci *Lecidea*-type (fig. 14f), *Porpidia*-type, or *Lecanora*-type (fig. 14e) 24

24. Spores 6.5-10 x 2.5-3.5 µm; paraphyses capitate at the tips; asci *Lecidea*-type; eastern North America, on dry rocks .... *Lecidea cyrtidia* Tuck.
24. Spores (10.3-)12-15(17.5) x 5-7.5 µm; paraphyses hardly expanded at the tips 25

25. Asci *Porpidia*-type; thallus greenish gray, continuous, K–; Pacific Northwest to central states, east to Quebec; often in or near freshwater .... *Bryobilimbia ahlesii* (Körber) Fryday et al. (syn. *Lecidea ahlesii*)
[Note: Young spores of *B. ahlesii* are halonate.]

25. Asci *Lecanora*-type; thallus pale, areolate to verrucose, K+ yellow (atranorin); Florida and California ....*Carbonea latypizodes* (Nyl.) Knopf & Rambold

26.(22) Thallus within substrate and not visible, or indistinct 27

26. Thallus clearly visible 30

27. Apothecia 0.15-0.4 mm in diameter; spores 6-10 x 2-5 µm; exciple pale within, dark at edge; on pebbles and boulders .... *Leimonis erratica* (Körber) R. C. Harris & Lendemer (syn. *Micarea erratica*)

27. Apothecia mostly over 0.5 mm in diameter 28

28. Spores 5-17 µm wide, usually halonate in an ink preparation, at least when young ....*Porpidia* key

28. Spores 2-5.5 µm wide, not halonate 29

29. Exciple pale within, dark at edge, distinct from hypothecium; thallus IKI+ blue (test white tissue under apothecia); apothecia usually clustered in groups, up to 2.5 mm in diameter; spores (5-)7.5-9(-12) x (2-)3-4 µm; containing confluentic acid ....*Lecidea auriculata* Th. Fr.

29. Exciple carbonaceous (very dark brown to black) throughout, confluent with hypothecium; thallus IKI–; apothecia usually separate, up to 1.0 mm in diameter; spores (8.5-)10.5-12(-15) x 3.5-5.5 µm; containing no lichen substances .... *Carbonea vorticosa* (Flörke) Hertel

30.(26) Asci narrowly cylindrical, walls uniformly IKI+ light blue, without an apical thickening (tholus) (*Schaereria*-type (fig. 14i); arctic-alpine 31

30. Asci club-shaped, with a distinct apical thickening (tholus) 32

31. Spores ellipsoid, 10-18 x 5-8 µm; thallus areolate, dark gray to brownish gray, prothallus black, conspicuous; medulla C+ pink (gyrophoric acid) ..... ***Schaereria fuscocinerea* (Nyl.) Clauzade & Cl. Roux**

31. Spores globose, 6-8(-10) µm in diameter; thallus areolate to subsquamulose, dark gray to brown; medulla C– or C+ faint pink (traces of gyrophoric acid) .... *Schaereria cinereorufa* (Schaerer) Th. Fr.

32. Paraphyses easily separated in a water mount; spores broadly ellipsoid 33

32. Paraphyses coherent in a water mount; spores ellipsoid 34

33. Thallus white, areolate to verrucose, K+ yellow, C– (atranorin); widespread .....*Lecidella carpathica* Körber

33. Thallus pale yellowish to greenish yellow, rimose to areolate; on rocks along the Pacific coast ....*Lecidella asema* (Nyl.) Knopf & Hertel (syn. *L. elaeochromoides*)

34. Apothecia less than 0.5 mm in diameter; typically on small stones in exposed habitats ....*Leimonis erratica* (Körber) R. C. Harris & Lendemer (syn. *Micarea erratica*)

34. Apothecia reaching more than 0.5 mm in diameter 35

35 Thallus thin or thick, white to pale or dark gray, not brownish or subsquamulose 36

35. Thallus thick, areolate to subsquamulose, usually brown or brownish gray 39

36. Asci *Lecanora*-type; thallus verrucose-areolate, K+ yellow; spores 12-15 x 5-7 µm; Florida and California ....*Carbonea latypizodes* (Nyl.) Knopf & Rambold

36. Asci *Lecidea*- or *Porpidia*-type; thallus continuous, rimose to rimose-areolate 37

37. Spores halonate, at least when young; asci *Porpidia*-type ....*Porpidia* key

37. Spores not halonate; asci *Lecidea*-type 38

38. Spores 4.5-8 µm broad; thallus K+ yellow or red (containing stictic or norstictic acid) ....***Lecidea lapicida* (Ach.) Ach.**

38. Spores 2-4 µm broad; thallus K– (containing confluentic acid) ....*Lecidea auriculata* Th. Fr.

39 (35). Thallus medulla C+ pink (gyrophoric acid), IKI–. Mostly western coastal ranges, but also Great Lakes region ....*Lecidea fuscoatra* (L.) Ach.

39. Thallus medulla C–, IKI+ blue 40

40. Medulla K+ red (norstictic acid); California and Rocky Mountains to arctic ....*Lecidea atrobrunnea* subsp. *saxosa* Hertel & Leukert (syn. *Lecidea syncarpa* Zahlbr)

40. Medulla K– or K+ yellow 41

41. Spores 10-17 x 5-9 µm; medulla K+ yellow, PD+ orange (stictic acid); thallus areolate, reddish brown; arctic-alpine ....*Lecidea paupercula* Th. Fr.

41. Spores mostly under 5.5 µm wide; medulla K– or K+ yellow; thallus becoming subsquamulose 42

42. Thallus dark brown, typically shiny, subsquamulose; spores 5-12 x 3-5 µm; medulla K– (confluentic acid with or without 2'-*O*-methylperlatolic acid or planaic acid, or has no substances; arctic-montane .... ***Lecidea atrobrunnea* (Lam. & DC.) Schaerer** *s. lat.*

42. Thallus yellowish brown to reddish brown, dull, areolate, rarely subsquamulose; spores 7-14.5 x 4-5.5 µm; containing no substances; Appalachian-Great Lakes .... *Lecidea brunneofusca* H. Magn.

43.(21) Epihymenium brown 44

43. Epihymenium olive, greenish, or blue-green 45

44. Thallus squamulose; asci similar to *Bacidia*-type, with K/I+ dark blue tholus and axial body; southern California .... *Miriquidica scotopholis* (Tuck.) B.D. Ryan & Timdal [also see *Psora* key]

44. Thallus continuous, areolate to smooth, not squamulose; asci *Fuscidea*-type .... *Fuscidea*

45. Thallus medulla IKI+ blue 46

45. Thallus medulla IKI– 51

46. Spores 11-18(-26) x (6-)7-9(-14) µm, halonate; asci *Porpidia*-type; thallus yellowish brown, areolate, areoles often with whitish margins; apothecia immersed in areoles (*Aspicilia*-like), often pruinose; western Great Lakes and Arizona .... *Immersaria athroocarpa* (Ach.) Rambold & Pietschm.

46. Spores 6-15 x 3.5-8 µm, not halonate; asci *Lecidea*-type 47

47. Thallus yellowish brown to red-brown, areolate to subsquamulose; mostly arctic-alpine, western 48

47. Thallus white to gray; continuous to areolate, or scanty 49

48. Thallus K+ red (norstictic acid; lacking planaic, confluentic, or 2'-*O*-methylperlatolic acid); California and Rocky Mountains to arctic ....*Lecidea atrobrunnea* subsp. *saxosa* Hertel & Leukert (syn. *Lecidea syncarpa* Zahlbr.)

48. Thallus K– (confluentic acid with or without 2'-O-methylperlatolic acid or planaic acid, or has no substances), or less frequently, K+ yellow (stictic acid) or K+ red (norstictic acid, together with either planaic, confluentic, or 2'-O-methylperlatolic acid); Arctic-montane .... ***Lecidea atrobrunnea* (Lam. & DC.) Schaerer** *s. lat.*

49. Thallus gray to brownish orange; lacking a prothallus; thallus K+ yellow or red (containing stictic or norstictic acid) .... ***Lecidea lapicida* (Ach.) Ach.**

49. Thallus pale gray to almost white; prothallus conspicuous, black; thallus K–, containing confluentic acid 50

50. Apothecia typically sunken between areoles and level with thallus surface, never pruinose; hypothecium colorless, confluent with exciple .... ***Lecidea tessellata* Flörke**

50. Apothecia sessile, not sunken, sometimes lightly pruinose; hypothecium pale brown, distinct from exciple .... *Lecidea confluens* (Weber) Ach.

51.(45) Paraphyses free in water or KOH mounts; thallus usually thin, white; lacking usnic acid 52

51. Paraphyses coherent in water and KOH 53

52. Hymenium lacking oil drops; growing on calcareous or non-calcareous rock; widespread temperate to boreal .... ***Lecidella stigmatea* (Ach.) Hertel & Leuckert**

52. Hymenium containing abundant tiny oil drops; growing on non-calcareous, siliceous rock, or less frequently on calcareous rock; western montane to arctic .... *Lecidella patavina* (A. Massal.) Knoph & Leuckert

53. Thallus yellow to yellowish white or chalky white, typically well developed; gray or black prothallus usually conspicuous; asci *Lecanora*-type; containing usnic acid; mostly arctic-alpine 54

53. Thallus gray to brownish or olive-brown, lacking usnic acid; prothallus present or absent 56

54. Hypothecium colorless throughout; thallus thin or thick, continuous to areolate, dull yellowish white or chalky white, with a blue-gray prothallus; spores 8-15 x 4-7.5 µm .... ***Lecanora marginata* (Schaerer) Hertel & Rambold**

54. Hypothecium yellowish to yellow-orange above, colorless below; thallus very thick, areolate to subsquamulose, pale to dark yellow; prothallus black 55

55. Thallus shiny, dark yellow to brownish; areoles fissured; spores 9-12 x 3.5-5 µm; contains alectorialic acid .... ***Calvitimela armeniaca* (DC.) Hafellner** (syn. *Tephromela armeniaca*)

55. Thallus typically dull, pale yellow; areoles smooth, not fissured; spores 11-15(-16) x (4.5-)6.0-8.5 µm; contains atranorin and usually usnic acid...... *Calvitimela aglaea* (Sommerf.) Hafellner (syn. *Tephromela aglaea*)

56.(53) Thallus pale brown to olive-brown, areolate to subsquamulose; spores mostly broadly ellipsoid, (7-)9-14(-16) x 5-8 µm; mainly southern California .... *Lecidea mannii* Tuck.

56. Thallus gray; spores narrow, less than 5 µm wide 57

57. Thallus typically well developed, continuous to areolate; planaic acid and 4-*O*-demethylplanaic acid present; spores 8.2-14 x 3.6-4.6 µm .... *Lecidea plana* (J. Lahm) Nyl.

57. Thallus typically endolithic or dispersed areolate; planaic acid absent, 4-*O*-demethylplanaic acid present; spores 6-12 x 2.5-4 µm .... *Lecidea laboriosa* Müll. Arg.

LEMPHOLEMMA Based in part on Gilbert et al. (2009).

1. Thallus with short, branched, cylindrical to flattened lobes up to 5 mm long, forming small umbilicate cushions on rock; apothecia absent but sometimes with granular goniocysts in disk- or cup-like structures at the branch tips; mainly western and Great Lakes region .... *Lempholemma cladodes* (Tuck.) Zahlbr.

1. Thallus more or less prostrate on soil, rock, or moss, not forming subfruticose cushions; typically fertile ...2

2. Lobes narrow, branched, 0.4-1.0 mm wide, 3-5(-15) mm long, folded or channeled when dry, sometimes developing globular to cylindrical isidia in older part of thallus; Montana and B.C. to Alaska .... *Lempholemma radiatum* (Sommerf.) Henssen

2. Lobes broad and membranous, thickened at the edges, with numerous marginal or laminal apothecia; widespread in the west and northeast but often overlooked 3

3. Spores broadly ellipsoid to almost globose, 9.5-13.5 x 8.5-11.5 µm .... *Lempholemma polyanthes* (Bernh.) Malme (syn. *L. myriococcum*)

3. Spores ellipsoid, 19-25 x 7.5-10.5 µm .... *Lempholemma chalazanum* (Ach.) B. de Lesd.

LEPRARIA (including *Leprocaulon* and "*Leproloma.*") Based on Lendemer (2013) and Lendemer and Hodkinson (2013).

1. Thallus producing slender or stout, solid stalks over which granules or granular soredia are deposited 2

1. Thallus entirely leprose, without stalks 5

2. Thallus slightly yellowish gray or yellowish green, covered with mealy granules that tend to be fuzzy with terpene crystals, PD–, K– or K+ yellow (usnic acid, zeorin, sometimes with rangiformic acid or atranorin); stalks slender, stiff, almost unbranched, 2-6 mm high, often dark brown to black at the base; Pacific coast and montane southwest

...................................*Leprocaulon americanum* Lendemer & Hodkinson (= *L. microscopicum* of N. Am. authors)

2. Thallus gray to bluish gray, lacking usnic acid or zeorin; stalks not darkening at the base 3

3. Thallus PD–, K+ pale yellow (atranorin, rangiformic acid); stalks sparsely to abundantly branched, with a central cartilaginous core; southern Rockies, Arizona, and Texas

....***Lepraria gracilescens* (Nyl.) Lendemer & Hodkinson** (syn. *Leprocaulon gracilescens*)

3. Thallus PD+ yellow or orange 4

4. Thallus PD+ orange, K+ yellow, or PD+ yellow, K– (atranorin with thamnolic, baeomycesic, or psoromic acid), UV–; stalks relatively stout, 0.25-0.5 mm wide, lacking a central cartilaginous core; montane Alaska and B.C. to Colorado

................................*Lepraria subalbicans* (I. M. Lamb) Lendemer & Hodkinson (syn. *Leprocaulon subalbicans*)

4. Thallus PD+ yellow, K–, UV+ white (baeomycesic and squamatic acids); stalks slender 0.2-0.3 mm wide; rare, southeast Alaska ....*Lepraria albicans* (Th. Fr.) Lendemer & Hodkinson (syn. *Leprocaulon albicans*)

5.(1) Thallus coarsely granular sorediate, not cottony, the soredia soon developing a thick pseudocortex giving it a firm appearance; margins irregular or, when on rocks, forming distinct, often concentric rings about 20-40 mm in diameter, pale to dark gray; hypothallus entirely absent; on siliceous rocks or moss, often exposed and in full sun; thallus K– or K+ yellow or red, PD– or PD+ yellow, orange, or red (many chemical races including those containing alectorialic, fumarprotocetraric, psoromic, stictic, norstictic acid, atranorin, and fatty acids); widespread ....***Lepraria neglecta* (Nyl.) Erichsen** (syn. *L. caesioalba, L. borealis*)

5. Thallus thin or thick, made up of fine, farinose soredia, never forming concentric rings; typically on shaded rock, bark, soil, or moss 6

6. Thallus forming round, adnate, or somewhat ascending lobes 3-6 mm across, often with thickened margins; brownish or white cottony hypothallus present 7

6. Thallus thick or thin, not forming well-developed, ascending, round lobes with thickened margins; hypothallus present or absent 9

7. On soil; thallus white or brownish white, superficial granular soredia often difficult to discern; PD+ yellowish, K+ yellow (atranorin, with either pannaric 6-methylester or norascomatic, roccellic and (or) rangiformic acids); coastal California .... *Lepraria xerophila* Tønsberg

7. On bark, wood, or rock; thallus blue-gray to greenish; PD+ red; eastern; hypothallus often dark gray to brown 8

8. Containing protocetraric acid; on shaded rock faces or tree trunks; common, East Temperate .... ***Lepraria normandinioides*** **Lendemer & R. C. Harris** [Plate 448 in LNA as *Leproloma membranaceum*]

8. Containing fumarprotocetraric acid; Appalachian-Ozark .... *Lepraria oxybapha* Lendemer

9.(6) Thallus developing a distinct white or brown cottony hypothallus that subtends the leprose upper layer 10

9. Thallus composed of a single layer of soredia, dispersed or aggregated 15

10. Thallus rather thick, greenish white, UV+ blue-white, KC+ pink (divaricatic and nordivaricatic acids (no atranorin); East Temperate, on rock .... *Lepraria cryophila* Lendemer

10. Thallus UV−, KC− 11

11. Thallus PD− or PD± yellowish, K+ yellowish (atranorin, zeorin, pallidic acid); hypothallus thin; East Temperate, on conifer bark, usually in wetlands ....*Lepraria harrisiana* Lendemer

11. Thallus PD+ deep yellow to orange or red-orange, K− or K+ yellow 12

12. Thallus PD+ deep yellow or yellow-orange, K−, KC+ red (alectorialic ± barbatolic or protocetraric acid, no atranorin); on tree bases, or calcareous or non-calcareous rocks; northeastern and western .... *Lepraria eburnea* J. R. Laundon

12. Thallus PD+ orange, K+ yellow, lacking alectorialic acid 13

13. Thallus pale green to yellowish green, rarely greenish gray; hypothallus brownish or white; containing atranorin, stictic acid, and zeorin, sometimes roccellic acid; widespread and very common .... ***Lepraria finkii*** **(Bouly d Lesd.) R. C. Harris** (= "*L. lobificans*" of N. Am authors)

13. Thallus yellowish gray; hypothallus white; containing dibenzofuranes (pannaric acid group) 14

14. Thallus containing pannaric and roccellic acids ± atranorin; on shaded rock faces; rare, northeastern .... *Lepraria membranacea* (Dickson) Vainio (syn. *Leproloma membranaceum*)

[Note: Plate 448 in LNA named *Leproloma membranaceum* actually shows *Lepraria normandinioides*.]

14. Thallus containing mainly pannaric 6-methylester ± roccellic acid, sometimes with yellowish pigments that are K+ purple; on bark and rock; pan-temperate to western boreal .... *Lepraria vouauxii* (Hue) R. C. Harris in Egan

15.(9) Thallus PD+ yellow-orange, red, or orange 16

15. Thallus PD− or PD± weak yellow 19

16. Thallus PD+ red, K− (fumarprotocetraric acid) 17
16. Thallus PD+ yellow to orange, K− or K+ yellow to reddish brown 18

17. Southeastern coastal plain, on conifer bark; lacking roccellic acid .... *Lepraria friabilis* Lendemer et al.
17. Pacific Northwest, rare in the east, on bark; containing roccellic acid .... *Lepraria torii* Pérez-Ortega & T. Sprib.

18. Thallus PD+ deep yellow, K− or K+ yellowish, KC+ red (alectorialic ± barbatolic acid); northeastern and western to the arctic .... *Lepraria eburnea* J. R. Laundon
18. Thallus PD+ orange, K+ yellow or red, KC− (atranorin, zeorin, and either stictic acid [widespread] or salazinic acid [Appalachian]) .... *Lepraria elobata* Tønsberg (syn. *L. salazinica*)

19.(15) Thallus UV+ blue-white (divaricatic acid and zeorin, lacking fatty acids) 20
19. Thallus UV− or dull yellowish (lacking divaricatic acid, containing fatty acids and atranorin) 21

20. West coast, on conifer bark; nordivaricatic acid present (minor) (KC+ pink) ....*Lepraria pacifica* Lendemer
20. Eastern, mainly on hardwoods; nordivaricatic acid absent (KC−) .... *Lepraria hodkinsoniana* Lendemer

21. Containing rangiformic and roccellic acids, lacking zeorin; thallus greenish to greenish gray; Pacific Northwest and Appalachian-Great Lakes region .... *Lepraria jackii* Tønsberg
21. Containing pallidic acid and zeorin; thallus bluish green when fresh; East Temperate, on bark, especially conifers, rarely rock 22

22. Thallus thin, without a trace of a hypothallus; not especially associated with wetlands .... *Lepraria caesiella* R. C. Harris
22. Thallus thin or thick, with a thin whitish hypothallus usually detectable; usually found in wetlands .... *Lepraria harrisiana* Lendemer

LEPTOGIUM (including *Leptochidium*)

[Note: Recent genetic work (Otálora et al. 2013) necessitates a reclassification of many species of *Leptogium*, placing them in the genus *Scytinium*. The nomenclatural ramifications, however, are still not resolved (Jørgensen et al. 2013).]

1. Growing submerged at least part of the year; fairly rare species 2

1. In dry or moist habitats, not submerged in water 4

2. Lobes 3-10 mm broad, ascending and fan-like from a single point; strongly veined below .... *Peltigera* key, couplets 1-2)

2. Lobes 0.2-1.5 mm broad; closely attached by many points, lacking veins below 3

3. On rocks; lobes elongate, generally sterile, 8 spores per ascus; in the western mountains .... *Leptogium rivale* Tuck.

3. On tree bases (especially *Fraxinus*) or rocks periodically submerged; almost always fertile, 4 spores per ascus; Manitoba to New England .... *Leptogium rivulare* (Ach.) Mont.

4. Lower surface tomentose with a mat of white hairs 5

4. Lower surface smooth, or wrinkled, not hairy (except for scattered tufts in a few species) 12

5. Isidia and lobules absent; Arizona and New Mexico 6

5. Isidia or lobules present on upper surface or margins of lobes 7

6. Surface conspicuously wrinkled .... ***Leptogium rugosum* Sierk**

6. Surface smooth .... *Leptogium burgessii* (L.) Mont. (rare morphs lacking lobules)

7. Thallus lobulate, not isidiate 8

7. Thallus isidiate, but isidia can be somewhat flattened in some species 9

8. Lobules confined to the lobe margins; tomentum short but conspicuous, 35-100 µm long, composed of branched threads with short cylindrical or barrel-shaped cells; southwestern .... *Leptogium burgessii* (L.) Mont.

8. Lobules on lobe surface as well as margins; tomentum consisting of a faint fuzz made up of threads 20-35 µm long with spherical cells; Appalachians to Ontario and New Brunswick .... *Leptogium laceroides* (B. de Lesd.) P. M. Jørg.

9. Fine, stiff, colorless hairs on the edges or upper surface of the lobes; isidia cylindrical to flattened and lobe-like when older, often bearing fine hairs .... ***Leptochidium albociliatum* (Desm.) M. Choisy**

9. Fine hairs absent or sparse and confined to lobe surface; isidia cylindrical to granular, never flattened or hairy 10

10. Surface of lobes clearly wrinkled; isidia cylindrical or slightly thicker at the tips .... ***Leptogium pseudofurfuraceum*** **P.M. Jørg. & Wallace** (= "*L. furfuraceum*" of N. Am. authors.)

10. Surface of lobes smooth, not wrinkled; isidia granular or cylindrical and branched 11

11. Thallus dark olive-brown or olive-gray when dry, never with fine colorless hairs; isidia mostly granular; common .... ***Leptogium saturninum*** **(Dickson) Nyl.**

11. Thallus slate gray, without an olive tint when dry, sometimes with scattered fine hairs on the surface; isidia mostly cylindrical; uncommon .... *Leptogium burnetiae* C. W. Dodge

12. (4) Thallus extremely small, gray to reddish brown, lobes 0.1- 0.2 mm wide, forming subfruticose clumps in moss over sandy soil, rock, or bark; thallus composed entirely of pseudoparenchyma 13

12. Thallus lobes usually more than 0.2 mm wide, not subfruticose; pseudoparenchyma only in upper and lower cortex 14

13. Apothecia often abundant; thallus lobes and sometimes apothecial margins with cylindrical, isidia-like branches usually less than 1 mm long; on bare or mossy soil, rarely on bark; common in west, rare elsewhere .... *Leptogium tenuissimum* (Wallr.) Körber. *s. lat.*

13. Apothecia very rare; entire thallus consisting of elongate, branched, cylindrical, isidia-like lobes up to 2 mm long; frequent on bark, especially tree bases, among mosses; Temperate to boreal .... *Leptogium teretiusculum* (Wallr.) Arnold

14. Thallus slate gray 15

14. Thallus brown at least when dry, or olive, black, or greenish gray 31

15. Isidia present, cylindrical or, in part, flattened, resembling narrow lobules; apothecia abundant or rare 16

15. Isidia absent (lobules present or absent); apothecia usually present and abundant 21

16. Upper surface of lobes smooth and even 17

16. Upper surface of lobes wrinkled 19

17. Lobes 1-3 mm wide, appearing almost squamulose; on shaded, mossy limestone; East Temperate, infrequent .... *Leptogium dactylinum* Tuck.

17. Lobes 2-6 mm wide, clearly foliose; on bark or rock 18

18. Lobes 2-4 mm wide; isidia mostly cylindrical, sometimes flattened and lobulate; on bark or sometimes mossy rock; mainly East Temperate, very common, rare in Pacific Northwest .... ***Leptogium cyanescens* (Rabenh.) Körber**

18. Lobes 3-6 mm wide; isidia all lobulate; on shaded siliceous rock; southern U.S., especially common in southwest, rare elsewhere in the west .... *Leptogium* "*denticulatum*" *sensu* Sierk

[Note: Jørgensen & Nash (2004) note that this species does not agree with the type of *L. denticulatum* Nyl., but no other name is yet available for it.]

19.(16) Isidia granular at first, later cylindrical in part; thallus more than 200 μm thick; southwestern; on rock or tree bases .... *Leptogium arsenei* Sierk

19. Isidia all cylindrical to flattened ; thallus less than 200 μm thick; southeastern, rare in southwest; on bark 20

20. Isidia flattened to cylindrical, not branched; wrinkles minute, not sharp or raised .... ***Leptogium austroamericanum* (Malme) C. W. Dodge**

20. Isidia branched and coralloid, mostly along ridges; wrinkles raised and sharp, forming a net-like pattern or running longitudinally with the lobes .... *Leptogium isidiosellum* (Riddle) Sierk

21.(15) Lobes curled inward and erect, forming tube-like tips .... ***Leptogium palmatum* (Hudson) Mont.** (syn. *Leptogium corniculatum*)

21. Lobes flat, convex, concave, or undulating and crisped at the margins, not forming tubes 22

22. Apothecial margins with abundant lobules; apothecia restricted to the thallus margins .... ***Leptogium marginellum* (Sw.) Gray**

22. Apothecial margins smooth and even or thickly wrinkled, with or without lobules; apothecia on the lobe surface 23

23. Apothecia constricted at the base and somewhat raised 24

23. Apothecia broadly attached, adnate or sunken into depressions 27

24. Spores 4 per ascus; northern Manitoba to southern Ontario ... *Leptogium rivulare* (Ach.) Mont. (see couplet 3)

24. Spores 8 per ascus 25

25. Lobes not wrinkled; mainly southeastern, scattered in southwestern mountains

.... *Leptogium azureum* **(Sw. *ex* Ach.) Mont.**

25. Lobes strongly wrinkled 26

26. Apothecial margins smooth; lobes separate, not anastomosing; East Temperate .... *Leptogium corticola* **(Taylor) Tuck.**

26. Apothecial margins with thick, frilled, or ruffled margins; lobes prostrate and anatomosing, with thin sharp wrinkles running the length of the lobes; Florida .... *Leptogium floridanum* Sierk

27.(23) Apothecia 2-7 mm in diameter, margins thickly wrinkled; southeastern U.S. .... *Leptogium phyllocarpum* **(Pers.) Mont.**

27. Apothecia 0.2-1.0 mm in diameter, margins smooth and even; western U.S. and Canada 28

28. Rounded, overlapping lobules on thallus surface; thallus 200-500 μm thick, often attached to substrate by tufts of pale hairs .... *Leptogium platynum* **(Tuck.) Herre**

28. Finely divided lobules on the lobe margins, or margins fairly even; thallus thin and membranous, less than 200 μm thick, attached directly to substrate, usually lacking hairs 29

29. Ascospores 4 per ascus; on bark .... *Leptogium polycarpum* **P. M. Jørg. & Goward**

29. Ascospores 8 per ascus; on mossy rock, soil, or tree bases 30

30. Lobes rounded, rarely indented or finely divided, 1-4 mm wide .... *Leptogium gelatinosum* **(With.) J. R. Laundon**

30. Lobes indented and finely divided, 0.5- 3 mm wide .... *Leptogium californicum* Tuck.

31.(14) Isidia present 32

31. Isidia absent 34

32. On bark; lobe margins scalloped; granular to cylindrical isidia on lobe margins and ridges; apothecia infrequent .... *Leptogium milligranum* Sierk

[Compare with *L. arsenei*; couplet 19]

32. On rocks; lobe margins finely divided 33

33. Isidia only on lobe margins (fig. 32); lobes wrinkled; apothecia often abundant; on mossy calcareous rock .... *Leptogium lichenoides* **(L.) Zahlbr.**

33. Cylindrical to somewhat flattened isidia on lobe surface; lobes smooth; on dry or mossy rocks in the Northwest .... *Leptogium subaridum* P. M. Jørg. & Goward

34.(31) Lobes curled inward, forming a tube .... ***Leptogium palmatum* (Hudson) Mont.** (syn. *Leptogium corniculatum*)

34. Lobes flat, concave, or undulating and crisped at the margins 35

35. Apothecia raised, especially on the crests of ridges and folds; mainly in the southeastern coastal plain .... ***Leptogium chloromelum* (Ach.) Nyl.**

35. Apothecia adnate or sunken into depressions; mainly western and northern 36

36. Rounded, overlapping lobules on thallus surface; thallus over 200 µm thick, often attached to substrate by tufts of hairs .... ***Leptogium platynum* (Tuck.) Herre**

36. Lobules, if present, marginal; thallus thin and membranous, less than 200 µm thick, lacking hairs on the lower surface 37

37. Margins very finely divided (fig. 32); thallus cushion forming; widespread, especially among mosses over limestone .... ***Leptogium lichenoides* (L.) Zahlbr.**

[Note: Compare with *L. californicum*, couplet 30.]

37. Margins usually smooth and even, or somewhat lobulate; Pacific Northwest 38

38. Ascospores 4 per ascus; on bark .... ***Leptogium polycarpum* P. M. Jørg & Goward**

38. Ascospores 8 per ascus; on mossy rock, soil, or bark .... ***Leptogium gelatinosum* (With.) J. R. Laundon**

LETHARIA See McCune and Altermann (2009).

1. Branches granular sorediate; apothecia rare .... ***Letharia vulpina* (L.) Hue**

1. Branches without soredia; apothecia abundant 2

2. Branches rough and ridged, dense, 0.5-3 mm in diameter, erect; widespread in Pacific Northwest .... ***Letharia columbiana* (Nutt.) J. W. Thomson**

2. Branches smooth, sparse, 0.5-2.5(-3.0) mm in diameter, drooping; California to Oregon .... *Letharia gracilis* Kroken *ex* McCune & Altermann

LETROUITIA

1. Spores brick-like muriform, many-celled, 27-34 x 13-16(-19) µm, 2 per ascus; apothecial disks dark reddish purple, or color obscured by heavy orange pruina ....*Letrouitia vulpina* (Tuck.) Hafellner & Bellem.
1. Spores transversely septate or with spiraled, subdivided locules, 4-8 per ascus; apothecial disks black, with or without pruina 2

2. Spores with spiraled, subdivided locules, 4-6 "turns" per spore, 22-41 x 11-16(-18) µm; apothecia sometimes with a light yellow pruina ....***Letrouitia parabola* Sant. & Hafellner**
2. Spores transversely septate, 6- to 8-celled, with lens-shaped locules, 24-30 x 9-13 µm; apothecia never with pruina ....*Letrouitia domingensis* (Pers.) Hafellner & Bellem.

LICHINELLA (including *Thyrea*)

1. Thallus thick, 230-500 µm thick when wet, deeply lobed and subfruticose, dull and often pruinose when dry, lobe margins not conspicuously thickened; ascomata buried in thallus and opening to surface by a tiny pit or hole; widely distributed on limestone, especially where it is periodically wet, but rare ....***Thyrea confusa* Henssen**
1. Thallus thin, 130-250 µm thick when wet, clearly flattened and foliose to strap-shaped, sometimes in umbilicate clumps, usually shiny, never pruinose; ascomata developing in bumps on thallus 2

2. Thallus sometimes becoming strap-shaped and subfruticose, 0.7-2.5(-3.5) mm in width, developing granules in older parts; widespread especially in central and southwestern regions ....***Lichinella nigritella* (Lettau) P. P. Moreno & Egea**
2. Lobes rounded and convex 1-4 mm in diameter, lacking granules on the surface; in central regions ....*Lichinella cribellifera* (Nyl.) P. P. Moreno & Egea

LICHENOMPHALIA (previously included within *Omphalina*) Based in part on Watling and Woods (2009).

1. Thallus squamulose, squamules green, not sorediate, 2-5 mm broad, often lobed, closely appressed to soil or peat; arctic-alpine to boreal; fruiting bodies beige to pale yellow mushrooms lacking a purplish-brown zone at the top of the stem ....*Lichenomphalia hudsoniana* (H. S. Jenn.) Redhead et al. (syns. *Omphalina hudsoniana*, *Coriscium viride*)
1. Thallus granular, consisting on minute, bright green globules 50-100 µm in diameter formed from a fungal envelope enclosing a number of cells of the green alga, *Coccomyxa*; on peaty wood and mosses, boreal to arctic-alpine 2

2. Cortical cells of fungal envelopes and hyphae among the granules (2.5-)3-4(-6) µm across; fruiting bodies brown to pale yellow, often with a purplish brown zone at the top of the stem; widespread .... ***Lichenomphalia umbellifera*** **(L.: Fr.) Redhead et al.** (syn. *Omphalina umbellifera, O. ericetorum*)

2. Cortical cells of fungal envelopes and hyphae among the granules less than 3 µm across 3

3. Cortical cells and hyphae among the granules 1-1.5 µm across; fruiting bodies bright lemon yellow to yellow-orange; western? .... *Lichenomphalia alpina* (Britzelm.) Redhead et al.

3. Cortical cells and hyphae among granules 2-3 µm across; fruiting bodies brown; western? .... *Lichenomphalia velutina* (Quélet) Redhead et al.

LITHOTHELIUM Based in part on Harris (1973), as *Plagiocarpa*.

1. Spores remaining colorless, with central cells larger than tip cells, 18-27(-30) x 7-10(-12) µm; northeastern .... *Lithothelium hyalosporum* (Nyl.) Aptroot (syns. *Arthopyrenia hyalospora, Plagiocarpa hyalospora, Pleurotrema solivagum*)

1. Spores brown when mature, 30-42(-45) x 12-16(-18) µm; eastern 2

2. Spores 4-celled .... *Lithothelium phaeosporum* R. C. Harris

2. Spores (4-)8-celled .... *Lithothelium septemseptatum* (R. C. Harris) Aptroot

LOBARIA

[Note: New genetic research (Moncada et al. 2013) has shown that the classification of species of *Lobaria* needs major realignment. Since not all North American *Lobaria* species were included or reevaluated, a traditional treatment is followed below, but is provisional.]

1. Photobiont blue-green; cephalodia absent 2

1. Photobiont green; cephalodia present 5

2. Thallus pale to dark brown, rather shiny, always distinctly scrobiculate (with depressions and ridges much like *Lobaria pulmonaria*); tomentum on lower surface black; rare, B.C. to Alaska 3

2. Thallus gray to yellow-gray, dull; smooth to somewhat scrobiculate; gray soredia produced on margins and on lobe surface; tomentum on lower surface pale; infrequent to fairly common 4

3. Lacking any soredia or isidia; medulla PD+ yellow to orange, K+ yellow to red (norstictic acid); Alaska .... *Lobaria pseudopulmonaria* Gyelnik

3. Soredia and (or) isidia on margins and ridges of upper surface; medulla PD–, K–; British Columbia .... *Lobaria retigera* (Bory) Trevisan

4. Stiff, tiny, colorless hairs on the lobe tips or on the upper surface of the lobes; rhizines rope-like, covered with perpendicular hairs; medulla PD–, K– (norstictic and stictic acids absent); infrequent, in Pacific Northwest .... ***Lobaria hallii*** **(Tuck.) Zahlbr.**

4. Stiff, tiny, colorless hairs absent; rhizines tufted; medulla PD+ yellow, K+ red (norstictic and stictic acids present); west and east coasts, Appalachians, and subarctic .... ***Lobaria scrobiculata*** **(Scop.) DC.**

5. Lobes with a network of depressions and ridges; thallus loosely attached; medulla KC–, C– (gyrophoric acid absent) 6

5. Lobes relatively smooth and even, without a network of ridges and depressions; thallus closely appressed to substrate; medulla KC+ red, C+ pink (gyrophoric acid) 7

6. Soredia, often mixed with isidia, present on ridges and lobe margins; medulla PD+ yellow to orange, K+ yellow to red (norstictic and stictic acids) .... ***Lobaria pulmonaria*** **(L.) Hoffm.**

6. Soredia and isidia absent; medulla PD–, K– .... ***Lobaria linita*** **(Ach.) Rabenh.**

7. Lobes 5-20 mm broad; Appalachian-Great Lakes region .... ***Lobaria quercizans*** **Michaux**

7. Lobes 2-4(-6) mm broad; southeastern coastal plain 8

8. Lobules abundantly produced on upper surface and margins of lobes; cephalodia always small and inconspicuous warts, seen best on lower thallus surface; pycnidia and apothecia rare .... ***Lobaria tenuis*** **Vainio**

8. Lobules rarely present; cephalodia sometimes forming small branched outgrowths on the upper thallus surface; pycnidia appearing as abundant black dots on the lobe surface; apothecia abundant .... ***Lobaria ravenelii*** **(Tuck.) Yoshim.**

LOBOTHALLIA Based in part on Ryan (2004c).

1. Apothecia remaining immersed in thallus, finally becoming broadly sessile; thallus dark gray to olive-gray or almost black, lobes flat and fanning at tips, closely appressed to substrate; spores ellipsoid, 11-15 x 6-8.5 µm; medulla and cortex PD–, K–on siliceous rocks; scattered in west .... *Lobothallia radiosa* (Hoffm.) Hafellner

1. Apothecia immersed only when immature, soon becoming sessile or even constricted at base; thallus pale to dark, gray to brown, typically free from substrate at tips 2

2. Cortex and medulla PD–, K–; thallus usually dark greenish gray; lobes convex, narrow, sometimes overlapping; apothecia typically black or very dark brown; spores broadly ellipsoid to subglobose, 10-18 x 6-10 µm; montane and arctic, siliceous rock, frequently in mountain streams .... *Lobothallia melanaspis* (Ach.) Hafellner

2. Cortex and often medulla PD+ yellow, K+ red (sometimes weak) (norstictic acid); lobes convex to somewhat flattened; apothecia medium to dark red-brown sometimes darkening to almost black; thallus gray to dark brown; widespread in west extending to Great Lakes region; on dry rock 3

3. Lobes flat or finally convex; spores broadly ellipsoid to subglobular, 8-13 x 7-9.5 µm; thallus yellowish to copper brown or gray to brownish gray; on siliceous or sometimes calcareous rock ....***Lobothallia praeradiosa* (Nyl.) Hafellner**

3. Lobes very convex, almost foliose in appearance; spores ellipsoid, 11-16 x 5-8(-10) µm; thallus shades of gray, or, less frequently, light brown; on siliceous rock ....***Lobothallia alphoplaca* (Wahlenb.) Hafellner**

LOPADIUM (including *Schadonia*)

1. Spores 1 per ascus, over 50 µm long; paraphyses mostly unbranched 2

1. Spores 2-8 per ascus, 20-40 x 9-18 µm; paraphyses branched and anastomosing; rare; arctic alpine, on peat 4

2. Thallus thick, verruculose to coralloid-isidioid; mostly on moss and peat, sometimes rock in arctic or subarctic ....*Lopadium coralloideum* (Nyl.) Lynge

2. Thallus thin, not verruculose or isidioid 3

3. On bark, arctic to boreal, rare in Appalachians ....***Lopadium disciforme* (Flotow) Kullhem**

3. On moss and peat, arctic-alpine ....*Lopadium pezizoideum* (Ach.) Körber

4. Spores 8 per ascus ....*Schadonia alpina* Körber

4. Spores 2-4 per ascus ....*Schadonia fecunda* (Th. Fr.) Vězda & Poelt

LOXOSPORA

1. Apothecia rare; thallus sorediate or producing masses of small, hollow pustules that break down into granular schizidia or soredia 2

1. Apothecia common, lecanorine, often with a torn or ragged margin; thallus without hollow pustules or soredia 3

2. Thallus with abundant farinose to granular soredia in the older central parts; pustules small if present, forming true soredia; almost exclusively on conifers; mainly Appalachian-Great Lakes, also Pacific Northwest .... *Loxospora elatina* (Ach.) A. Massal.

2. Thallus without true soredia, only with conspicuous pustules that break down into granular schizidia; almost exclusively on deciduous trees; eastern temperate
.... ***Variolaria pustulata*** **(Brodo & W. L. Culb.) Lendemer et al.** (syn. *Loxospora pustulata*)

3. Apothecial disks red-brown, not pruinose or lightly pruinose; thallus rather thick and irregular; apothecia constricted at base; Appalachian-Great Lakes region .... ***Loxospora ochrophaea*** **(Tuck.) R. C. Harris**

3. Apothecial disks purplish, heavily pruinose, thallus very thin, more or less smooth; apothecia sessile; eastern boreal, rare in Appalachians .... ***Loxospora cismonica*** **(Beltr.) Hafellner**

MEGALARIA

1. Spores 23-30 x 10-15 µm; exciple dark greenish at outer edge, paler within; thallus K– (lacking atranorin); on hardwood bark in the northeast .... *Megalaria grossa* (Pers. *ex* Nyl.) Hafellner

1. Spores mostly 13-20 x 5-11 µm 2

2. Thallus blue-gray, granular; epihymenium blue-green, K+ green, HNO_3 + red; spores broadly ellipsoid, (15-)17-21(-26) x (7.5-)9-11(-13.5) µm; Pacific Northwest .... *Megalaria brodoana* Ekman & Tønsberg

2. Thallus thin, continuous; exciple brown-purple or reddish purple; epihymenium greenish or red-purple; spores ellipsoid to narrow, 12-19(-24) x 5-7(-8) µm 3

3. Hypothecium red-brown to orange, K+ purple, HNO_3–; epihymenium black to greenish; thallus K+ yellow (atranorin); Appalachian-Great Lakes distribution, on hardwoods, especially beech and maple
....***Megalaria laureri*** **(Hepp *ex* Th. Fr.) Hafellner**

3. Hypothecium gray to blue-green, K– or green, HNO_3+ red; epihymenium greenish or red-purple; thallus K– (lacking substances); mostly Pacific Northwest ....*Megalaria columbiana* (G. K. Merr.) Ekman

MELANELIA (including *Cetrariella* [p.p.], *Melanelixia, Melanohalea, Montanelia, Tuckermanella,* and "*Neofuscelia*" group of *Xanthoparmelia*)

1. Soredia present 2

1. Soredia absent (but see also couplet 13) 9

2. Medulla C+ red or pink 3

2. Medulla C– 6

3. Pseudocyphellae abundant and conspicuous; lobes elongated; gyrophoric acid present; white or brown soralia laminal or partly marginal; soredia granular
....***Montanelia tominii* (Oxner) Divakar et al.** (syn. *Melanelia tominii*)

3. Pseudocyphellae absent; lobes rounded; lecanoric acid present 4

4. Soredia entirely marginal, farinose, forming powdery crescents; isidia absent
....***Melanelixia albertana* (Ahti) O. Blanco et al.** (syn. *Melanelia albertana*)

4. Soredia at least partly laminal; soredia granular, often mixed with or arising from isidia 5

5. Soredia entirely laminal, arising from a disintegration of the cortex or isidia, leaving yellowish green patches; thallus surface not pruinose, without cortical hairs
....***Melanelixia subaurifera* (Nyl.) O. Blanco et al.** (syn. *Melanelia subaurifera*)

5. Soredia both laminal and marginal, arising from pustules or breakdown of the cortex, brown to whitish; thallus surface commonly pruinose, and often with a fuzz of minute, colorless hairs on the lobe tips
....***Melanelixia subargentifera* (Nyl.) O. Blanco et al.** (syn. *Melanelia subargentifera*)

6.(2) Soredia mostly on upper surface of lobes, dark brown (or whitish where abraded); rhizines abundant, brown or black; lower surface dark brown or black; medulla UV+ blue-white (perlatolic acid) 7

6. Soredia mostly marginal; rhizines sparse, very pale beige or tan; lower surface pale brown; medulla UV– (perlatolic acid absent) 8

7. Soredia fine, in rounded mounds mostly near the lobe tips
....***Montanelia sorediata* (Ach.) Divakar et al.** (syn. *Melanelia sorediata*)

7. Soredia coarse, granular, in irregular patches on the older parts of the thallus
....***Montanelia disjuncta* (Erichsen) Divakar et al.** (syn. *Melanelia disjuncta*)

8. On bark and wood, rarely rock, in western North America; medulla K–, PD– (fatty acids); lobes concave, undulating or crisped at the margins ....***Tuckermanopsis chlorophylla* (Willd.) Hale**

8. On rock in the Appalachian region; medulla K+ red, PD+ yellow (norstictic and stictic acids); lobes flat
....***Melanelia culbersonii* (Hale) Thell**

9.(1) Isidia or isidioid pustules present 10
9. Isidia absent 21

10. Lobes 0.4-0.7 mm wide; rhizines mostly marginal; on wood or bark
.... ***Tuckermanella coralligera* (W. A. Weber) Essl.** (syn. *Tuckermanopsis coralligera*)
10. Lobes mostly more than 1 mm wide; rhizines all over lower surface; on rock, wood, or bark 11

11. Isidia spherical or globular 12
11. Isidia cylindrical, flattened, club-shaped, or conical, unbranched or branched 15

12. On bark or wood, rarely rock; isidia globular to conical or cylindrical, solid
.... ***Melanohalea elegantula* (Zahlbr.) O. Blanco et al.** (syn. *Melanelia elegantula*)
12. On rock; isidia granular or pustule-like, hollow, sometimes breaking into soredia-like schizidia 13

13. Medulla K+ yellow or red, PD+ orange, KC– (unidentified substance); mass of small pustules often breaking into schizidia ......***Xanthoparmelia subhosseana* (Essl.) O. Blanco et al.** (syn. *Neofuscelia subhosseana*)
13. Medulla K–, PD–, KC–, KC+ pinkish violet or pink 14

14. Medulla KC+ pinkish violet (glomelliferic, glomellic and perlatolic acids); pustular isidia 0.1-0.5 mm in diameter, rarely breaking up and becoming schizidioid
.... ***Xanthoparmelia loxodes* (Nyl.) O. Blanco et al.** (syn. *Neofuscelia loxodes*)
14. Medulla KC– or KC+ pink (divaricatic acid and a trace of gyrophoric acid); pustular isidia 0.1-0.2 mm in diameter, commonly breaking into masses of soredia-like schizidia
.... *Xanthoparmelia verruculifera* (Nyl.) O. Blanco et al. (syn. *Neofuscelia verruculifera*)

15.(11) Medulla C+ red or pink (lecanoric acid) 16
15. Medulla C– (lecanoric acid absent) 17

16. Isidia unbranched, mostly shorter than 0.2 mm; abrading to leave yellowish green patches ....
.... ***Melanelixia subaurifera* (Nyl.) O. Blanco et al.** (syn. *Melanelia subaurifera*)
16. Isidia frequently branched and longer than 0.2 mm (fig. 33c); not forming yellowish patches where abraded
.... ***Melanelixia fuliginosa* (Fr. *ex* Duby) O. Blanco et al.** (syn. *Melanelia fuliginosa*)
[Note: *Melanelixia glabratula* (Lamy) Sandler & Arup is included within *M. fuliginosa* by Esslinger (2014), but European

studies show that *M. fuliginosa s. str.* is dark, shiny, and largely saxicolous and can be distinguished from *M. glabratula*, which is usually pale, dull to shiny, and largely corticolous (Arup & Berlin 2011).]

17. Isidia cylindrical or conical when mature (fig. 33a); upper surface dull, often pruinose 18

17. Isidia flattened or club-shaped when mature; upper surface shiny or dull 19

18. Isidia developing from conical warts (i.e., broad at the base); on bark and wood, rarely rock .... ***Melanohalea elegantula* (Zahlbr.) O. Blanco et al.** (syn. *Melanelia elegantula*)

18. Isidia developing from globular warts constricted at the base; only on rock .... *Melanohalea infumata* (Nyl.) O. Blanco et al.

19. Growing on rock; isidia solid; medulla UV+ blue-white (perlatolic acid) .... ***Montanelia panniformis* (Nyl.) Divakar et al.** (syn. *Melanelia panniformis*)

19. Growing on bark or wood, rarely on rock; isidia solid or hollow; medulla UV– 20

20. Isidia hollow, club-shaped to flattened, but rarely branched or cylindrical and never bearing rhizines (fig. 33b); upper surface of lobes shiny; growing on bark, rarely wood or rock; widespread in the west ....***Melanohalea exasperatula* (Nyl.) O. Blanco et al.** (syn. *Melanelia exasperatula*)

20. Isidia solid, cylindrical when young but soon becoming flattened and lobe-like, sometimes branched, developing tiny rhizines on one side (fig. 33d); upper surface of lobes dull; growing exclusively on bark and wood, central California to southern B.C. east to the Canadian Rockies .... *Melanohalea subelegantula* (Essl.) O. Blanco et al. (syn. *Melanelia subelegantula*)

21.(9) Growing on bark... 22

21. Growing on rock... 30

22. Medulla C+ red, KC+ red (lecanoric acid); California........... ***Melanelixia californica* A. Crespo & Divikar** (= North American records of *Melanelia (Melanelixia) glabra*; Plate 498 in LNA)

22. Medulla C–, KC– 23

23. Rhizines sparse, mostly marginal and resembling cilia; lobes 0.5-1.6 mm wide .... ***Tuckermanella fendleri* (Nyl.) Essl.** (syn. *Tuckermanopsis fendleri*)

23. Rhizines abundant all over lower surface, never resembling cilia; lobes (0.5-)1-5 mm wide 24

24. Lower surface with a network of plates and ridges; spores spherical, 12-32 per ascus; lobes long and

narrow often convex; medulla K+ red (norstictic acid)

.... *Melanohalea trabeculata* (Ahti) Blanco et al. (syn. *Melanelia trabeculata*)

24. Lower surface lacking ridges and plates; spores 8-32 per ascus; lobes mostly short; lacking norstictic acid 25

25. Medulla PD+ red (fumarprotocetraric acid) 26

25. Medulla PD– 28

26. Spores 15-20 x 8-12.5 µm (Hinds & Hinds 2007); pseudocyphellae inconspicuous or absent, often lobulate on older parts of the thallus; mainly Appalachians, rarer into Great Lakes region

.... *Melanohalea halei* (Ahti) O. Blanco et al. (syn. *Melanelia halei*)

26. Spores 9-16 x 5-9 µm; boreal to arctic 27

27. Pseudocyphellae sparse or inconspicuous on the thallus and apothecial margins; lobes smooth and even

.... ***Melanohalea septentrionalis* (Lynge) O. Blanco et al.** (syn. *Melanelia septentrionalis*)

27. Pseudocyphellae abundant and conspicuous especially on the apothecial margins; lobes wrinkled or bumpy (rugose) .... ***Melanohalea olivacea* (L.) O. Blanco et al.** (syn. *Melanelia olivacea*)

28.(25) Spores 12-32 per ascus; lobes often with lobules, at least in center of thallus

....***Melanohalea multispora* (A. Schneider) O. Blanco et al.** (syn. *Melanelia multispora*)

28. Spores 8 per ascus; lobules absent 29

29. Pseudocyphellae absent or rare; western

....***Melanohalea subolivacea* (Nyl.) O. Blanco et al.** (syn. *Melanelia subolivacea*)

29. Pseudocyphellae on apothecial margins, lobes, and thallus warts; East Temperate

....*Melanohalea exasperata* (De Not.) O. Blanco et al. (syn. *Melanelia exasperata*)

30.(21) Pseudocyphellae abundant and conspicuous; rhizines mostly on or close to the margins 31

30. Pseudocyphellae absent or inconspicuous; rhizines more or less uniformly distributed 34

31. Pseudocyphellae marginal or close to margins; pycnidia prominent, on lobe margins; medulla PD– or PD+ orange 32

31. Pseudocyphellae laminal; pycnidia scattered on lobe surface, entirely buried in thallus with just the ostiole showing; medulla PD– or PD+ red, K– 33

32. Margins of the lobes slightly raised (rimmed); lower surface typically black, at least in part; medulla PD+ orange, K+ yellow, KC–, UV– (stictic acid); widespread arctic-alpine .... ***Melanelia hepatizon* (Ach.) Thell**

32. Margins of lobes not raised (rimmed); lower surface pale to dark brown; medulla PD–, K–, KC+ pink, UV+ white (alectoronic acid); western montane .... *Cetrariella commixta* (Nyl.) A. Thell & Kärnefelt (syn. *Melanelia commixta*)

33. Medulla C+ pink, KC+ red, PD– (gyrophoric acid); thallus thick and stiff; pycnidia sparse or very inconspicuous; conidia rod-shaped ....***Montanelia tominii* (Oxner) Divakar et al.** (syn. *Melanelia tominii*)

33. Medulla C–, KC–, PD+ red or PD– (fumarprotocetraric acid sometimes present); thallus moderately thick; pycnidia abundant and conspicuous; conidia dumbbell-shaped ....***Melanelia stygia* (L.) Essl.**

34.(30) Medulla C+ red (lecanoric acid); infrequent in southwest .... *Melanelixia glabroides* (Essl.) O. Blanco et al. (syn. *Melanelia glabroides*)

34. Medulla C– 35

35. Thallus brown to olive-brown, usually producing abundant, overlapping, flattened lobules; lower surface wrinkled; conidia rod-shaped ....***Montanelia panniformis* (Nyl.) Divakar et al.** (syn. *Melanelia panniformis*)

35. Thallus reddish brown or yellowish brown, without abundant, overlapping lobules; lower surface smooth; conidia dumbbell-shaped 36

36. Lower surface and rhizines dark brown or black; pycnidia sparse or very inconspicuous; medulla K–, PD–, KC+ pinkish violet, UV+ white (perlatolic acid) .... ***Xanthoparmelia loxodes* (Nyl.) O. Blanco et al.** (syn. *Neofuscelia loxodes*) [morphotype lacking isidioid pustules]

36. Lower surface and rhizines pale brown; pycnidia abundant and conspicuous, UV– 37

37. Medulla K+ yellow or red-orange, PD+ orange, KC– (stictic and sometimes norstictic acids); Arizona and New Mexico ....***Xanthoparmelia atticoides* (Essl.) O. Blanco et al.** (syn. *Neofuscelia atticoides*)

37. Medulla K–, PD–, KC– (containing only fatty acids); mainly Arizona 38

38. Lobes less than 1.7 mm wide .... *Xanthoparmelia brunella* (Essl.) O. Blanco et al. (syn. *Neofuscelia brunella*)

38. Lobes 1-3.5 mm wide .... *Xanthoparmelia ahtii* (Essl.) O. Blanco et al. (syn., *Neofuscelia ahtii*)

[Note : *Xanthoparmelia infrapallida* (Essl.) O. Blanco et al. (syn. *N. infrapallida*) is similar to *X. ahtii* but less frequent; it differs in fatty acid chemistry (see Esslinger 1977, 2002).]

MELASPILEA

1. Parasitic on crustose lichens, especially *Pyrenula*; ascomata script-like, straight or curved, clustered, 0.2-0.5 x 0.1-0.2 mm; spores 15-21 x 6-9 µm; southeastern .... *Melaspilea tribuloides* (Tuck.) Müll. Arg.
1. Growing directly on tree bark, not parasitic 2

2. Ascomata elongate, straight or curved, up to 2 mm long, 0.3 mm wide 3
2. Ascomata round to irregular in shape, up to 1 mm in diameter 4

3. Ascomata straight to curved, rarely branched, 0.3-1.0 mm long, opening by a slit or broad; spores 15-19(-23) x 6-9 µm, pale brown, upper cell round and larger than lower cell; northeastern .... *Melaspilea demissa* (Tuck.) Zahlbr.
3. Ascomata mostly flexuose or curved, sometimes branched once, rarely straight, 0.5-0.8 mm long, thin margins usually prominent; spores 11-17 x 5-6.6 µm, oblong, slightly constricted, colorless or finally brownish; Appalachian .... *Melaspilea deformis* (Tuck.) Zahlbr.

4. Spores (18-)20-27 x (8-)10-13.5 µm; hypothecium pale orange-brown; thallus just a gray stain or disappearing; southeastern .... *Melaspilea maculosa* (Fr.) Müll. Arg.
4. Spores 12-18 x 6-8.5 µm; hypothecium colorless; thallus white, very thin or disappearing; rare in New England, mainly central U.S. .... *Melaspilea arthonioides* (Fée) Nyl.

MICAREA (including *Frutidella* and *Leimonis*) Based in part on Czarnota (2007), Coppins (2009).

1. On soil, moss, and dead plants, arctic-alpine; spores 1- to 2-celled; epihymenium greenish, HNO_3+ red......2
1. On wood, bark, rock, occasionally bryophytes 5

2. Thallus pale, with corticate granules; spores 1-celled, length to width ratio 2-3:1, (11-)15-19(-26) x 5-7(-9) µm, photobiont cells chlorococcoid, 6-12 µm in diameter; asci *Lecanora*-type; thallus KC+ orange, UV+ white (thiophanic acid, sphaerophorin); infrequent .... *Frutidella caesioatra* (Schaerer) Kalb (syn. *Lecidea caesioatra*) [see also *Biatora* key]
2. Thallus pale or dark; spores 1- to 2(-3)-celled, length to width ratio 3-4:1, (10-)12-20(-25) x 3-5(-6) µm; photobiont cells micareoid, 4-7 µm in diameter; thallus KC–, UV– ....3 [See also *Toninia* key]

3. Thallus very dark gray to black, without cephalodia; hypothecium red-brown, K– ; rare .... *Micarea turfosa* (A. Massal.) Du Rietz

3. Thallus white to pale gray, with lumpy brown cephalodia (sometimes hidden between areoles) 4

4. Hypothecium purple-brown, K+ purple .... *Micarea assimilata* (Nyl.) Coppins
4. Hypothecium red-brown, K– .... *Micarea incrassata* Hedl.

[Note: *Bilimbia lobulata* is similar but has 2- to 4-celled spores and no cephalodia.]

5.(1) Spores mostly 1-celled, ellipsoid, mostly 6-11 x 2.5-4.5 µm 6
5. Spores mostly 2- or more celled, elongate, usually over 10 µm long 10

6. Growing on wood; epihymenium olive-gray to olive brown, K+ violet, C+ violet, red or pink-brown; hypothecium colorless; B.C., Manitoba to Maritimes 7
6. Growing on rock; thallus thin or well developed; epihymenium or hymenium greenish or blue-green, or almost colorless, K–, C– 8

7. Thallus thin, almost imperceptible .... *Micarea misella* (Nyl.) Hedl. (syn. *M. globularis*)
7. Thallus consisting of green soredia-like granules (goniocysts) ....*Micarea prasina* Fr. *s. lat.* [see couplet 12]

8. Hypothecium colorless or vaguely greenish; northeastern ....*Brianaria bauschiana* (Körber) S. Ekman & M. Svensson (syn. *Micarea bauschiana*)
8. Hypothecium brown to black 9

9. Exciple well developed; apothecia flat to convex often with a persistent, thin apothecial margin; hypothecium brown .... *Leimonis erratica* (Körber) R. C. Harris & Lendemer (syn. *Micarea erratica*)
9. Exciple not developed at all; apothecia convex to hemispherical, lacking any margin; hypothecium greenish black .... *Brianaria sylvicola* (Flotow *ex* Körber) S. Ekman & M. Svensson (syn. *Micarea sylvicola*)

10.(5) Spores not more than (1-)2-celled, 8-14(-17) µm long; epihymenium olive-gray, K+ purple, C+ purple 11
10. Spores 2- or more celled, mostly over 12 µm long; epihymenium various 13

11. Thallus gray to gray-green, areolate, without goniocysts; spores 8-14 x 2.0-3.5(-4) µm, somewhat curved; apothecia and thallus C+ pink (gyrophoric acid); rather rare, west and east coasts .... *Micarea denigrata* (Fr.) Hedl.
11. Thallus appearing dark green leprose, consisting of tiny green soredia-like goniocysts; spores 8-14(-17) x 2.5-4(-5), straight; apothecia and thallus C–; common and widespread 12

12. Containing methoxymicareic acid .... *Micarea micrococca* (Körber) Gams *ex* Coppins

12. Containing micareic acid .... *Micarea prasina* Fr.

[Note: See Barton & Lendemer (2014) for TLC methodology to distinguish methoxymicareic acid from micareic acid.]

13.(10) Hypothecium dark brown to purple-brown; apothecial sections C– 14

13. Hypothecium colorless; apothecial sections C+ pink to violet, or C– 16

14. Spores 12-20 x 4-5.5 µm; thallus C+ pink; apothecia always black; hypothecium purple-brown, K+ purple; widespread .... *Micarea melaena* (Nyl.) Hedl.

14. Spores 20-40 x 1.5-3.0 µm; thallus C–; apothecia gray to black; eastern 15

15. Hypothecium brown-black, turning green in K; spores 4- to 8-celled; Wisconsin to North Carolina .... *Micarea endocyanea* (Tuck.) R. C. Harris

15. Hypothecium purple-brown, K– or intensifying purple; spores 2- to 4-celled (septa sometimes obscure); coastal plain, Nova Scotia to Florida .... *Micarea chlorosticta* (Tuck.) R. C. Harris

[Note: This is the first report of this species from Canada: Nova Scotia, Kejimkujic National Park, on dead spruce, *A.E. Roland 140* (CANL).]

16. Spores curved ("comet-shaped"), strongly tapered at one end, 18-35(-40) x 3-5 µm, (3-)5- to 8-celled; asci *Lecanora*-type; apothecial sections C–; thallus granular, dark green; apothecia dark brown to black .... ***Scoliciosporum chlorococcum* (Stenh.) Vězda**

16. Spores not, or only slightly curved, not tapered at one end; asci *Micarea*-type (like *Porpidia*); apothecial sections (especially epihymenium) C+ pink (gyrophoric acid) or C+ violet (olive-gray pigment) 17

17. Spores 25-40 x 4.5-6 µm, (4-)6- to 8-celled; thallus and apothecial section C+ pink (gyrophoric acid); apothecia very pale gray to brownish or dark gray; probably oceanic, B.C. and Newfoundland .... *Micarea cinerea* (Schaer.) Hedl.

17. Spores under 25 µm long, 2- to 4(-8)-celled 18

18. Thallus PD+ red (fumarprotocetraric acid), C–; apothecia white to pale gray, 0.1-0.22 mm in diameter; spores 17-25 x 1.7-2.5 µm, (2-)4-celled; apothecial tissues, if pigmented, K+ violet, C+ violet; infrequent, New England to North Carolina .... *Micarea neostipitata* Coppins & P. May

18. Thallus PD–, C+ pink (gyrophoric acid); apothecia pale gray to black; eastern 19

19. Epihymenium olive-gray to olive-brown, K+ violet, C+ violet or pink; spores 2- to 4(-8)-celled, (10-)12-24 x 2.0-3.0 µm; infrequent .... *Micarea globulosella* (Nyl.) Coppins

19. Epihymenium greenish to aeruginose, at least in part, K–, C– or C+ pink; spores, mostly 4-celled, (12-)14-20(-24) x 3.5-5.5 µm; common .... ***Micarea peliocarpa* (Anzi) Coppins & R. Sant.**

MULTICLAVULA

[Note: Similar non-lichenized club fungi such as those belonging to *Clavaria, Clavariadelphus, Clavuliopsis,* and *Ramariopsis,* all of them almost invariably have larger fruiting bodies (2-8 cm tall) and do not emerge from an algal mat as do species of *Multiclavula*, which are typically 0.5-2 cm tall.]

1. Fruiting bodies (basidiocarps) dark yellow to orange; on rotting logs; boreal, common .... *Multiclavula muscida* (Fr.) R. Peterson

1. Fruiting bodies very pale yellow; on soil or peat 2
[Note: Distribution uncertain; possibly boreal-montane, widespread.]

2. Spores up to 12 µm long; fruiting bodies fleshy, opaque throughout .... *Multiclavula vernalis* (Schwein.) R. Peterson

2. Spores 5-8 x 2-3.5 µm; fruiting bodies hard and brittle when dry, often translucent at base .... ***Multiclavula corynoides* (Peck) R. Peterson**

MYCOBLASTUS: (include *Violella*) Based in part on Spribille et al. (2011), James and Watson (2009).

1. Thallus thin, gray, usually with a darker prothallus, always sorediate, apothecia rarely seen, on bark.......... 2

1. Thallus thick, rarely sorediate, apothecia black, hemispherical, usually present; on bark, peat, and rarely rock 3

2. Thallus and prothallus bluish-gray, PD–, UV+ white (perlatolic acid) .... *Mycoblastus caesius* (Coppins & P. James) Tønsberg

2. Thallus and prothallus gray, PD+ red, UV– (fumarprotocetraric acid) .... *Violella fucata* (Stirton) T. Sprib. (syn. *Mycoblastus fucatus*)

3. Spores (1-)2 per ascus, usually under 70 µm long; apothecial base always colorless; thallus occasionally producing yellowish soralia; western .... *Mycoblastus affinis* (Schaerer) T. Schauer (syn. *M. alpinus*)

3. Spores 1 per ascus, 70-100 x 25-45 µm; base of apothecia (as seen in section) typically bright red, rarely colorless; thallus rarely sorediate 4

4. Epihymenium and upper parts of hymenium containing tiny crystals that are bi-refringent in polarized light; mainly northeastern .... *Mycoblastus sanguinarioides* Kantvilas

4. Epihymenium and upper parts of hymenium lacking any crystals (polarized light negative); mainly western .... ***Mycoblastus sanguinarius* (L.) Norman** *s. str.*

MYRIOTREMA (including *Leptotrema* and *Leucodecton*) Based on Harris (1995).

1. Thallus greenish gray, smooth and continuous, with white granular soredia in irregular soralia; ascomata hard to see, with tiny, brown, shallow ostioles; Florida to Georgia .... *Myriotrema erodens* R. C. Harris

1. Thallus not sorediate 2

2. Spores colorless, muriform or submuriform, 9-20 x 4-6 µm, 8 per ascus; thallus PD+ yellow, K– (psoromic acid) .... ***Myriotrema rufigerum* (Harm.) Hale**

2. Spores brown, muriform 3

3. Spores 1 or 2 per ascus, 70-130 x 20-30 µm; thallus PD+ orange, K+ yellow (stictic acid) .... *Myriotrema reclusum* (Krempelh.) Hale

3. Spores 8 per ascus 4

4. Spores 45-55 x 10-13 µm, many-celled; ascomata round, not crowded; medulla PD+ orange, K+ yellow (stictic acid complex); throughout southeast .... *Leucodecton subcompunctum* (Nyl.) A. Fritsch (syn. *Myriotrema subcompunctum*)

4. Spores under 30 µm long, few-celled 5

5. Ascomata round, with black, pore-like openings, without pruina; thallus with a cortex; spores 14-28 x 9-14 µm; medulla PD–, K–, but containing large colorless crystals and pockets of red pigment (K+ purple); on trees and mosses .... *Leptotrema wightii* (Taylor) Müll. Arg. (syn. *Myriotrema wightii*)

5. Ascomata with angular, irregular to elongate openings, with blue-gray pruinose disks, often in whitish areas without a cortex; spores 13-16 x 7-10 µm; medulla PD+ orange, K+ yellow (stictic acid complex); Florida and Georgia .... *Leucodecton glaucescens* (Nyl.) A. Fritsch (syn. *Myriotrema glaucescens*)

NEPHROMA

1. Photobiont green; cephalodia present; on the ground and mossy rocks and logs mostly in northern boreal, arctic, and alpine habitats 2

1. Photobiont blue green; cephalodia absent; on mossy bark or rock 3

2. Thallus pale green, becoming browned especially at the margins; cortex KC– (usnic acid absent); lobes undulating or crisped at the margins; cephalodia forming small, inconspicuous warts on lower thallus surface; lobules present .... ***Nephroma expallidum*** **(Nyl.) Nyl.**

2. Thallus yellow-green; cortex KC+ orange-yellow (usnic acid); lobes flat; cephalodia forming internal swellings visible externally as rounded dark bumps on the thallus surface; lobules absent .... ***Nephroma arcticum*** **(L.) Torss.**

3. Soredia present 4

3. Soredia absent 5

4. Thallus brown; soredia mainly on the lobe margins and sometimes on the surface; lobes mostly smooth; widespread, common .... ***Nephroma parile*** **(Ach.) Ach.**

4. Thallus yellow-green (with usnic acid); soredia mainly on the thallus surface; lobes with a network of depressions and sharp ridges; western, rare .... ***Nephroma occultum*** **Wetmore**

5. Isidia present 6

5. Isidia absent 7

6. Isidia cylindrical, unbranched or branched; lobes with a network of depressions and sharp ridges, or wrinkled; lobules absent; apothecia absent .... ***Nephroma isidiosum*** **(Nyl.) Gyelnik**

6. Isidia flattened; lobes smooth and even; lobules present; apothecia typically abundant .... ***Nephroma helveticum*** **Ach.**

7. Medulla yellow, K+ red or deep purple (anthraquinones) .... ***Nephroma laevigatum*** **Ach.**

7. Medulla white, K– (anthraquinones absent) 8

8. Lower surface tomentose, with scattered pale bumps .... ***Nephroma resupinatum*** **(L.) Ach.**

8. Lower surface smooth .... ***Nephroma bellum*** **(Sprengel) Tuck.**

OCELLULARIA (including *Stegobolus*) Based on Harris (1995).

[Note: All three species below are found mainly in Florida.]

1. Spores muriform, many-celled, 1(-2) per ascus, more than 100 μm long; thallus C–, PD– (no substances); fruiting bodies in small bumps (verrucae), 0.9-1.5 mm across, one per verruca

....*Ocellularia sanfordiana* (Zahlbr.) Hale

1. Spores only transversely septate 2

2. Spores 6- to 8-celled, 16-24 x 5-7 µm, 8 per ascus; fruiting 0.3-0.7 mm in diameter; thallus C–, PD+ yellow (psoromic acid) ....***Stegobolus granulosus* (Tuck.) A. Fritsch** (syn. *Ocellularia granulosa*)
2. Spores 19- to 23-celled,100-130 x 20-35 µm, 1 per ascus; fruiting bodies 0.7-1.5 mm in diameter; thallus C+ pink, PD– (gyrophoric acid) ....*Ocellularia americana* Hale

OCHROLECHIA Revisions based in part on Brodo & Lendemer (2012) and Kukwa (2011).

[Note: All spot tests with C must be done on the soralia or on thick sections of the apothecia or thallus, not on the thallus surface.]

1. Growing on rock, soil, peat, or moss 2
1. Growing on bark or wood 9

2. On rock; thallus thick 3
2. On soil, peat, or moss 6

3. Thallus sorediate ....***Ochrolechia androgyna* (Hoffm.) Arnold**
3. Thallus without soredia 4

4. In temperate eastern U.S.; thallus with thick isidia ....***Ochrolechia yasudae* Vainio**
4. Western; thallus lacking isidia 5

5. Apothecia lecanorine, with a broad disk; thallus and apothecial cortex C+ pink (gyrophoric acid); medulla UV– ....*Ochrolechia tartarea* (L.) A. Massal. *s. lat.* [Note: needs confirmation.]
5. Apothecia opening by a narrow ostiole, which is slightly depressed, not forming a hole; thallus area around ostiole C+ pink (gyrophoric acid), medulla UV+ white (alectoronic acid); on maritime or alpine rocks near Pacific coast; rare .... *Ochrolechia subplicans* (Nyl.) Brodo

6.(2) Thallus cortex and medulla C– (lacking gyrophoric acid); apothecia abundant, disks coarsely pruinose-scabrose, C– or C+ yellow; thallus thin or, more commonly, thick and granulose to verruculose, never sorediate; arctic-alpine and western interior ....***Ochrolechia upsaliensis* (L.) A. Massal.**

6. Thallus cortex and often medulla C+ pink (gyrophoric acid); apothecia abundant or rare, disks smooth or cracked, not pruinose or scabrose, C+ pink; thallus smooth to verrucose or coarsely isidiate; with or without discrete round or irregular and spreading soralia 7

7. Thallus typically producing elongate, often branched or "frayed," fruticose filaments or spines at the margins; excipular ring usually visible and well developed around the disk and sometimes spreading over the apothecial margin; variolaric acid absent; widespread arctic-alpine .... ***Ochrolechia frigida*** **(Sw.) Lynge** *s. lat.*

[Note: Kukwa (2011) includes within this polymorphic species: *O. gyalectina*, *O. inequatula*, *O. gonatodes*, and *O. frigida* f. *lapuensis*.]

7. Thallus marginal projections or filaments rare or absent 8

8. Thallus lacking variolaric acid, always sorediate with irregular patches of yellowish soredia; apothecia very rare; widespread boreal to arctic-alpine .... ***Ochrolechia androgyna*** **(Hoffm.) Arnold**

8. Thallus containing variolaric acid, usually without soredia; apothecia commonly produced, up to 8 mm in diameter, lacking an excipular ring in mature apothecia; arctic-boreal, from Alaska and Newfoundland (not between) .... *Ochrolechia alaskana* (Verseghy) Kukwa

9.(1) Thallus sorediate 10

9. Thallus without soredia 16

10. Soredia C–; apothecia rare; thallus containing variolaric acid; rare species 11

[Note: See also couplet 26 below.]

10. Soredia C+ pink to red (gyrophoric acid); apothecia rare or common 12

11. Soralia at first discrete, later becoming irregular and confluent; thallus shiny and smooth; lacking fatty acids; on acid bark, especially conifers .... *Ochrolechia turneri* (Sm.) Hasselrot

11 Soredia generally widespread on thallus, almost leprose; containing lichesterinic and sometimes protolichesterinic acids; generally on deciduous trees .... *Ochrolechia microstictoides* Räsänen

12. Thallus very thin, membranous, C–; containing variolaric acid; boreal ....*Ochrolechia gowardii* Brodo

12. Thallus thin or thick, cortex or medulla C+ red; lacking variolaric acid 13

13. Thallus and (or) sorala UV+ bright yellow (lichexanthone); thallus relatively thin; widespread .... *Ochrolechia arborea* (Kreyer) Almb.

13. Thallus and soralia UV–; thallus relatively thick 14

14. Soredia coarsely granular, originating from the breakdown of isidia; cortex C– or weakly C+ pink, medulla C+ red; rare in Florida .... *Ochrolechia antillarum* Brodo

14. Soredia powdery, fine to coarse, not from isidia; cortex C+ red, medulla C– or C+ weak pink; north temperate to boreal or arctic 15

15. Fatty acids present .... ***Ochrolechia androgyna* (Hoffm.) Arnold**

15. Fatty acids absent .... *Ochrolechia mahluensis* Räsänen

16.(9) Eastern, but some species occasionally in the southwest 17

16. Western 22

17. Thallus thick, more or less covered with thick isidia; cortex C+ pink .... ***Ochrolechia yasudae* Vainio**

17. Thallus thin or thick, smooth to verrucose, but not isidiate; cortex C+ pink or C– 18

18. Thallus and medulla C–; apothecia with pruinose or rough, scabrose, C+ pink disks; thallus rather thick, rough to verruculose; very common in Appalachian-Great Lakes region and southwest; on conifer bark and wood; apothecia large and flat, 1.3-5 mm in diameter .... *Ochrolechia pseudopallescens* Brodo

18. Thallus cortex or medulla and apothecial disks C+ pink or red; mostly on deciduous trees; apothecia 0.6-2(-3) mm in diameter, pruinose or not, rough or smooth 19

19. Spores 1 per ascus, over 70 µm long ....
... ***Varicellaria velata* (Turner) Schmitt & Lumbsch** (syn. *Pertusaria velata*) [see also couplet 33 in *Pertusaria* key]

19. Spores 4-8 per ascus, under 70 µm long 20

20. Cortex of apothecial margin C–, medulla C+ red; apothecia usually pruinose; apothecial margins smooth and even; thallus UV+ yellow or UV–; mainly southeastern coastal plain, infrequent in southwest
.... ***Ochrolechia africana* Vainio**

20. Cortex of apothecial margin C+ pink, medulla C+ pink or C–; apothecia with or without pruina; apothecial margins smooth or rough 21

21. Thallus UV– (lacking lichexanthone); medulla C–; apothecia usually without pruina; apothecial margins verrucose; Appalachian-Great Lakes .... ***Ochrolechia trochophora* (Vainio) Oshio**

21. Thallus UV+ yellow (lichexanthone); medulla C+ pink (rarely C–); apothecia with or without pruina; apothecial margins usually smooth; Appalachian-Great Lakes and southwest .... *Ochrolechia mexicana* Vainio

22.(16) Thallus with granular isidia; cortex C–, medulla C+ pink (variolaric and gyrophoric acids); Arizona .... *Ochrolechia subisidiata* Brodo

22. Thallus lacking isidia; Pacific Northwest 23

23. Cortex and medulla of thallus and apothecial margin C– (containing variolaric acid) 24

23. Cortex of thallus and apothecial margin C+ pink to red (lacking variolaric acid) 27

24. Apothecial disks C+ pink (scratch the surface of the disk before testing) 25

24. Apothecial disks C– 26

25. Apothecia remaining small and pore-like except in oldest apothecia, which become broad, with a thallus-colored margin; thallus thin or well developed and verruculose; on conifer bark, common .... ***Ochrolechia juvenalis* Brodo**

25. Apothecia broadening relatively early in their development, with a pinkish disk-colored margin; thallus very thin, often endophloeodal; on very hard wood, rarely bark, of conifers; infrequent .... *Ochrolechia subathallina* H. Magn

26. Thallus thick, verrucose, chalky white to pale gray, rarely with patches of granular soredia .... *Ochrolechia farinacea* Howard

26. Thallus extremely thin .... *Ochrolechia szatalaënsis* Verseghy

27. Thallus very thin and smooth; apothecial margins smooth, lacking a pinkish ring next to the disk, containing few algae in the amphithecium; on deciduous trees, especially alder .... ***Ochrolechia laevigata* (Räsänen) Verseghy**

27. Thallus thick and often verrucose, or rather thin and verruculose, rarely smooth; apothecial margins smooth to verrucose, often or always with an inner ring of pinkish tissue next to the disk 28

28. Algae abundant in the apothecial margin, but absent or spotty below the hypothecium; inner ring always present on apothecial disks; hymenium 320-410 μm high; on conifer bark and wood .... ***Ochrolechia oregonensis* H. Magn.**

28. Algae forming a continuous layer under the hypothecium, and also present in the apothecial margin; inner ring sometimes present on apothecial disks; hymenium 180-280 µm high; mostly on deciduous trees, rarely on conifers .... *Ochrolechia subpallescens* Verseghy

OPEGRAPHA (including *Alyxoria, Dimidiographa, Zwackhia*, and species of *Arthonia* with carbonaceous exciple)

Based in part on Ertz & Egea (2007) and Tønsberg & Brodo (1992).

[Note: This genus includes over 40 North American species and is in need of a thorough review on this continent.]

1. On rock 2
1. On bark or wood 5

2. Punctiform soralia present containing fine white soredia; on shaded, humid rock faces; infrequent, but probably overlooked; widespread 3
2. Soredia absent 4

3. Thallus usually C+ pink (gyrophoric acid often with, or replaced by, schizopeltic acid); often fertile, with round apothecia having rough disks with sterile bumps (gyrose); spores 4-celled, 20-25(-27) x 4-5 µm .... *Opegrapha gyrocarpa* Flotow
3. Thallus C– (confluentic and 2'-*O*-methylmicrophyllinic acids); rarely fertile, round to conical apothecia; spores 6-celled, 32-34 x 4-5 µm .... *Opegrapha zonata* Körber (syn. *Enterographa zonata*)

4. On calcareous rock; thallus thin or endolithic, often parasitic on other lichens, C–; spores 4-celled, 22-29 x 5.5-8 µm; California, Alaska, and Newfoundland .... *Opegrapha rupestris* Pers.
4. On siliceous rock; thallus distinct, not parasitic, C+ pink (erythrin and lecanoric acids); spores (3-)4-celled, 13-17 x 3-4.5 µm; California ...... *Lecanographa brattiae* (Egea & Ertz) Ertz & Tehler (syn. *Opegrapha brattiae*)

5.(1) Spores 4-celled 6
5. Spores 5- or more celled 8

6. Spores (17.5-)19-25 x 5.5-10 µm; lirellae fusiform, straight or curved, sometimes pruinose; thallus creamy white to pale brown, forming small discrete patches; western, on bark and wood of various kinds .... *Opegrapha herbarum* Mont.
6. Spores under 5 µm wide; widely distributed 7

7. Thallus white in delimited patches; lirellae long and narrow with thick walls, opening by a narrow slit; spores 13-18 (-20) µm long .... *Arthonia atra* (Pers.) A. Schneid. (syn. *Opegrapha atra*)

7. Thallus brownish and indistinct, mostly within the bark; lirellae short, frequently branched or star-shaped, scattered, with thin walls, opening broadly; spores 15-27 µm long .... *Opegrapha rufescens* Pers.

8.(5) Spores 12- to 16-celled, 25-60 x 6-9 µm; lirellae short and broad, elliptical, rarely forked; widely distributed .... *Zwackhia viridis* (Pers. *ex* Ach.) Poetsch. & Scheid. (syn. *Opegrapha viridis*)

8. Spores 5- to 8-celled 9

9. Lirellae with a red-brown surface, long, branched, opening broadly with a very thin black margin that is sometimes hard to see; spores 16-21 x 4-6 µm; southeastern coastal plain .... *Dimidiographa longissima* (Müll. Arg.) Ertz & Tehler (syn. *Opegrapha longissima*)

9. Lirellae entirely black, sometimes pruinose, long or short; spores 20-40 µm long 10

10. Spores 2.5-4.5 wide; lirellae long, very narrow (less than 0.25 mm wide), opening by a slit, often branched; eastern .... *Opegrapha vulgata* Ach. (syn. *O. cinerea*)

10. Spores 6-9 µm wide; lirellae fusiform, sometimes forked, opening broadly; very widely distributed .... ***Alyxoria varia* (Pers.) Ertz & Tehler** (syn. *Opegrapha varia*)

OPHIOPARMA

1. On bark and wood, rarely rocks, in the Pacific Northwest south to California; spores straight .... ***Ophioparma rubricosa* (Müll. Arg.) S. Ekman**

1. On rocks in arctic-alpine regions 2

2. Spores tapered and curved, more than 30 µm long................................. ***Ophioparma ventosa* (L.) Norman**

2. Spores fusiform, straight, less than 30 µm long .... *Ophioparma lapponica* (Räsänen) Hafellner & R. W. Rogers

PANNARIA (including *Coccocarpia, Fuscopannaria, Massalongia, Parmeliella, Pectenia, Psoroma, Vahliella,* and *Vestergrenopsis*) Based in part on Goward et al. (1994) and Jørgensen (2000). See McCune & Geiser (2009) for Pacific Northwest species.

1. Soredia present 2

1. Soredia absent; isidia present or absent 8

2. Upper surface with an erect or appressed tomentum .... *Erioderma* key

2. Upper surface of lobes without hairs or tomentum 3

3. Lobes 2-5 mm wide, clearly foliose, with coarse soredia along the lobe margins and sometimes take over the entire thallus; medulla PD+ orange (pannarin); on bark, Great Lakes to Maritimes, southern Rockies .... ***Pannaria conoplea* (Ach.) Bory**

3. Lobes smaller, closely appressed squamulose to areolate, with soredia at the edges of the squamules, less commonly on the lobe surface, often coalescing into a continuous sorediate crust; medulla PD–; uncommon 4

4. On soil or rock; thallus olive-brown, with blue-gray soredia; Pacific Northwest .... *Fuscopannaria cyanolepra* (Tuck.) P. M. Jørg.

4. On bark 5

5. Thallus surface blue-gray, thin, with very coarse gray soredia forming along the lobe margins and on the lobe surface and finally dominating the thallus center; prothallus absent; lacking fine needles in old specimens (no triterpenoids); an oceanic species on both coasts .... *Parmeliella parvula* P.M. Jørg.

5. Thallus surface yellowish brown to yellowish gray or olive, although soredia can be blue-gray; often covering larger patches; triterpenoids often present producing fine colorless needles on old specimens 6

6. Thallus yellowish brown, surface scabrid (crusty and rough), squamulose, lobes 1-3 mm wide; soredia on lobe margins, coarsely granular; conifer forests of west and east coasts and Great Lakes region, rare at high elevations in Appalachians .... *Fuscopannaria ahlneri* (P. M. Jørg.) P. M. Jørg.

6. Thallus yellowish brown to dark brown or olive brown; surface not scabrid 7

7. Western; thallus consisting of small, thin, flat to imbricate squamules less than 0.7 mm wide, pale- to olive-brown, with lead-gray soredia forming at the margins, on a thin black prothallus .... *Fuscopannaria mediterranea* (Tav.) P. M. Jørg.

7. Appalachians north to the Maritime Provinces; thallus consisting of thick, spreading squamules up to 2 mm across; soredia forming under lobe tips (labriform), granular to farinose .... *Fuscopannaria sorediata P.M. Jørg.*

8.(1) Thallus foliose, either minute (with lobes less than 1 mm wide), or relatively broad, generally spreading to form rosettes, not shingled at the thallus margins 9

8. Thallus squamulose, with lobes about as long as they are broad, under 2 mm wide, commonly overlapping at least in part 28

9. Upper surface of lobes tomentose .... *Erioderma* key
9. Upper surface of lobes lacking tomentum 10

10. Lower surface smooth; thallus olive to brown or yellow-brown 11
10. Lower surface blue-black with a tomentum or hypothallus, or with a conspicuous prothallus; thallus gray to blue-gray or yellowish gray to yellow-brown 17

11. Lobes 2-4(-5) mm wide .... ***Nephroma helveticum*** **Ach.**
11. Lobes 0.1-1.5 mm wide, elongate or branching 12

12. Growing on soil, moss, or moss-covered rock; thallus chocolate brown, with an upper cortex of pseudoparenchyma; lobes 0.5-1.5 mm wide; apothecia frequent, biatorine, red-brown; spores 2-celled, 8 per ascus; photobiont *Nostoc* 13
12. Growning on bare rock or bark; thallus olive to yellowish brown, lacking an upper cortex; lobes under 1.0 mm wide; on bare rock; apothecia rare or common; spores 1- to 2-celled; photobiont *Scytonema* 14

13. Lobes often 2X longer than they are wide, usually more than 2 mm long; on moss or mossy rock; humid western forests and northern boreal region .... ***Massalongia carnosa*** **(Dickson) Körber**
13. Lobes short, about as long as wide, less than 2 mm long; on dry soil in arid regions in the west; rare .... *Massalongia microphylliza* (Nyl. *ex* Hasse) Henssen

14. On bark in the southwest; thallus isidiate; apothecia biatorine, common; spores 1-celled, (32-)43-55 x 1.5-2(-3) µm, 8 per ascus .... *Koerberia biformis* A. Massal.
14. On bare siliceous rock; thallus with or without isidia; apothecia lecanorine or biatorine; spores 1- to 2-celled, under 35 µm long 15

15. Cylindrical isidia on thallus surface; on wet rocks, arctic-alpine and northwest coast; apothecia lecanorine, rare; spores 1- to 2-celled, 7-10 x 4-6 µm, up to 16 per ascus .... ***Vestergrenopsis isidiata*** **(Degel.) Dahl**
15. Cylindrical isidia absent or rare; on wet or dry rocks 16

16. Lobes spreading, not contiguous, 0.1-0.3(-1.0) mm wide, but margins divided into flattened lobules; spores narrowly ellipsoid, 8 per ascus; on dry rocks in shaded canyons mainly in southwestern U.S., just reaching southern British Columbia; rare
.... *Vestergrenopsis sonomensis* (Tuck.) T. Sprib. & Muggia (syn. *Koerberia sonomensis*)

16. Lobes forming radiate rosettes of typically contiguous lobes 0.5-1.2 mm at the tips; not at all lobulate on the margins; spores 1-celled, broadly ellipsoid to kidney-shaped, 12-16 per ascus; on moist rocks in the open; locally common in coastal Alaska on glacial outwash, rarer in mountains to California
.... *Vestergrenopsis elaeina* (Wahlenb.) Gyelnik

17.(10) Isidia or isidia-like, elongate lobules present 18

17. Isidia absent, but rounded or globular lobules sometimes present 21

18. Isidia on thallus surface; photobiont *Scytonema*; lower surface with a cortex, at least in part
.... ***Coccocarpia palmicola* (Sprengel) Arv. & D. J. Galloway**

18. Isidia or lobules marginal; photobiont *Nostoc*; lower surface entirely without a cortex 19

19. Prothallus very thick, brown-black, projecting like a fringe all around the thallus; lobes 0.7-2 mm wide; coastal plain, Florida to North Carolina .... *Parmeliella pannosa* (Sw.) Nyl.

19. Prothallus thin, conspicuous or not, or prothallus absent; mostly north of southeastern coastal plain 20

20. Thallus with thick lobes 1-3(-4) mm wide, slightly ascending, without a prothallus; apothecia abundant; cortex and medulla PD+ orange (pannarin) .... ***Pannaria tavaresii* P. M. Jørg.**

20. Thallus with thin lobes 0.3-1 mm wide, thin, flat, with a conspicuous blue-black prothallus; apothecia rare; medulla PD– 35

21.(17) Lobes narrow, linear and branched, 0.1-0.5 mm wide; apothecia fringed with hairs; rare, southeast; on bark and leaves .... *Coccocarpia stellata* Tuck.

21. Lobes wider than 0.5 mm, usually rounded, not linear; apothecia not fringed 22

22. Apothecia biatorine, convex with thin or disappearing margins 23

22. Apothecia lecanorine, more or less flat or concave, with persistent raised margins 24

23. Lower surface pale brown to black with a cortex; lobes smooth or with concentric ridges; spores (6-)7-14(-16) x (2-)3-5 µm; mainly southern coastal plain; common

.... ***Coccocarpia erythroxyli* (Sprengel) Swinscow & Krog**

23. Lower surface blue to black, cortex absent; lobes smooth with longitudinal radiate ridges; spores 17-25 x 7-10 µm; northeastern maritime coast; rare ....*Pectenia plumbea* (Lightf.) P. M. Jørg et al. (syn. *Degelia plumbea*)

24. Lobes mostly 2-6 mm wide, with branching veins or ridges on the upper surface, PD+ orange or PD– (with or without pannarin); apothecial margins bumpy; rare, Appalachians and coastal plain to Maritime Provinces .... ***Pannaria lurida* (Mont.) Nyl.** *s. lat.*

24. Lobes mostly 0.2-3 mm wide, more or less smooth; apothecial margins smooth and even, or toothed 25

25. Lobes 0.2-0.7 mm wide, more squamulose than foliose; spores subglobular, 9-12 x 8-11 µm; medulla PD–; eastern and southwestern .... *Pannaria subfusca* P. M. Jørg.

25. Lobes 0.7-3 mm wide, with foliose marginal lobes 26

26. Spores ellipsoid, 15-19(-24) x 8-10 µm; lobes very thick, up to 3 mm broad; medulla PD+ orange (pannarin); rare (Nova Scotia) .... *Pannaria rubiginosa* (Thunb.) Delise
[Note: The description and figure under *P. rubiginosa* in LNA represent a distinctive taxon I am provisionally calling *P. "subrubiginosa"* (see couplet 27). Genetic work is needed on the European, Appalachian, and South African populations.]

26. Spores ellipsoid to subglobular, (10.5-)11.5-15.5(-16.5) µm long; lobes relatively thin, rarely over 1.5 mm wide; medulla PD–, cortex PD+ faint orange or frequently PD– 27

27. Western; round or globular pruinose lobules frequent on lobe margins and tips; thalli small, forming compact rosettes; lower surface with short tomenum rarely visible from above
.... *Pannaria rubiginella* P. M. Jørg. & Sipman

27. Eastern, Appalachians and southern Rockies; round lobules only on older central parts of thallus; thalli irregular in shape, forming large rosettes; blue-black tomentum on lower surface usually very thick, often extending out and visible from above as a fringe
............................***Pannaria "subrubiginosa"* Brodo, *ined.*** [Plate 558 in LNA as "*P. rubiginosa*"; see couplet 26]

28.(8) Thallus grass-green to brownish green when wet; photobiont green 29

28. Thallus dark gray to brownish gray when wet; photobiont blue-green 30

29. Spores 19-28 x 8-10 µm; thallus abundantly squamulose with finely lobed squamules; widespread arctic-alpine ....***Psoroma hypnorum* (Vahl) Gray**

29. Spores 15-19(-22) x (7-)8.5-12.5 µ; thallus with minute squamules, almost granular; western montane to Alaska........ *Psoroma tenue* Henssen

30. Thallus with isidia or cylindrical, isidia-like outgrowths or lobules 31
30. Thallus without isidia, isidia-like outgrowths or lobules 37

31. On soil or moss, rarely rocks; prothallus absent 32
31. On wood or bark, rarely rock; temperate to boreal; prothallus present or absent; edges of squamules not white or felty 34

32. Squamules dark brown throughout, deeply incised and becoming coralloid-branched, without erect cylindrical isidia; California to Oregon .... *Fuscopannaria coralloidea* P. M. Jørg.
32. Edges of squamules gray to bluish white and felty; thick 33

33. Lobe margins with cylindrical to granular outgrowths, or entire thallus a mass of short, erect isidia; arctic-alpine .... ***Fuscopannaria praetermissa* (Nyl.) P. M. Jørg.**
33. Lobes developing rounded, sometimes globular isidia-like lobules on the squamule edges, conspicuously white pruinose at the margins; west coast .... *Fuscopannaria aurita* P.M. Jørg.

34.(31) Thallus blue-gray, not brownish; black prothallus well developed; lacking atranorin 35
34. Thallus brownish gray to dull brownish, prothallus present or absent 36

35.(20) With slender, cylindrical isidia on the lobe margins and sometimes almost covering the entire thallus; coastal and southern boreal .... ***Parmeliella triptophylla* (Ach.) Müll. Arg.**
35. With flattened lobules on lobe surface and margins; Appalachians ....*Parmeliella appalachensis* P. M. Jørg.
[Note : *Parmeliella parvula*, couplet 5, is very similar but has granular soredia instead of lobules or isidia.]

36. Apothecia almost always present; spore walls rough and sculptured; spores subglobose, under 12 µm long; lacking atranorin .... *Pannaria subfusca* P. M. Jørg. (see couplet 24)
36. Apothecia sparse; spore walls smooth; spores ellipsoid, over 12 µm long; thallus pale brown, dull to "satiny," developing narrow, elongated but usually flattened lobules on the lobe margins, especially in older parts of the thallus; containing atranorin; west coast .... *Fuscopannaria laceratula* (Hue) P.M. Jørg.
[Note: Compare also with *Fuscopannaria leucostictoides*, couplet 43, which usually has rough, scabrose surface.]

37.(30) Growing directly on rock 38

37. Growing on bark, moss, or soil (sometimes on moss over rock) 40

38. On maritime rocks splashed with sea water; edges of squamules bluish white; apothecia appearing biatorine; Pacific Northwest .... *Fuscopannaria maritima* P.M. Jørg.

38. On non-maritime rocks; edges of squamules not white or with whitish areas 39

39. Apothecia red-brown to dark brown or black, lacking well-developed thalline margins; spores narrow, (12.5-)14-17 x 5-6.5 µm; growing especially on moist rock walls in shade; Appalachian-Great Lakes and west coast .... ***Vahliella leucophaea* (Vahl) P. M. Jørg.** (syn. *Fuscopannaria leucophaea*)

39. Apothecia black, with prominent thalline margins; spores broad, 10-12 x 6-7 µm; arctic-alpine to boreal .... *Vahliella hookerioides* (P. M. Jørg.) P. M. Jørg.

40.(37) Growing on moss, mossy rock, or tree bases, or soil; spore walls conspicuously rough or sculptured; prothallus black and membranous or absent; apothecia broad, lecanorine .... ***Protopannaria pezizoides* (P.M. Jørg.) S. Ekman** (syn. *Pannaria pezizoides*)

40. Growing directly on bark; spore walls smooth 41

41. Apothecia biatorine with smooth, pale brown margins, 0.4-0.75(1.0) mm in diameter; squamules rarely edged in white; squamules less than 1.0 mm wide, often little more than areoles but overlapping (shingled) when well developed, lobed but usually not deeply incised, typically on a thin black prothallus; spores 16-20(-22) x (7-)8-9(-10) µm; common, wet coastal areas from northern California to British Columbia .... ***Fuscopannaria pacifica* P. M. Jørg** [Plate 343 in LNA as "*F. saubinetii*"]

41. Most apothecia lecanorine with a bumpy, whitish margin; black prothallus usually conspicuous; squamules 0.6-3 mm across, edges white and felty 42

42. Spore walls conspicuously rough and sculptured; spores 9-12 x 8-11 µm, subglobose .... *Pannaria subfusca* P. M. Jørg. (see couplet 25)

42. Spore walls smooth; spores more than 14 µm long 43

43. Spores tapering to a point at one or both ends, 23-27 x 9-11 µm; thallus reddish brown; without atranorin .... ***Fuscopannaria leucosticta* (Tuck.) P. M. Jørg.**

43. Spores ellipsoid, not pointed at the ends, (14-)16-20(-22) x (7-)8-10 µm; thallus yellowish brown to yellowish gray, usually scabrose; containing atranorin (not detectable with a K test) .... ***Fuscopannaria leucostictoides* Ohlsson**

PARMELIA (including *Canoparmelia, Myelochroa, Paraparmelia, Parmelina, "Parmelinopsis," Pseudoparmelia*)

[Note : For yellow-green usnic acid-containing parmelioid lichens (e.g., *Flavoparmelia, Flavopunctelia*), see Key J.]

1. Marginal cilia or cilia-like rhizines present on lobe margins or in axils of lobes, often sparse and visible only with a hand lens 2
1. Marginal cilia or cilia-like rhizines absent 14

2. Lobes broad, 4-20 mm wide; thallus usually loosely attached over entire surface, ascending .... *Parmotrema* key
2. Lobes narrow to moderate, mostly under 5 mm wide; thallus closely appressed to substrate 3

3. Cilia bulbous at base, very short and stiff (noticeable only with a hand lens) .... *Bulbothrix*
3. Cilia not bulbous at base, short or long 4

4. Isidia present 5
4. Isidia absent 8

5. Rhizines regularly and abundantly forked .... *Hypotrachyna*
5. Rhizines unbranched 6

6. Medulla C+ red or pink, KC+ pink (gyrophoric acid); isidia without black cilia growing from the tips .... ***Hypotrachyna minarum*** **(Vainio) Krog & Swinscow** (syn. *Parmelinopsis minarum*)
6. Medulla C–, KC– or KC+ orange-yellow or pinkish violet 7

7. Medulla white, PD–, K–, KC+ pinkish violet (4,5 di-*O*-methylhiascic acid and 5-*O*-methylhiascic acids); isidia with black cilia at the tips .... ***Hypotrachyna horrescens*** **(Taylor) Krog & Swinscow** (syn. *Parmelinopsis horrescens*)
7. Medulla pale yellow, PD+ orange, K+ yellow to red, KC– or KC+ orange-yellow (secalonic and galbinic acids); isidia without cilia at the tips .... ***Myelochroa obsessa*** **(Ach.) Elix & Hale**

8.(4) Rhizines abundantly forked .... *Hypotrachyna*
8. Rhizines not branched 9

9. Soredia or schizidia present; medulla pale yellow, at least close to algal layer or below pustules; apothecia absent or rare 10

9. Soredia or schizidia absent; medulla white or pale yellow; apothecia abundant 12

10. Medulla C+ pink, KC+ red (gyrophoric acid); lobes square or truncated .... ***Hypotrachyna spumosa* (Asahina) Krog & Swinscow** (syn. *Parmelinopsis spumosa*)

10. Medulla C–, KC– or C+ yellow, KC+ yellow (secalonic acid derivatives); lobes rounded 11

11. Medulla PD–; soredia developing from pustules on the thallus surface; common, East Temperate .... ***Myelochroa aurulenta* (Tuck.) Elix & Hale**

11. Medulla PD+ orange (galbinic acid); soredia in hemispherical soralia close to lobe tips; infrequent in Appalachian Mountains .... *Myelochroa metarevoluta* (Asahina) Elix & Hale

12. Rhizines absent from a broad or narrow zone close to the margin; medulla PD+ red (protocetraric acid) .... ***Parmotrema submarginale* (Michaux) DePriest & B. Hale** (syn. *Parmotrema michauxianum*)

12. Rhizines more or less uniformly distributed; medulla PD– or PD+ orange 13

13. More or less restricted to California; medulla white, PD–, K–, C+ red (lecanoric acid) .... ***Parmelina coleae* Argüello & A. Crespo** (= *Parmelina quercina* of N. Am. authors; Plate 566 in LNA)

13. Eastern U.S. and adjacent Canada; medulla pale yellow, PD+ orange, K+ yellow or red, C+ yellow (galbinic acid) .... ***Myelochroa galbina* (Ach.) Elix & Hale**

14.(1) Rhizines absent or sparse at least in a broad or narrow zone close to the lobe margin; thallus usually loosely attached, and (or) with lobes 4-20 mm wide 15

14. Rhizines abundant, uniformly distributed; thallus usually broadly attached, with lobes mostly under 4 mm wide 18

15. Pseudocyphellae conspicuous on the lobe surface, appearing as white dots 16

15. Pseudocyphellae absent or inconspicuous 17

16. Thallus with marginal soredia; lobes with a smooth, even surface; medulla KC+ red and (or) UV+ white .... *Cetrelia*

16. Thallus lacking soredia or isidia; lobes with a network of ridges and depressions; medulla KC–, UV– .... *Platismatia*

17. Thallus surface usually very uneven, with ridges and depressions or otherwise wrinkled; rhizines sparse throughout; pycnidia along the lobe margins .... *Platismatia*

17. Thallus generally smooth and at most folded, without sharp ridges and depressions; rhizines absent from a zone near the margin but usually abundant in the thallus center; pycnidia on the lobe surface ...*Parmotrema*

18.(14) Pseudocyphellae dot-shaped or on net-like ridges, abundant and conspicuous, or sparse (but sometimes absent in young, underdeveloped specimens) 19

18. Pseudocyphellae absent at all stages of development 29

19. Pseudocyphellae round or irregular, not forming a reticulate pattern of white markings; medulla K–, often C+ red or pink .... *Punctelia*

19. Pseudocyphellae irregular, forming a reticulate pattern of white markings; medulla K+ red in most species, never C+ pink or red 20

20. Isidia present 21

20. Isidia absent 24

21. Isidia globular, mostly without a cortex and dull, sometimes resembling soredia .... ***Parmelia hygrophila* Goward & Ahti**

21. Isidia cylindrical or flattened, unbranched or branched, shiny, with a continuous cortex 22

22. Medulla K– or yellowish, PD+ red-orange, KC+ pink (protocetraric acid); isidia marginal and often laminal; rhizines unbranched or forked at the tips, not squarrose; on bark; California to Alaska; infrequent .... *Parmelia pseudosulcata* Gyelnik (= *P. kerguelensis* of N. Am. authors)

22. Medulla K+ red, PD+ yellow (salazinic acid); common 23

23. Rhizines slender, squarrose except close to lobe margins; isidia especially abundant along margins but also laminal .... ***Parmelia squarrosa* Hale**

23. Rhizines thick, unbranched or rarely forked; isidia mostly laminal .... ***Parmelia saxatilis* (L.) Ach.**

24.(20) Soredia absent 25

24. Soredia present 27

25. On bark; rhizines squarrose in older parts of thallus; apothecia almost always present; northern New England and adjacent maritime Canada .... *Parmelia fertilis* Müll. Arg.

25. On rock; rhizines unbranched or forked, not squarrose 26

26. Medulla K+ red, PD+ yellow (salazinic acid); montane, boreal, and arctic .... ***Parmelia omphalodes* (L.) Ach.**

26. Medulla K–, PD+ red (fumarprotocetraric acid); Appalachian .... *Parmelia neodiscordans* Hale

27.(24) Soredia coarsely granular, mostly limited to lobe margins; mature rhizines unbranched or sometimes forked; on rock; uncommon, boreal to arctic; containing usnic acid in soralia .... *Parmelia fraudans* (Nyl.) Nyl

27. Soredia powdery, on lobe surface and margins; lacking usnic acid 28

28. Mature rhizines squarrose; on bark, wood, rock, and sometimes soil; extremely common and widely distributed .... ***Parmelia sulcata* Taylor**

28. Mature rhizines unbranched; on bark; California to Pacific Northwest; infrequent .... *Parmelia barrenoae* Divikar et al.

29.(18) Lower surface white or pale to dark brown 30

29. Lower surface black 34

30. Medulla pale lemon yellow; thallus without soredia, isidia, or lobules; cortex and medulla KC+ orange (secalonic acid) .... ***Pseudoparmelia uleana* (Müll. Arg.) Elix & T. H. Nash**

30. Medulla white, cortex KC– or difficult to interpret due to strong K+ yellow reaction 31

31. Lower surface white or pale brown; cortex and medulla PD+ orange, K+ deep yellow (thamnolic acid) .... *Imshaugia*

31. Lower surface medium to dark brown; cortex PD–, K+ pale yellow (atranorin) 32

32. Medulla PD+ yellow, K+ red (salazinic acid) .... *Canoparmelia salacinifera* (Hale) Elix & Hale (syn. *Pseudoparmelia salacinifera*)

32. Medulla PD–, K– 33

33. Thallus surface with isidia; lobe tips with a network of white maculae

....*Canoparmelia caroliniana* **(Nyl.) Elix & Hale**

33. Thallus surface without isidia, but with granular soredia; maculae sparse and inconspicuous ....*Canoparmelia texana* **(Tuck.) Elix & Hale**

34.(29) Rhizines dichotomously branched; medulla white .... *Hypotrachyna*

34. Rhizines unbranched or sometimes squarrose; medulla white or pale yellow 35

35. Soredia or fragments originating from the breakdown of pustules or hollow warts present 36

35. Soredia and soredia-like fragments absent 39

36. Medulla pale yellow, at least close to algal layer, K–, or K+ yellow, PD–, UV–; common, East Temperate ....*Myelochroa aurulenta* **(Tuck.) Elix & Hale**

36. Medulla entirely white; southeastern U.S. 37

[Note: Poorly developed or immature specimens of *Punctelia* and *Parmelia* may key out here.]

37. On rocks; lobes 1-2 mm wide, smooth; soralia discrete, in hemispherical mounds, containing dark gray granular soredia; medulla PD+ red-orange, KC+ pink (protocetraric acid) .... *Canoparmelia alabamensis* (Hale & McCull.) Elix (syn. *Paraparmelia alabamensis)*

37. On bark of hardwoods; lobes 2-5 mm broad, wrinkled or ridged; soralia discrete or running together, especially along ridges, remaining pale; medulla PD+ orange or PD–, KC– or KC+ faint purple 38

38. Medulla K+ yellow, PD+ orange, UV– (stictic acid) ....*Crespoa crozalsiana* (B. de Lesd. *ex* Harm.) Lendemer & Hodkinson (syn. *Canoparmelia crozalsiana*)

38. Medulla K–, PD–, UV+ blue-white (divaricatic acid) ....*Canoparmelia texana* **(Tuck.) Elix & Hale**

39.(35) Isidia absent; medulla pale yellow, at least near algal layer ....*Myelochroa galbina* **(Ach.) Elix & Hale**

39. Isidia present; medulla white or pale yellow 40

40. Lobes with a network of whitish maculae, wrinkled and rough; medulla white, PD–, K–, UV+ white (perlatolic acid) ....*Canoparmelia caroliniana* **(Nyl.) Elix & Hale**

40. Lobes without maculae, smooth and uniform (where there are no isidia); medulla PD+ red or yellow, K+ or K– 41

41. Medulla pale yellow, at least close to algal layer, PD+ orange, K+ orange to red, KC– (galbinic acid); Appalachian-Ozark and western Great Lakes .... ***Myelochroa obsessa* (Ach.) Elix & Hale**

41. Medulla white, PD+ red, K–, KC+ pink (protocetraric acid); Florida .... *Canoparmelia amazonica* (Nyl.) Elix & Hale (syn. *Pseudoparmelia amazonica*)

PARMELIOPSIS

1. Thallus gray, cortex K+ yellow, KC– (atranorin); sorediate in irregular patches on the upper surface of the lobes; lower surface dark brown; boreal and western ....***Parmeliopsis hyperopta* (Ach.) Arnold**

1. Thallus yellowish green, K–, KC+ gold (usnic acid) 2

2. Soralia in hemispherical mounds on lobe tips; lower surface pale to dark brown; Great Lakes to Maritime Provinces and Maine .... *Parmeliopsis capitata* R. C. Harris *ex* J. W. Hinds & P. L. Hinds

2. Soredia in irregular patches or diffuse on upper surface of lobes 3

3. Lower surface yellowish white; soredia diffuse, arising in part from pustules; southeastern coastal plain .... *Parmeliopsis subambigua* Gyelnik

3. Lower surface dark brown; soredia in patches, not from pustules; boreal-montane and western .... ***Parmeliopsis ambigua* (Wulfen) Nyl.**

PARMOTREMA (including "*Rimelia*" and "*Canomaculina*")

1. Rhizines abundant to the edge of the lobes; white maculae usually conspicuous at least on lobe tips, often in a reticulate pattern 2

1. Rhizines absent in a broad zone close to the lobe margin; maculae not common and usually inconspicuous, never reticulate 7

2. Soredia and isidia absent; pycnidia and apothecia abundant .... ***Parmotrema cetratum* (Ach.) Hale** (syn. *Rimelia cetrata*)

2. Isidia or soredia present 3

3. Thallus isidiate; medulla K+ red (salazinic acid) 4

3. Thallus sorediate, with soredia on the upper surface, especially on the lobe tips and margins 5

4. Maculae scattered, not reticulate nor associated with a network of cracks; isidia cylindrical, developing on the lobe surface; containing norlobaridone as well as salazinic acid

.... ***Parmotrema subtinctorium* (Zahlbr.) Hale** (syn. *Canomaculina subtinctoria*)

4. Maculae creating a network of fine white cracks at least on lobe tips; lobes with cylindrical to branched isidia, occasionally ciliate at the tips, sometimes granular and grading into soredia, on thallus surface and margins .... ***Parmotrema subisidiosum* (Müll. Arg.) Hale** (syn. *Rimelia subisidiosa*)

5. Medulla PD+ yellow, K+ red (salazinic acid); mainly southeastern and into New England .... ***Parmotrema reticulatum* (Taylor) M. Choisy** (syn. *Rimelia reticulata*)

5. Medulla PD–, K–; southern Appalachians 6

6. Medulla CK+ orange, UV+ yellow (diffractaic acid and lichexanthone) .... *Parmotrema diffractaicum* (Essl.) Hale (syn. *Rimelia diffractaica*)

6. Medulla CK–, UV– (caperatic acid) .... *Parmotrema simulans* (Hale) Hale (syn. *Rimelia simulans*)

7.(1) Marginal cilia or cilia-like rhizines present (although sometimes sparse), on lobe margins or in axils of lobes 8

7. Marginal cilia or cilia-like rhizines absent 29

8. Soredia absent 9

8. Soredia present 20

9. Isidia present 10

9. Isidia absent 16

10. Medulla bright yellow .... ***Parmotrema sulphuratum* (Nees & Flotow) Hale**

10. Medulla white 11

11. Thallus yellow-green; cortex K–, KC+ yellow (usnic acid); medulla K–, PD–; southeastern coastal plain 12

11. Thallus gray to yellowish gray; cortex K+ yellow, KC– (atranorin); medulla K+ yellow or red, PD+ yellow or orange, or K–, PD– 13

12. Medulla C–, KC– (containing protolichesterinic acid) .... ***Parmotrema xanthinum* (Müll. Arg.) Hale**

12. Medulla C+ pink and (or) KC+ pink (gyrophoric acid) .... *Parmotrema madagascariaceum* (Hue) Hale

13. Maculae present; medulla K+ red, UV+ yellow-orange (mostly in lower half of medulla close to lower cortex) (salazinic acid and lichexanthone); isidia without cilia .... ***Parmotrema ultralucens* (Krog) Hale**

13. Maculae absent; medulla K– or K+ yellow, UV+ white or UV– (lichexanthone absent); isidia sometimes with short black cilia growing out of the tips 14

14. Isidia partly breaking down into granular soredia; medulla K–, KC+ red, UV+ white (alectoronic acid) .... *Parmotrema mellissii* (C. W. Dodge) Hale

14. Isidia not breaking down into soredia; medulla K+ yellow, KC–, UV– (stictic acid) 15

15. Lobes 4-12 mm wide; isidia mostly ciliate; medulla KC– (norlobardone absent); common, East Temperate and west coast .... ***Parmotrema crinitum* (Ach.) M. Choisy**

15. Lobes 2-7 mm wide; isidia not ciliate; medulla KC+ rose (containing norlobaridone); southeast, infrequent .... *Parmotrema internexum* (Nyl.) Hale

16.(9) Apothecial disks not perforated; maculae absent; lower surface brown, never white at edge; medulla PD+ red, K–, KC+ red (protocetraric acid); common throughout the southeast .... ***Parmotrema submarginale* (Michaux) DePriest & B. Hale** (syn. *Parmotrema michauxianum)*

16. Apothecial disks perforated with an irregular hole through the center; maculae present on upper surface; lower surface uniform or splotched with white 17

17. Medulla PD+ orange, K+ red or K+ yellow, KC– 18

17. Medulla PD–, K–, KC+ pink, UV+ blue-white (alectoronic acid); southeastern coastal plain .... *Parmotrema subrigidum* Egan (= "*P. rigidum*" of N. Am. authors.)

18. Medulla K+ yellow (stictic acid, major product); eastern Texas and Louisiana .... *Parmotrema preperforatum* (W. L. Culb.) Hale

18. Medulla K+ red (norstictic or salazinic acid) 19

19. Lower surface usually splotched with white near the margins; containing norstictic acid; mainly southeastern coastal plain and Appalachians .... ***Parmotrema perforatum* (Jacq.) A. Massal.**

19. Lower surface pale to dark brown, not splotched with white; containing salazinic acid; arid central states, Texas to North Dakota .... *Parmotrema eurysacum* (Hue) Hale

20.(8) Medulla K–, PD–, KC+ red (alectoronic acid) ....21 (See also couplet 37)

20. Medulla K+ yellow, orange, or red, PD+ yellow, orange, or red, KC– or KC+ pink (alectoronic acid absent) 24

21. Soredia developing from short isidia, many of which are ciliate (with short black hairs) .... *Parmotrema mellissii* (C. W. Dodge) Hale

21. Soredia developing from breakdown of thallus cortex, not isidia 22

22. Soredia on the upper surface of lobe tips; west coast, Appalachian and Great Lakes regions ....***Parmotrema arnoldii* (Du Rietz) Hale**

22. Soredia along the lobe margins; southeastern coastal plain 23

23. Lower surface with conspicuous white blotches near the lobe margins ...*Parmotrema louisianae* (Hale) Hale

23. Lower surface brown near the margins, not white-blotched ....***Parmotrema rampoddense* (Nyl.) Hale**

24.(20) Cilia long or short, often abundant, scattered along lobe margins; medulla PD+ yellow or orange 25

24. Cilia short, sparse, and usually confined to lobe axils 33

[Note: Species with sparse cilia will key out at 33 together with species lacking cilia (from couplet 29).]

25. Medulla K+ yellow or orange (stictic acid as the major product) 26

25. Medulla K+ red (salazinic or norstictic acid as the major product) 27

26. Upper surface of thallus uniform, lacking white maculae; lower surface brown at the edges; Appalachian-Great Lakes, west coast ....***Parmotrema perlatum* (Hudson) M. Choisy** (syn. *Parmotrema chinense*)

26. Upper surface with a network of white maculae; lower surface blotched with white areas near the margins; south central and southeastern areas and southern California ....*Parmotrema hypoleucinum* (Steiner) Hale

27. Lower surface with blotches of ivory white close to the margins; medulla PD+ yellow (norstictic acid); common throughout southeast ....***Parmotrema hypotropum* (Nyl.) Hale**

27. Lower surface uniformly black to dark brown, rarely with white areas; medulla PD+ orange (salazinic acid) 28

28. Soredia narrowly restricted to the lobe margins; Appalachians, California ....***Parmotrema stuppeum* (Taylor) Hale**

28. Soredia on upper surface of lobe tips close to the margins; eastern U.S. .... ***Parmotrema margaritatum* (Hue) Hale**

29.(7) Soredia absent, but sometimes isidiate 30

29. Soredia present 32

30. Medulla pale orange-yellow, K+ yellow, C+ pink (gyrophoric acid); thallus isidiate; Gulf coast to Georgia .... ***Parmotrema endosulphureum* (Hillm.) Hale**

30. Medulla white, K–, C+ red or C– (gyrophoric acid absent) 31

31. Upper surface dull; medulla C+ red (lecanoric acid); thallus isidiate; common throughout the Southeast .... ***Parmotrema tinctorum* (Delise *ex* Nyl.) Hale**

31. Upper surface rather shiny; medulla C–; thallus with or without isidia; mostly northern and western species, not in the coastal plain .... *Platismatia*

[Note: See also couplet 15, *Parmotrema internexum*, which can lack marginal cilia.]

32.(29) Medulla K+ yellow or red, PD+ yellow, orange or red 33

32. Medulla K–, PD– 37

33.(24) PD+ red, KC+ pink, K– or K+ yellow (protocetraric acid) 34

33. PD+ yellow to orange, K+ red (norstictic or salazinic acid) 36

34. Medulla K+ yellow (echinocarpic acid); soredia granular; southeastern coastal plain .... ***Parmotrema dilatatum* (Vainio) Hale**

34. Medulla K– (lacking echinocarpic acid) 35

35. Soredia yellowish when fresh (containing usnic acid), granular, on lobe surface close to or on margins; lower surface black close to margins with occasional white blotches; southeastern coastal plain .... *Parmotrema dominicanum* (Vainio) Hale

35. Soredia white (lacking usnic acid), fine, mostly on short, crescent-shaped lobes on older parts of thallus; lower surface uniformly shiny brown near edge, lacking white blotches; southeast north to Pennsylvania .... *Parmotrema gardneri* (C. W. Dodge) Sérus.

36.(33) Containing salazinic acid; soredia farinose; Gulf Coast to Georgia, frequent

.... ***Parmotrema cristiferum* (Taylor) Hale**

36. Containing norstictic acid; soredia granular; Florida, rare .... *Parmotrema rubifaciens* (Hale) Hale

37.(32) Soralia on the thallus surface; thallus closely appressed to substrate, lobes flat; lower surface more or less uniform in color; medulla UV+ blue-white (divaricatic acid) .... ***Canoparmelia texana* (Tuck.) Elix & Hale**

37. Soralia marginal; thallus loosely attached over entire surface, ascending; lobes undulating or crisped at the margins; lower surface pale to dark brown or blotched with white close to the margin; medulla UV– 38

38. Medulla C+ red or pink (lecanoric acid) .... ***Parmotrema austrosinense* (Zahlbr.) Hale**

38. Medulla C– 39

39. Soredia in crescent-shaped soralia on older lobes; containing caperatic acid; on Gulf coastal plain .... ***Parmotrema praesorediosum* (Nyl.) Hale**

39. Soredia scattered over lobe margins, not in crescent-shaped soralia; boreal and western, not on Gulf coastal plain .... ***Platismatia glauca* (L.) W. L. Culb. & C. F. Culb.**

PELTIGERA

1. Thalli translucent when wet, growing underwater, attached to rocks by a single point and fanning out; strongly veined below; containing cyanobacteria; in mountain streams 2

1. Thalli opaque even when wet; usually attached by many rhizines (if fan-like, then containing green algae) 3

2. Eastern (Appalachians to Quebec); containing methyl gyrophorate .... *Peltigera hydrothyria* Miadl. & Lutzoni (syn. *Hydrothyria venosa*)

2. Western mountains; containing no lichen substances ....***Peltigera gowardii* Lendemer & H. O'Brien** [Plate 369 in LNA as "*Hydrothyria venosa*"]

3. Photobiont green; cephalodia present 4

3. Photobiont blue-green; cephalodia absent 8

4. Thallus attached by a single point at one edge, fanning out; rhizines absent; cephalodia forming tiny nodular lobules on the lower surface mostly on the dark veins; disks more or less flat .... ***Peltigera venosa* (L.) Hoffm.**

4. Thallus attached by numerous rhizines except at the lobe margin; cephalodia forming scales on the upper surface of the lobes; disks saddle-shaped 5

5. Cephalodia in the form of lobed scales loosely attached to the upper thallus surface, easily detached ....***Peltigera britannica*** **(Gylenik) Holt.-Hartw. & Tønsberg**

5. Cephalodia in the form of round to slightly scalloped scales closely appressed, firmly fixed to the thallus surface 6

6. Lower surface black in the center changing abruptly to white at the margins, or with a regular pattern of white and dark areas near the margins, but lacking well-defined veins; lobes more or less flat, not crisped or undulating at the margins; rhizines forming an intricately branched and anastomosing mat; lower surface of apothecia with a continuous, bumpy cortex ....***Peltigera aphthosa*** **(L.) Willd.**

6. Lower surface with conspicuous veins 7

7. Lobes crisped and undulating at the margins; lower surface of apothecia with green, scale-like, patches; widespread, especially associated with calcareous soils ....***Peltigera leucophlebia*** **(Nyl.) Gyelnik**

7. Lobes rounded and even, not crisped; lower surface of apothecia with a continuous cortex; western, associated with deep moss and persistent snow patches ....*Peltigera chionophila* Goward & Goffinet

8.(3) Soredia present 9

8. Soredia absent 12

9. Soredia in irregular gray patches on thallus surface 10

9. Soredia marginal or on the lower surface of the lobe tips 11

10. Lobes typically deeply concave, up to 10 mm in diameter; outermost rhizines distinct and pointed; medulla KC– (usually lacking methyl gyrophorate or gyrophoric acid); on bare soil or thin moss cover in dry open habitats ....***Peltigera didactyla*** **(With.) J. R. Laundon**

10. Lobes typically flat or somewhat concave, often lobed and up to 15 mm in diameter; lower surface covered with flocculent rhizines; soralia often KC+ pink (which quickly disappears) (methyl gyrophorate and gyrophoric acid); in deep mosses typically in forests .... *Peltigera extenuata* (Vainio) Lojka

11. Upper surface smooth to scabrous, not tomentous; lower surface with conspicuous veins; photobiont *Nostoc* ....***Peltigera collina*** **(Ach.) Schrader**

11. Upper surface tomentose at least near lobe margins; lower surface more or less uniform; photobiont *Scytonema* .... *Erioderma* key (including *Leioderma*)

12.(8) Isidia present on the thallus surface 13

12. Isidia absent; lobules present or absent 14

13. Lobes mostly 5-10 mm broad, concave; isidia flat and scale-like, peltate .... ***Peltigera lepidophora*** **(Nyl. *ex* Vainio) Bitter**

13. Lobes 7-25 mm broad, turned down at the margins; isidia granular to erect and flattened .... ***Peltigera evansiana*** **Gyelnik**

14. Lobules present on margins and along cracks 15

14. Lobules absent 18

15. Upper surface dull or tomentose, at least at lobe tips ...... ***Peltigera praetextata*** **(Flörke *ex* Sommerf.) Zopf**

15. Upper surface rather shiny 16

16. Lower surface with a regular pattern of white and dark areas; rhizines arising in more or less concentric bands; disks rather flat .... ***Peltigera elisabethae*** **Gyelnik**

16. Lower surface with conspicuous veins; rhizines distributed along the veins; disks saddle-shaped 17

17. Apothecia red-brown; veins on lower surface pale, with a stretched, elongate appearance; containing peltidactylin and dolichorrhizin; Pacific Northwest .... ***Peltigera pacifica*** **Vitik.**

17. Apothecia black; veins dark, fused; lacking peltidactylin and dolichorrhizin; southern Appalachians .... *Peltigera phyllidiosa* Goffinet & Miadl.

[Note: This species is much like *P. neckeri*, but with lobules.]

18.(14) Lower surface without distinct veins or patterning 19

18. Lower surface with distinct veins, or with a regular pattern of white and dark areas; rhizines abundant 21

19. Tiny lichen, lobes 3-7 mm across, tiny brown apothecia on the upper surface; on trees, oceanic woods from Newfoundland to Maine; very rare .... *Erioderma pedicellatum* (Hue) P. M. Jørg.

19. Large lichen, lobes 7-20 mm across, apothecia, if present, marginal 20

20. Lower surface black in the center grading to white at the margins; rhizines sparse; thallus thick, dark green when wet .... ***Peltigera malacea* (Ach.) Funck**

20. Lower surface pale; thallus thin, not turning dark green when wet; east coast (rare on west coast) .... *Peltigera hymenina* (Ach.) Delise

21.(18) Upper surface rather shiny 22

21. Upper surface dull, scabrose, or tomentose 27

22. Apothecial disks black; maculae present on thallus surface .... ***Peltigera neckeri* Hepp *ex* Müll. Arg.**

22. Apothecial disks brown; maculae absent 23

23. Veins pale, narrow and conspicuously raised, with discrete, ropy rhizines 24

23. Veins dark (at least in older parts of thallus), broad or narrow, relatively flat, fuzzy or smooth; thallus not membranous and relatively smooth 25

24. Thallus thin and membranous, often with a blistered appearance, lobes 10-30 mm across; veins distinctly fuzzy with an erect tomentum; close examination of lobe tips usually reveals a thin tomentum; common .... ***Peltigera membranacea* (Ach.) Nyl.**

24. Thallus smooth, not blistered, lobes 5-10(-14) mm across; veins not fuzzy; lobes entirely shiny with no trace of tomentum; infrequent but widely distributed .... *Peltigera degenii* Gyeln.

25. Rhizines arising in more or less concentric bands; apothecia flat; spores mostly under 45 µm long .... ***Peltigera horizontalis* (Hudson) Baumg.**

25. Rhizines distributed along the veins; apothecia saddle-shaped; spores mostly over 45 µm long 26

26. Lobes 7-15(-20) mm across, with crisped margins; upper surface shiny throughout; veins reach edge of lobes where they fuse into a continuous brown margin .... ***Peltigera polydactylon* (Necker) Hoffm.**

26. Lobes 20-40 mm across, round and not crisped; surface rather dull; veins typically do not reach margin, which remains pale .... ***Peltigera neopolydactyla* (Gyelnik) Gyelnik** *s. lat.*
[Note: A west coast population that turns dark green when wet and has dark, fused veins may be distinct and can be called *P. occidentalis* (E. Dahl) Kristinsson. Typical *P. neopolydactyla* turns dark gray when wet and has paler veins barely reaching the margin.]

27.(21) Small species, lobes up to 15 mm wide 28

27. Large species, lobes 10-40 mm wide 32

28. Upper surface with erect tomentum (sometimes confined to extreme lobe tips) and often scabrose at the lobe edges; veins very dark, contrasting with white spaces between them, distinctly fuzzy with an erect tomentum .... ***Peltigera kristinssonii*** **Vitik.**

28. Tomentum appressed to lobe surface, usually abundant 29

29. Lobe margins entire; lobes strongly erect; veins low, not prominent ....***Peltigera didactyla*** **(With.) Laundon** [Note: See also couplet 10.]

29. Lobe margins scalloped; lobes mostly adnate; veins prominent 30

30. Rhizines tufted, running together; lobe surface heavily tomentose; widely distributed ....***Peltigera rufescens*** **(Weiss) Humb.**

30. Rhizines unbranched or brush-like, separate and distinct; lobe surface heavily tomentose or dull but almost without tomentum 31

31. Lower surface pale, with pale more or less naked veins; rhizines pale, unbranched, typically smooth; widespread .... ***Peltigera ponojensis*** **Gyelnik**

31. Lower surface with a network of raised, overlapping, very fuzzy veins, pale at the edge and darker toward the center of the thallus; rhizines dark brown, thick and fuzzy at the base and brush-like at the tips; western boreal ....*Peltigera retifoveata* Vitik.

32.(27) Upper surface scabrose; veins distinct, pale near margins, more or less smooth; rhizines short, tufted, dark brown throughout; cortex usually C+ pink (gyrophoric acid and tenuiorin) ....***Peltigera scabrosa*** **Th. Fr.**

32. Upper surface tomentose, at least at the lobe tips, not at all scabrose 33

33. Rhizines fibrous and smooth, tufted, or brush-like; upper surface tomentose, dull; lower surface with raised, pale veins, sometimes dark in the center of the thallus ....***Peltigera canina*** **(L.) Willd.**

33. Rhizines rope-like, discrete, unbranched, often long 34

34. Thallus very thin, with a blistered appearance; veins narrow, strongly raised, pale; rhizines with a fuzzy surface; tomentum on upper surface often confined to extreme lobe tips with the remainder of the surface rather shiny; widespread .... ***Peltigera membranacea*** **(Ach.) Nyl.**

34. Thallus thin or thick, smooth, not blistered; veins broad, dark; mainly western 35

35. Veins smooth to fibrous and raised, reddish brown; spaces between veins broader than veins; rhizines slender, long; boreal to western montane .... *Peltigera cinnamomea* Goward

35. Veins spongy or tomentose, broader than spaces between them, pale near margin, darker brown in thallus center, not red-brown; rhizines thick, very fuzzy especially at base; western ....*Peltigera retifoveata* Vitik.

PELTULA (including *Heppia*) Based in part on Büdel & Nash (2002).

1. Sorediate on the margins of the squamules 2

1. Not sorediate 3

2. Individual thalli under 2 mm in diameter, with undulating margins; spores broadly ellipsoid to subglobose, 5-7.5 x 3-4.5 µm; southwestern to Dakotas .... *Peltula bolanderi* (Tuck.) Wetmore

2. Individual thalli 1.5-3.0(-10) mm in diameter, with downturned margins; spores ellipsoid, 6-9 x 3-5 µm; widespread central to western temperate .... ***Peltula euploca* (Ach.) Poelt**

3. Thallus with erect more or less cylindrical or spherical lobes 4

3. Thallus squamulose; mostly southwestern or interior 7

4. Lobes almost globose, mostly hollow with a webby medulla, sometimes with granular isidia; southwestern .... *Peltula clavata* (Kremp.) Wetmore

4. Lobes solid, lacking isidia 5

5. Spores ellipsoid, 4.5-7.5 x 3-4.5 µm; lobes mostly flattened, appearing areolate; California to Texas .... *Peltula zahlbruckneri* (Hasse) Wetmore

5. Spores globose or almost globose, 3.5-4.5 µm in diameter 6

6. Most cylindrical lobes rounded at summit; fertile lobes about the same size as sterile lobes mainly Appalachian, rare in central California .... ***Peltula cylindrica* Wetmore**

6. Most cylindrical lobes flattened at summit; fertile lobes larger than sterile lobes; southeastern, rare in Arizona ....*Peltula tortuosa* (Nees) Wetmore

7.(3) Spores ellipsoid, 8 per ascus 8

7. Spores globose or broadly ellipsoid, many per ascus 10

8. Thallus gelatinous, without a clearly defined algal layer or cortices; widely distributed, especially in west .... *Heppia lutosa* (Ach.) Nyl.

8. Thallus squamulose, with a distinct algal layer and upper or lower cortex; mainly southwestern to Midwestern 9

9. Thallus grayish pruinose, rough, with deeply concave lobes; lower cortex absent; hymenium reddish in IKI .... ***Heppia conchiloba* Zahlbr.**

9. Thallus brown to olive, without pruina, smooth, with flat lobes that are upturned at the edges; lower cortex well developed; hymenium mostly blue in IKI, although can be reddish in epihymenium .... *Heppia adglutinata* (Kremp.) A. Massal.

10.(7) Spores broadly ellipsoid; growing directly on rock or soil 11

10. Spores globose; growing on soil; squamules round; epihymenium K+ red to violet or purple 12

11. Squamules scalloped (with rounded lobes); epihymenium K– .... ***Peltula obscurans* var. *hassei* (Zahlbr.) Wetmore**

11. Squamules unlobed; epihymenium K+ red-purple .... ***Peltula obscurans* (Nyl.) Gyelnik var. *obscurans***

12. Squamules with edges turned down, upper surface rough (rugose), lower surface blackish; apothecial margin usually absent .... ***Peltula richardsii* (Herre) Wetmore**

12. Squamules with edges turned up, upper surface smooth, lower surface pale brown; apothecial margin prominent and usually persistent .... ***Peltula patellata* (Bagl.) Swinscow & Krog**

PERTUSARIA (includes *Varicellaria* and *Variolaria*) Some spore measurements from Dibben (1980) or Harris & Ladd (2005).

1. Apothecia buried in verrucae with straight or sloping sides, or that are slightly constricted at the base, having small ostiole-like openings 2

1. Apothecia lecanorine, with narrow or broad disks, or becoming partly or entirely "sorediate" or pruinose at the summit 27

[Note: Lendemer (pers. comm.) points out that the powdery granules on the disks of species of *Variolaria* and related lichens rarely, if ever, contain algae and are not true soredia.]

2. Spores (4-)8 per ascus 3

2. Spores 2-4 per ascus 11

3. Thallus yellowish green or sometimes yellow-gray; cortex C+ yellow-orange, KC+ orange, UV+ dark orange (thiophaninic acid) 4
3. Thallus greenish gray, not yellowish; cortex C–, KC–, UV– or UV+ yellow 6

4. Ostiole area depressed, dark brown to black, often pruinose, C–; epihymenium K+ violet; medulla K+ red (norstictic acid); Appalachian-Great Lakes distribution .... *Pertusaria rubefacta* Erichsen
4. Ostiole area raised and yellow, C+ orange; epihymenium K–; southeastern 5

5. Medulla K+ yellow, PD+ orange (stictic acid) .... ***Pertusaria texana* Müll. Arg.**
5. Medulla K–, PD– (containing variolaric acid) .... *Pertusaria epixantha* R. C. Harris

6.(3) Spores in two irregular rows within the ascus 7
6. Spores lined up in a single row within the ascus 8

7. Epihymenium K–; medulla K–, PD–; thallus and hemispherical to subglobose verrucae rather smooth; ostioles dot-like; on bark, Appalachian-Ozark distribution .... *Pertusaria ostiolata* Dibben
7. Epihymenium K+ purple; medulla K+ red, PD+ yellow (norstictic acid); thallus and steep-sided, discrete verrucae usually rough and rugose; ostioles sometimes coalescing; on bark and rocks; East Temperate .... *Pertusaria propinqua* Müll. Arg.

8. Spores 50-130 x 25-45 µm; thallus and verrucae UV+ bright yellow (lichexanthone), smooth or rough and warty; ostioles often with a pale border; throughout the southeast ..... ***Pertusaria paratuberculifera* Dibben**
8. Spores 20-60 x 12-32 µm; thallus and verrucae UV– or UV+ dull pinkish orange or pale yellow, smooth to rough or verruculose 9

9. Ostioles with a white border; thallus rough; spores 50-95 x 25-40 µm, radially roughened walls; Florida .... *Pertusaria iners* R. C. Harris
9. Ostioles without a white border; thallus smooth; northern 10

10. Epihymenium K–; spores 36-60 x 18-32 µm; Great Lakes region east to Maritimes and Massachusetts, also Alaska .... *Pertusaria alpina* Hepp *ex* Ahles
10. Epihymenium K+ red-violet; spores 20-36 x 12-24 µm; western boreal montane, east to Great Lakes region .... *Pertusaria sommerfeltii* (Flörke *ex* Sommerf.) Fr.

11.(2) Growing on bark, moss, peat, or soil 12
11. Growing directly on rock 25

12. On soil or peat, mostly arctic-alpine (rarely on hardwood bark on east coast); spores 2 per ascus; abundant black ostioles in separate or frequently confluent verrucae; epihymenium K+ violet; medulla K+ yellow (stictic acid) .... *Pertusaria subobducens* Nyl.
12. On bark or moss, not arctic-alpine 13

13. Spores mostly 4 per ascus 14
13. Spores 2(-3) per ascus 19

14. Thallis isidiate, usually on moss or moss-covered rocks, rarely bark; medulla PD–, K– (containing 2'-*O*-methylperlatolic acid); eastern, rare .... *Pertusaria globularis* (Ach.) Tuck.
14. Thallus not isidiate; on bark 15

15. Medulla K–, PD–; epihymenium K+ violet or K–; southeastern 16
15. Medulla K+ yellow or red; PD+ yellow or orange; epihymenium K– 17

16 Cortex UV–; epihymenium K+ violet; fertile verrucae mostly with steep sides, ostioles black, coalesced into a flat area at the summit; southeastern coastal plain .... *Pertusaria sinusmexicani* Dibben
16. Cortex UV+ yellow (lichexanthone); epihymenium K–; verrucae smooth with sloping sides, ostioles pale, remaining discrete; Appalachian-Ozark .... *Pertusaria valliculata* Dibben

17. Medulla K+ red, PD+ yellow (norstictic acid); cortex C+ yellow-orange, UV+ orange-red (thiophaninic acid); fertile verrucae steep-sided or becoming lecanorine with dark, pruinose disks; southwestern, on conifers or rarely rock .... *Pertusaria wulfenioides* B. de Lesd.
17. Medulla K+ yellow, PD+ pale orange (stictic acid); cortex C–, UV+ pale pink-orange (2,7-dichlorlichexanthone); fertile verrucae mostly with sloping sides 18

18. Spores 75-130 μm long, with rough inner walls; East Temperate .... *Pertusaria tetrathalamia* (Fée) Nyl.
18. Spores 50-100 μm long, with smooth inner walls; East Temperate and west coast .... *Pertusaria leioplaca* DC. (syn. *P. leucostoma*)

19.(13) Thallus pale greenish yellow to yellow-gray or greenish gray; cortex C+ yellow-orange, KC+ orange, UV+ orange-red (xanthone); thallus medulla PD+ orange, K+ persistently yellow (stictic acid); spores mostly 50-150 x 30-50 µm 20

19. Thallus gray; cortex C–, KC–, UV– or UV+ yellow or orange-pink; spores mostly (90-)100-200 x 30-65 µm 21

20. Inner spore walls rough and grooved when mature; ostioles pale brown, typically separate; mainly on coastal plain .... ***Pertusaria xanthodes* Müll. Arg.**

20. Inner spore walls smooth; ostioles black, sometimes clustered or fused; East Temperate .... *Pertusaria pustulata* (Ach.) Duby

21. Medulla PD–, K– (containing variolaric acid); verrucae obscure, rarely well developed; southeastern .... *Pertusaria obruta* R. C. Harris

21. Medulla PD+ yellow or red, K+ yellow or red, or K– (lacking variolaric acid); verrucae well developed [but see couplet 24] 22

22. Ostioles black; epihymenium K+ violet; medulla K+ yellow, PD+ orange (stictic acid); Great Lakes to Maritime Provinces .... *Pertusaria consocians* Dibben

22. Ostioles pale to dark, sometimes black; epihymenium K– 23

23. Inner wall of spores with radiating channels; spores sometimes greenish and K+ dull violet; medulla K+ yellow, PD+ orange (stictic acid); Appalachian-Great Lakes .... ***Pertusaria macounii* (I. M. Lamb) Dibben**

23. Inner wall of spores smooth, not channeled, always colorless and K– 24

24. Medulla PD+ red, K– or brownish (fumarprotocetraric acid); fertile warts usually distinct; Southeastern .... ***Pertusaria subpertusa* Brodo**

24. Medulla PD+ yellow, K+ red (norstictic acid); fertile warts often fused; East Temperate .... *Pertusaria neoscotica* I. M. Lamb

25.(11) Thallus gray, thin; spores narrowly ellipsoid (75-)85-150(-245) x 25-70 µm, with rough inner walls; thallus cortex C–, KC–; medulla K+ red (norstictic acid); eastern North America ..... ***Pertusaria plittiana* Erichsen**

25. Thallus sulphur yellow, thick; spores broad, (50-)60-90(-114) x 30-50(-60) µm, with smooth walls; cortex C+ orange, KC+ orange (thiophaninic acid); medulla K– 26

26. Spores 2(-3) per ascus, southern California .... ***Pertusaria flavicunda*** **Tuck.**

26. Spores 4 per ascus; southern Arizona to western Texas .... *Pertusaria arizonica* Dibben

27.(1) Growing on mosses or dead vegetation; arctic-alpine 28

27. Growing on bark or rock; not arctic-alpine 29

28. Thallus white to bluish gray, verrucose, but verrucae not resembling isidia, black, pruinose apothecia immersed in the verrucae; all spot tests negative .... ***Pertusaria panyrga*** **(Ach.) A. Massal.**

28. Thallus composed of white, cylindrical isidia with black, sometimes lightly pruinose apothecia at the summits; thallus cortex and medulla PD+ red, K+ brownish (fumarprotocetraric acid) .... ***Pertusaria dactylina*** **(Ach.) Nyl.**

29. On rock; thallus sulphur yellow, cortex C+ yellow-orange; spores 2(-4) per ascus .... ***Pertusaria flavicunda*** **Tuck.** (see also couplet 26)

29. Usually on bark; thallus gray to yellowish green, not sulphur yellow, cortex C–, C+ yellow-orange, or C+ red; spores 1-8 per ascus 30

30. Thallus UV+ yellow or orange-red 31

30. Thallus UV–; spores 1, 2, or 8 per ascus 36

31. Thallus cortex UV+ orange, C+ yellow, KC+ orange (thiophaninic acid); spores 4-8 per ascus; medulla K+ red (norstictic acid) 32

31. Thallus cortex UV+ yellow, C–, KC– (lichexanthone); spores 1(-2) per ascus, but often absent; medulla K– or K+ yellow (lacking norstictic acid); eastern 33

[Note: *Varicellaria velata* (couplet 33) has a thin cortex and may reveal the C+ red, KC+ red reactions of the medulla.]

32. Spores mostly 4 per ascus; southwestern, on conifers or rarely rock .... *Pertusaria wulfenioides* B. de Lesd.

32. Spores 8 per ascus; Appalachian-Great Lakes and southern California, mainly on hardwoods .... lecanorine form of *Pertusaria rubefacta* Erichsen (see couplet 4)

33. Medulla C+ red, KC+ red, PD– (lecanoric acid); verrucae usually expanded into lecanorine apothecia; Appalachians .... ***Varicellaria velata*** **(Turn.) Schmitt & Lumbsch** (syn. *Pertusaria velata*)

[Note: This is a rare chemotype with lichexanthone.]

33. Medulla C–, KC– or KC+ pink to violet 34

34. Medulla KC+ pink becoming violet, PD– (hypothamnolic acid); verrucae becoming granular ("sorediate") at the summit; throughout southeast .... "*Pertusaria*" *hypothamnolica* Dibben

34. Medulla KC–, PD+ yellow to orange 35

35. Medulla K+ yellow (haemathamnolic acid); verrucae pruinose or granular, somewhat raised, not expanded; southern coastal plain ...."*Pertusaria*" *commutata* Müll. Arg. (syn. *Pertusaria copiosa*)

35. Medulla K– (baeomycesic acid); verrucae small, coarsely pruinose to granular; Florida ...."*Pertusaria*" *floridana* Dibben

36.(30) Medulla C+ red or pink, KC+ red; apothecia pruinose or not, with smooth margins, never rough or appearing sorediate 37

36. Medulla C–, KC– or KC+ violet; apothecia pruinose with rough and ragged margins, or dissolved into soredia-like granules 39

37. Spores 1(-2) per ascus, over 100 µm long; thallus usually thick, medulla C+ red (lecanoric acid); thallus usually thick; fertile verrucae very *Lecanora*-like, up to 1.2 mm in diameter, with heavily pruinose disks and thick, smooth margins; common, East Temperate .... ***Varicellaria velata* (Turn.) Schmitt & Lumbsch** (syn. *Pertusaria velata*)

37. Spores 8 per ascus, under 60 µm long; medulla C+ pink (mainly gyrophoric acid) 38

38. Disk black or almost black, usually pruinose; thallus thin, pale greenish-gray, often with white dots (pseudocyphellae); apothecia *Lecanora*-like, sessile, 0.5-1.2 mm in diameter, with fairly smooth margins; spores (10-)15-20(-23) x (7-)10-13.5 µm; west coast, on conifer bark ....*Pertusaria glaucomela* (Tuck.) Nyl.

38. Disk pale pinkish or orange, pruinose or not; thallus very thin, often patchy, lacking white dots; apothecia bursting through thallus leaving a thin, irregular margin, 0.3-0.8 mm in diameter; spores (16-)20-35(-42) x (10-)12-16(-20) µm; on west and east coasts ....*Pertusaria carneopallida* (Nyl.) Anzi *ex* Nyl.

39. Soralia or sorediate verrucae KC+ purple (picrolichenic acid); spores 1 per ascus but rarely found .... ***Variolaria amara* Ach.** (syn. *Pertusaria amara*)

39. Soralia (if present), verrucae, and thallus medulla KC–; spores 1-8 per ascus 40

40. Thallus medulla K+ bright yellow, PD+ orange (thamnolic acid); spores 2 per ascus but sometimes hard to find ....

Variolaria trachythallina **(Erichson) Lendemer, Hodkinson & R. C. Harris** (syn. *Pertusaria trachythallina*)

[Note: See also *Loxospora* key: *Variolaria pustulata,* in which apothecia are extremely rare.]

40. Thallus medulla K–, PD–, K+ red, PD+ yellow, or K± brownish, PD+ red; spores not 2 per ascus 41

41. Thallus medulla PD–, K– or K+ red, PD+ yellow; spores 1 per ascus 42

41. Thallus medulla PD+ red, K+ brown (fumarprotocetraric acid); spores 1-8 per ascus 43

42. Medulla PD–, K– (no substances); widely distributed

.... ***Variolaria ophthalmiza*** **(Nyl.) Darb.** (syn. *Pertusaria ophthalmiza*)

42. Medulla PD+ yellow, K+ red (norstictic acid); Appalachian-Great Lakes

.... *Variolaria waghornei* (Hulting) Darb. (syn. *Pertusaria waghornei*)

43. Apothecial disks usually distinct, various colors from yellow to green or black, usually pruinose; apothecial margins ragged, often in concentric layers; spores 8 per ascus; Pacific Northwest

.... ***"Pertusaria" subambigens*** **Dibben**

43. Apothecial disks usually obscured by "soredia" and aborted, rarely black and pruinose with "sorediate" margins; spores 1 per ascus; Appalachian-Great Lakes region

.... *Variolaria multipunctoides* (Dibben) Lendemer et al. (syn. *Pertusaria multipunctoides*)

PHAEOGRAPHIS (including *Leiorreuma*; see also *Platygramme* key) Based on Harris (1995). See that publication for a more complete key.

[Note: All species in this key are found mainly in the southern parts of the southeastern coastal plain except *Ph. inusta*, which occurs only north of Florida and into New England.]

1. Spores submuriform, 27-35 x 10-13 µm, but often not developed; lirellae white caused by a dense pruina, branched or star-like; containing no lichen substances .... *Phaeographis asteroides* (Fink) Lendemer

1. Spores transversely septate 2

2. Lirellae covered with deep red pruina, K+ yellow-green; spores 6- to 10-celled, 21-35 x 9-11 µm

.... *Phaeographis haematites* (Fée) Müll. Arg.

2. Lirellae white, brown, or black 3

3. Spores 4- to 6(-8)-celled; lirellae brown or black, sometimes lightly pruinose, flat 4

3. Spores 6- to 10-celled 6

4. Lirellae clustered on small white patches, barely or not raised; hymenium without oil droplets; spores 4-celled; thallus medulla K+ red (norstictic acid); Florida ....*Phaeographis intricans* (Nyl.) Staiger (syn. *Sarcographa intricans*)

4. Lirellae not clustered on small white patches; hymenium containing oil droplets; thallus medulla usually K– (no substances); eastern coastal plain 5

5. Walls of lirellae thin, not prominent, carbonized at the sides but not at the base; spores (15-)17-25 x (6-)7-9(-10) µm, 4- to 6-celled; coastal plain mostly north of Florida ......***Phaeographis inusta* (Ach.) Müll. Arg.**

5. Walls of lirellae completely carbonized at the base as well as at the sides; spores (13-)15-18 x 5-6(-7) µm, 4-celled; coastal plain including Florida ...... *Leiorreuma sericeum* (Eschw.) Staiger (syn. *Phaeographis sericea*)

6. Lirellae elongate, sometimes branched, black, pruinose, wall carbonized, flat, emerging from the bark and leaving "bark flaps" as part of the wall; spores 6- to 7-celled, 23-32 x 7-9 µm; thallus often reddish; containing no lichen substances ....*Phaeographis erumpens* (Nyl.) Müll. Arg.

6. Lirellae almost round, somewhat lobed or irregular in shape 7

7. Spores 6-celled, 18-24 x 6-7 µm; margin of lirellae not at all prominent; lirellae 0.1-0.3 mm in diameter, immersed in and pulling away from surrounding bark; thallus UV+ yellow (lichexanthone) .... *Phaeographis punctiformis* (Eschw.) Müll. Arg.

7. Spores (6-)8- to 10-celled; margin of lirellae black and prominent; lirellae 0.4-1.0 mm in diameter; thallus UV– (no lichen substances) .... *Phaeographis lobata* (Eschw.) Müll. Arg.

PHAEOPHYSCIA (including *Physciella*)

1. Soredia absent 2

1. Soredia present 8

2. Rhizines absent; thallus extremely closely appressed, almost crustose (but with a lower cortex) .... ***Hyperphyscia syncolla* (Tuck. *ex* Nyl.) Kalb.**

2. Rhizines present, sparse or abundant; thallus easily detached from substrate, clearly foliose 3

3. Thallus very loosely attached, growing over soil and moss; lower surface white to pale brown; rhizines sparse, white to very pale brown .... ***Phaeophyscia constipata* (Norrlin & Nyl.) Moberg**

3. Thallus closely to somewhat loosely attached, growing on bark or rock; lower surface black; rhizines abundant, black 4

4. Lobes 1-3(-4) mm across; cilia-like rhizines long .... ***Phaeophyscia hispidula* (Ach.) Essl.**
4. Lobes 0.2-1 mm across; cilia-like rhizines very short or absent 5

5. Growing on rock .... ***Phaeophyscia decolor* (Kashiw.) Essl.**
5. Growing on bark, moss, or wood 6

6. Erect lobules abundant on lobe margins; East Temperate .... *Phaeophyscia squarrosa* Kashiwadani
6. Lobules absent 7

7. Stiff, tiny, colorless hairs on the upper surface of the lobes; uncommon, northeast, central and southwest .... *Phaeophyscia hirtella* Essl.
7. Stiff, tiny, colorless hairs absent; common, East Temperate and southwest, Alberta and British Columbia .... ***Phaeophyscia ciliata* (Hoffm.) Moberg**

8.(1) Rhizines absent; thallus attached directly to substrate, almost crustose .... *Hyperphyscia*
8. Rhizines sparse or abundant; thallus attached by rhizines, clearly foliose 9

9. Lower surface and rhizines white to pale brown 10
9. Lower surface and rhizines black 14

10. Cortex K+ yellow, atranorin present; rhizines sparse; conidia rod-shaped ...***Physcia dubia* (Hoffm.) Lettau**
10. Cortex K–, atranorin absent; rhizines abundant; conidia ellipsoid 11

11. Soredia on lobe surface or lobe margins 12
11. Soredia entirely marginal or on lobe tips 13

12. Lobes 0.1-0.3(-0.5) mm wide; soralia round, not marginal, sometimes wider than lobe; mostly Ozarks and mid-west, rare elsewhere in the east .... *Phaeophyscia insignis* (Mereschk.) Moberg
12. Lobes 0.3-1.0 mm wide; soralia irregular in shape, sometimes marginal as well as on lobe surface; northeastern to Montana and Colorado .... *Physciella melanchra* (Hue) Essl.

13. Lobes mostly 0.1-0.3 mm broad, short or elongate, ascending, with coarse soredia or isidioid granules at the lobe tips; mainly western from Canadian prairies to southwestern U.S. .... *Phaeophyscia nigricans* (Flörke) Moberg

13. Lobes mostly 0.3-1.0 mm broad, flat, with powdery soredia in lip-shaped soralia; mainly Eastern Temperate .... ***Physciella chloantha* (Ach.) Essl.**

14.(9) Medulla red-orange, sometimes only in spots .... ***Phaeophyscia rubropulchra* (Degel.) Essl.**

14. Medulla white 15

15. Stiff, tiny, colorless hairs on the upper surface of the lobes close to margins 16

15. Stiff, tiny, colorless hairs absent 17

16. Soredia fine or granular, greenish or brownish where fresh, not isidioid, rarely producing colorless hairs; lobes 0.5-2 mm wide, typically pale brown and often maculate; black rhizines not extending as a "hairy" fringe around the lobes; on bark of hardwoods; common and widespread across U.S., but rare in Canada .... ***Phaeophyscia hirsuta* (Mereschk.) Essl.** (syn. *Ph. cernohorskyi*)

16. Soredia very coarse, dark brown to blackish, almost isidioid, often producing fine colorless hairs; lobes 0.5-1.3(-1.5) mm wide, typically dark brown without maculae; long black rhizines often extend out from lower surface as a fringe; on rock or on moss over rock, or on bark, especially in the north; mainly western montane, but scattered in the east .... *Phaeophyscia kairamoi* (Vainio) Moberg

17. Soredia black, very coarse (almost isidia), along the lobe margins; on rock .... ***Phaeophyscia sciastra* (Ach.) Moberg**

17. Soredia greenish, granular to farinose, on the margins and tips of the lobes or on the thallus surface; on bark or rock 18

18. Lobes 1-3(-4) mm across; cilia-like rhizines long .... ***Phaeophyscia hispidula* (Ach.) Essl.**
[Note: subsp. *hispidula*: fine soredia on thallus surface; subsp. *limbata* Poelt: coarse or isidioid soredia on lobe margins.]

18. Lobes 0.5-1.5(-2) mm across; cilia-like rhizines very short 19

19. Soredia on the upper surface of the lobe tips, or on expanded, turned-back lobe tips (labriform soralia) ...20

19. Soredia on the lobe margins, close to the margins, or on thallus surface 21

20. Soredia farinose, forming small hemispherical greenish mounds on the upper surface of the lobe tips; lobes mostly less than 0.7 mm wide; on bark, or rarely on calcareous rock (e.g., limestone) .... ***Phaeophyscia pusilloides* (Zahlbr.) Essl.**

20. Soredia granular, mostly in lip-shaped soralia at the lobe tips; lobes 0.5-1(-2) mm wide; on siliceous rock (e.g., granite), or occasionally on bark .... ***Phaeophyscia adiastola* (Essl.) Essl.**

21. Soredia farinose, mainly on lobe surface in depressed patches, but some marginal .... ***Phaeophyscia orbicularis* (Necker) Moberg** (see also couplet 12)

21. Soredia granular, mainly on lobe tips, occasionally on surface in mounds .... ***Phaeophyscia adiastola* (Essl.) Essl.**

PHAEORRHIZA

1. Thallus clearly squamulose throughout; apothecia immersed to convex and marginless, rarely appearing lecanorine, and then with only a thin margin; thallus sections (under microscope) producing a K+ yellow wash, lower cortex sometimes becoming dark reddish purple in K; arid interior regions of the west .... *Phaeorrhiza sareptana* (Tomin) Mayerh. & Poelt

1. Thallus areolate to squamulose at the thallus margins; apothecia sunken in areoles one per areolae and appearing lecanorine with a thick margin; thallus sections and lower cortex K–; mainly arctic-alpine .... ***Phaeorrhiza nimbosa* (Fr.) Mayerh. & Poelt**

PHLYCTIS Based in part on an unpublished key by R. C. Harris (pers. comm.); also see Joshi et al. (2012).

1. Thallus sterile, granular or sorediate 2

1. Thallus with ascomata buried in thalline warts or barely visible mounds 4

2. Thallus blue-gray, covered with granules, K–, PD+ yellow (psoromic acid); southeastern .... *Phlyctis boliviensis* Nyl. (syn. *P. ludoviciensis*)

2. Thallus almost white; K+ red, PD+ yellow (norstictic acid) 3

3. On bark; thallus rather thin, soraliate to almost leprose; common in the Appalachian-Great Lakes region and Pacific Northwest .... ***Phlyctis argena s. lat.***

[*Phlyctis argena* (Sprengel) Flotow in the strict sense appears to be a species with a smooth to shiny thallus having discrete rimmed soralia and is found only in a few localities in the Maritime Provinces.]

3. On non-calcareous rock; thallus very thick, rimose, with punctiform clusters of very coarse soredia or granules; mostly Appalachians and Ozarks .... *Phlyctis petraea* R. C. Harris, *ined.*

4. Spores muriform, 2 per ascus, mostly under 85 µm long; rare (Nova Scotia) .... *Phlyctis agelaea* (Ach.) Flotow

4. Spores muriform or transversely septate, 1 per ascus, mostly over 80 µm long 5

5. Thallus sorediate, containing norstictic acid; spores muriform .... ***Phlyctis argena* (Sprengel) Flotow**

5. Thallus without soredia, but sometimes granular 6

6. Thallus containing psoromic acid (K–, PD+ yellow); spores transversely septate, not muriform; common in southeastern U.S. .... *Phlyctis boliviensis* Nyl. (syn. *P. ludoviciensis*)

6. Thallus containing norstictic acid (K+ red, PD+ yellow); spores muriform; uncommon in Pacific Northwest, the southwest, and Appalachian-Great Lakes region .... *Phlyctis speirea* G. Merr.

PHYLLOPSORA

1. Thallus distinctly squamulose, at least at the thallus margins, with lobate squamules 0.3-0.5(-1.0) mm across, often finely divided, having a white tomentose lower surface; the older parts of the thallus producing finely divided phyllidia resembling isidia or coarse granules; apothecia common, red-brown, flat to convex, margins thin and finally disappearing; spores 9-12 x 1.8-2.3 µm; common throughout southeast .... ***Phyllopsora parvifolia* (Pers.) Müll. Arg.**

1. Thallus typically crustose, or forming obscure squamules, isidiate; apothecia common or not, spores 7.5-12 x 2-3 µm; coastal plain 2

2. Thallus with long, cylindrical, mostly unbranched isidia, commonly reaching 0.4-0.7 mm long .... *Phyllopsora corallina* (Eschw.) Müll. Arg.

2. Thallus with short to long, somewhat flattened, and prostrate, frequently branched isidia-like phyllidia, 0.1-0.3 mm long .... *Phyllopsora confusa* Swinscow & Krog

PHYSCIA (including *Heterodermia, Imshaugia,* and *Pyxine*) For *Heterodermia*, based largely on Moberg & Nash (2002) and Lendemer (2009).

1. Soredia absent 2

1. Soredia present 27

2. Isidia present 3

2. Isidia absent 5

3. Rhizines more or less uniformly distributed, short and mostly unbranched; cortex and medulla K+ deep yellow, PD+ orange (thamnolic acid); on conifers and birch in northern and montane regions .... ***Imshaugia aleurites* (Ach.) S. F. Meyer**

3. Rhizines mostly on or close to the margins; cortex K+ yellow; medulla PD– or PD+ yellow (lacking thamnolic acid); on deciduous trees in southeastern U.S. 4

4. Lower surface yellow or orange, wobby, lacking a cortex; medulla K+ yellow (or deep purple where pigmented) .... ***Heterodermia crocea* R. C. Harris**

4. Lower surface white to tan, smooth, with a cortex; medulla K+ red (salazinic acid) .... *Heterodermia granulifera* (Ach.) W. L. Culb.

5.(2) Thallus attached by only a few points, almost fruticose, with long narrow lobes 6

5. Thallus closely or loosely attached; lobes rarely long and narrow 8

6. On soil and mosses in the dry interior regions; lobes less than 0.5 mm wide .... ***Phaeophyscia constipata* (Norrlin & Nyl.) Moberg**

6. On bark or rarely rock; southern U.S.; lobes 0.5-2 mm wide 7

7. Apothecial margins smooth and even; southwestern U.S. .... ***Heterodermia erinacea* (Ach.) W. A. Weber**

7. Apothecial margins toothed or lobulate; south central or southeastern U.S. .... ***Heterodermia echinata* (Taylor) W. L. Culb.**

8. Medulla uniformly yellow; apothecia lecideine, black; cortex UV+ yellow-orange (lichexanthone) .... ***Pyxine berteriana* (Fée) Imshaug**

8. Medulla white, sometimes with pigmented spots; apothecia lecanorine, brown to black; cortex UV– (lichexanthone absent) 9

9. Cortex K– (atranorin absent) .... ***Phaeophyscia hispidula* (Ach.) Essl.**

9. Cortex K+ yellow (atranorin, except for *Imshaugia*) 10

10. Medulla with patches of orange-yellow pigment (anthraquinones), K+ yellow (atranorin), or K+ purple where pigmented; thallus heavily white pruinose; uncommon, in extreme southwest .... *Heterodermia rugulosa* (Kurok.) Wetmore

10. Medulla entirely white, K+ yellow (atranorin, or thamnolic acid in *Imshaugia*) or K–; thallus usually without pruina 11

11. Lobules present at least on the inrolled apothecial margins and usually the lobe margins, sometimes also on the thallus surface (occasionally absent in *H. hypoleuca*, which has brown squarrose rhizines) 12

11. Lobules absent; rhizines never squarrose 14

12. Lobules finely divided, on margins and upper surface of lobes; lower surface becoming purplish black in the center .... ***Heterodermia squamulosa* (Degel.) W. L. Culb.**

12. Lobules rounded or strap-shaped, on the margins of the lobes and apothecia; lower surface white to gray throughout 13

13. Lower surface cottony, entirely without a cortex; rhizines squarrose, well developed and often forming a mat; thallus not pruinose; mainly eastern, although also in southwestern U.S. .... ***Heterodermia hypoleuca* (Ach.) Trevisan**

13. Lower surface smooth, with a cortex, at least in part; rhizines unbranched or forked, occasionally squarrose, not forming a mat; thallus somewhat pruinose on lobe tips; southwestern U.S. .... ***Heterodermia diademata* (Taylor) D. D. Awasthi**

14.(11) On rock 15

14. On bark or wood 18

15. Thallus heavily pruinose .... ***Physcia biziana* (A. Massal.) Zahlbr.**

15. Thallus entirely without pruina 16

16. Lobes 0.3-0.5 mm wide; maculae absent; zeorin absent .... ***Physcia halei* J. W. Thomson**

16. Lobes 0.7-1.5(-2.5) mm wide; zeorin present 17

17. White spots (maculae) abundant on upper surface; apothecia black, without pruina; common, Appalachian-Great Lakes, boreal, western .... ***Physcia phaea* (Tuck.) J. W. Thomson**

17. Upper surface uniform, not spotted with white maculae; apothecia often pruinose; uncommon, western montane .... *Physcia cascadensis* H. Magn. [= *Ph. phaea* fide Moberg (1997)]

18.(14) Lower surface black; very tightly attached to substrate (almost crustose in appearance) .... *Dirinaria*

18. Lower surface white to pale brown; thallus closely or loosely attached to substrate 19

19. Spores colorless, single-celled; cortex PD+ orange, K+ deep yellow (thamnolic acid; atranorin absent); Appalachian-Great Lakes, southwest .... ***Imshaugia placorodia*** **(Ach.) S. F. Meyer**

19. Spores brown, 2-celled; cortex PD– or PD+ pale yellow, K+ yellow (atranorin present, thamnolic acid absent) 20

20. Lobes loosely attached, ascending, with long white cilia at the tips; surface with maculae; southern California .... *Physcia leptalea* (Ach.) DC.

20. Lobes closely attached, not ascending, lacking marginal cilia; surface with or without maculae 21

21. Medulla K+ yellow (atranorin) 22

21. Medulla K–; upper surface of lobes typically without maculae (rarely white spotted) 25

22. Spores with rounded cells (*Pachysporaria*-type); upper surface white spotted on older lobes but uniform on young lobes; apothecial margins sometimes with a fringe of rhizines on the lower side; zeorin absent; southeastern coastal plain .... ***Physcia neogaea*** **R. C. Harris**

22. Spores with angular cells (*Physcia*-type); upper surface distinctly maculate on young and old lobes; apothecial margins without rhizines on lower side; zeorin present 23

23. Lobes convex, 0.3-0.5 mm wide; southeastern .... ***Physcia pumilior*** **R. C. Harris**

23. Lobes flat or somewhat concave at the tips, 1-2(-3) mm wide 24

24. Lobes (0.5-)1-2(-3) mm wide, often elongate; apothecia mainly confined to central parts of thallus, ranging from large to small; temperate region .... ***Physcia aipolia*** **(Ehrh.** ***ex*** **Humb.) Fürnr.**

24. Lobes 0.2-0.6(-1.2) mm wide and rather short; apothecia appearing almost to the lobe tips, more or less uniform in size; boreal region .... *Physcia alnophila* (Vainio) Loht. et al.

25.(21) Thallus entirely without pruina .... ***Physcia stellaris*** **(L.) Nyl.**

25. Thallus typically pruinose 26

26. Lobes less than 1 mm across, convex; thallus under 3 cm in diameter; North Temperate .... *Physcia* sp. (cfr. *stellaris*)

26. Lobes more than 1 mm across, flat; thallus often more than 3 cm in diameter; western and Ontario .... ***Physcia biziana* (A. Massal.) Zahlbr.**

27.(1) Lower surface black or purplish, at least in center of thallus 28

27. Lower surface white to dark brown, gray, yellow, or orange 39

28. Rhizines absent, thallus attached directly to substrate; medulla UV+ blue-white (divaricatic acid) .... *Dirinaria*

28. Rhizines sparse or abundant; medulla UV– or UV+ yellow-orange or red (divaricatic acid absent) 29

29. Medulla yellow to pale orange 30

29. Medulla white 35

30. Soredia entirely marginal, granular or coarse 31

30. Soredia mostly laminal (but see *P. eschweileri* below), farinose, greenish white or green; thallus closely appressed; southeastern 33

31. Thallus with ascending lobe tips that are mostly 1-2.5 mm wide; widely distributed, East Temperate; soredia granular, bluish gray; thallus cortex K+ yellow, UV– (atranorin); medulla mustard yellow, K –; East Temperate .... ***Pyxine sorediata* (Ach.) Mont.**

31. Thallus appressed throughout; lobes mostly under 1 mm wide; restricted to southeastern coastal plain; soredia developing from fragments of pustules (schizidia), not bluish gray; medulla white to orange 32

32. Thallus cortex K–, UV+ yellow (lichexanthone); medulla salmon orange, K+ purple (anthraquinone) .... *Pyxine caesiopruinosa* (Nyl.) Imshaug

32. Thallus cortex K+ yellow, UV– (atranorin); medulla white to yellow, K– .... ***Pyxine eschweileri* (Tuck.) Vainio**

33.(30) Thallus UV–, K+ yellow (atranorin); medulla PD+ orange; some forms with marginal isidia-like granules (schizidia) from the breakdown of pustules .... ***Pyxine eschweileri* (Tuck.) Vainio**

33. Thallus UV+ yellow, K– (lichexanthone); medulla PD–, never with marginal schizidia 34

34. Medulla salmon orange, K+ purple (anthraquinone); restricted to southeastern coastal plain .... *Pyxine albovirens* (G. Meyer) Aptroot

34. Medulla pale yellow, K–; common southeast Temperate, rarely in southwest ....*Pyxine subcinerea* Stirton

35.(29) Thallus with crowded, overlapping lobes; cilia-like rhizines abundant along the margins; Appalachian-Great Lakes and western .... ***Phaeophyscia hispidula* (Ach.) Essl.** (see *Phaeophyscia* key, couplet 17)

35. Thallus forming flat rosettes; marginal cilia or cilia-like rhizines present or absent; mainly southeastern coastal plain 36

36. Cilia-like rhizines present along the lobe margins ....*Heterodermia casarettiana* (A. Massal.) Trevisan

36. Cilia-like rhizines or cilia absent 37

37. Medulla K+ yellow; rhizines sparse; zeorin present ....***Physcia sorediosa* (Vainio) Lynge**

37. Medulla K–; rhizines abundant; zeorin absent 38

38. Cortex K+ yellow, UV– (atranorin); medulla PD+ orange (testacein); spores 4-celled ....***Pyxine eschweileri* (Tuck.) Vainio**

38. Cortex K–, UV+ yellow-orange (lichexanthone); medulla PD–; spores 2-celled ...***Pyxine cocoes* (Sw.) Nyl.**

39.(27) Cortex K– (atranorin absent) 40

39. Cortex K+ yellow (atranorin) 42

40. Lobes pruinose at least at the tips; western ... ***Anaptychia elbursiana* (Szatala) Poelt** (syn. *Physconia thomsonii*)

40. Lobes without pruina; mostly northeastern 41

41. Soredia entirely marginal or on lobe tips; East Temperate to central U.S. ....***Physciella chloantha* (Ach.) Essl.**

41. Soredia on lobe surface or lobe margins; northeastern to Montana and Colorado ....*Physciella melanchra* (Hue) Essl.

42.(39) Soredia laminal or on the upper surface of the lobe tips 43

42. Soredia mostly marginal, on the lower surface of the lobe tips, or on expanded lobe tips 46

43. Marginal cilia common and abundant; thallus loosely attached and often ascending; lobes 0.2-0.5(-1) mm wide ....***Physcia tenella* (Scop.) DC.**

43. Marginal cilia absent; thallus closely attached to substrate; lobes mostly 0.5-1.5(-2.5) mm wide 44

44. Lower surface and rhizines pale brown; soredia blue-gray to white; thallus moderately thick; mainly on rock, rarely wood or bark .... ***Physcia caesia* (Hoffm.) Fürnr.**

44. Lower surface and rhizines white or almost white; soredia greenish or white; on trees or rocks 45

45. Soredia granular, in discrete, round, hemispherical soralia not developing from pustules; thallus thin and membranous; lobes 0.6-1.5(-2.5) mm wide; mainly on bark of deciduous trees, rarely on limestone; East Temperate .... ***Physcia americana* G. Merr.**

45. Soredia coarsely granular to isidioid, developing from the breakdown of pustules that cover the older parts of the thallus; lobes usually under 0.5 mm wide; on bark and rocks; southern U.S. to California .... *Physcia clementei* (Sm.) Lynge

46.(42) Marginal cilia or cilia-like rhizines common and abundant 47

46. Marginal cilia or cilia-like rhizines absent or sparse 58

47. Thalli closely appressed to substrate; soralia crescent-shaped; marginal cilia or cilia-like rhizines branched (fig. 10b) 48

47. Thalli loosely attached over entire substrate, or ascending and almost fruticose; soralia not crescent-shaped; cilia branched or unbranched 54

48. Lower surface smooth to fibrous (with fibers running in one direction), with a cortex, at least in part, white to pale brown, K– or K+ yellow .... ***Heterodermia speciosa* (Wulfen) Trevisan**

48. Lower surface fibrous throughout, entirely without a cortex, white orange or yellow 49

49. Lower surface pale orange, at least in spots (white to pale brown elsewhere), K+ purple where pigmented (anthraquinones) 50

49. Lower surface white or yellowish (but K–) 51

50. Orange pigment more or less covering lower surface; medulla K–, PD–; widespread eastern lichen, mainly at low to middle elevations .... ***Heterodermia obscurata* (Nyl.) Trevisan**

50. Orange pigment confined to some lobe tips; medulla usually K+ red, PD+ yellow (norstictic acid), less often K–, PD–; mainly high elevations in Appalachians, extending to Maritime provinces; infrequent .... *Heterodermia neglecta* Lendemer, R. C. Harris & E. Tripp

51. Lower surface white to pale brown, lacking yellow pigments 52

51. Lower surface partly or entirely yellowish (yellow pigment K–) 53

52. Soralia on broadened, reflexed lobe tips (labriform), with short, white, unbranched cilia; rhizines with irregular (not perpendicular) side branches; containing zeorin; uncommon but widespread in east and reported as far west as Alberta .... *Heterodermia galactophylla* (Tuck.) W. L. Culb.

52. Soralia mostly marginal, not on expanded, reflexed lobe tips; cilia black, sometimes branched; rhizines squarrose (like a bottle brush); medulla sometimes K+ red (± norstictic acid); on hardwood trees in southeast, lower elevations to coastal plain, but reported from as far north as Alberta .... *Heterodermia japonica* (Satô) Swinscow & Krog

53. Yellow pigment at center of thallus spreading toward the margin; medulla K+ red, PD+ yellow (norstictic acid); Appalachian Mountains .... *Heterodermia casarettiana* (A. Massal.) Trevisan

53. Yellow pigment only on lobe tips; with or without norstictic acid; coastal plain .... *Heterodermia japonica* (M. Satô) Swinscow & Krog

54.(47) Cilia short, 1-2(-3) mm long, unbranched; lobes relatively short, forming crowded clumps; lower surface smooth, with a cortex; soralia on expansions of lobe tips 55

54. Cilia 2-5 mm long, branched or unbranched; lobes broad or very long and narrow; lower surface cottony or webby, often pruinose, lacking a cortex 56

55. Soredia within hood-like expansions of lobe tips (between upper and lower cortices) .... ***Physcia adscendens* (Fr.) H. Olivier**

55. Soredia on widened, turned-back lobe tips (lip-shaped soralia) .... ***Physcia tenella* (Scop.) DC.**

56. Lobes broadening, hood-like, not long and narrow, with soredia under expanded "hoods"; rare, California coast .... *Heterodermia namaquana* Brusse

56. Lobes long and narrow, not broadening or hood-like; soredia on lower surface of lobe tips, which are not expanded or hood-like 57

57. Lower surface usually pale sulphur yellow, at least close to lobe tips (K+ yellow, not an anthraquinone); thallus prostrate; cilia white or brown; medulla PD–, K+ yellow (atranorin); Appalachians and southwestern Texas .... ***Heterodermia appalachensis* (Kurok.) W. L. Culb.**

57. Lower surface white; cilia and rhizines mostly black; thallus often ascending; medulla PD+ yellow, K+ yellow changing to red (salazinic acid); west coast and Appalachians, rarer elsewhere .... ***Heterodermia leucomela*** **(L.) Poelt**

[Note: Linda in Arcadia (2012) recommends conserving the spelling "leucomelos."]

58.(46) Soredia produced all along lobe margins 59

58. Soredia confined to lobe tips or on crescent-shaped marginal lobes 65

59. Lower surface fibrous, more or less streaked with black or dark gray; thallus pruinose, at least at lobe tips; rhizines brown to black; medulla K+ yellow .... ***Physcia atrostriata*** **Moberg**

59. Lower surface smooth, more or less uniformly white; thallus with or without pruina; rhizines white to pale brown, sometimes darkening at the tips; medulla K–, K+ yellow, or K+ red 60

60. Soredia formed in soralia emerging from the medulla, not by budding off at the lobe margins (blastidia); soredia largely confined to older central parts of thallus 61

60. Soredia formed as blastidia (fig. 19e) at the lobe margins 64

61. Medulla K+ yellow changing to red (salazinic acid); lobes elongated, up to 3 mm long, moderately thick, not finely divided; marginal cilia sometimes present ......***Heterodermia albicans*** **(Pers.) Swinscow & Krog**

61. Medulla K+ yellow or K– (lacking salazinic acid); lobes thin, short and rounded; marginal cilia never produced 62

62. Lobes smooth, without pruina, thin; medulla K–; southeastern .... *Physcia crispa* Nyl.

62. Lobes distinctly pruinose 63

63. On bark; medulla K+ yellow (atranorin); Florida, Texas, Arizona; infrequent ...... *Physcia undulata* Moberg

63. On rocks or hard wood; medulla K–; western, common .... *Physcia dimidiata* (Arnold) Nyl.

64.(60) Soredia all around margins including lobe tips; lobes 0.3-1.0 mm wide, finely divided; spores 13.5-18 x 6.5-9.0 µm; medulla K–; East Temperate .... ***Physcia millegrana*** **Degel.**

64. Soredia produced mainly on older central lobes, not on lobe tips; lobes (0.5-)1.0-2.0 mm wide; spores 19-25(-28) x 8-11 µm; medulla K+ yellow; central Texas .... *Physcia solistella* Essl. & Egan

65.(58) Rhizines black or, less frequently, pale; soralia mostly on marginal lobes; upper cortex prosoplectenchyma (fig. 5e) 66

65. Rhizines white or pale tan, sometimes darkening at the tips, or rarely dark gray, not squarrose; soralia usually on lobe tips or all around the margins; upper cortex pseudoparenchyma (fig. 5f) 68

66. Medulla K+ red (norstictic and salazinic acids); rhizines sparse; southwest and southeast as well as the western Great Lakes; infrequent........ *Heterodermia pseudospeciosa* Kurok.

66. Medulla K+ yellow or K–, lacking norstictic and salazinic acids; rhizines usually abundant, often squarrosely branched; common in the east, rare elsewhere........ 67

67. Lower surface white to pale brown, smooth to fibrous, with a cortex (at least in part) .... ***Heterodermia speciosa*** **(Wulfen) Trevisan**

67. Lower surface pale orange, at least in part (K+ purple, anthraquinones), webby, entirely without a cortex .... ***Heterodermia obscurata*** **(Nyl.) Trevisan**

68. Lobes very narrow and convex, 0.2-0.5 mm wide, hardly widening at the tips; soredia and blastidia very coarse, forming under and at lobe tips; medulla made up of pseudoparenchyma, not loose hyphae; on siliceous rock .... ***Physcia subtilis*** **Degel.**

68. Lobes 0.3-1.2(-3) mm, usually widening or fanning out at the tips; soredia fine or coarse; medulla with loosely organized hyphae; on various substrates 69

69. Lobes distinctly upturned with soredia forming in lip-shaped soralia; soredia fine or coarse 70

69. Lobe tips mostly flat or downturned; soralia not lip-shaped; soredia coarsely granular 71

70. Soredia very fine, greenish; on rocks, wood, or sometimes bark, exposed or partly shaded; very widespread and common .... ***Physcia dubia*** **(Hoffm.) Lettau**

70. Soredia coarsely granular, blue-gray to white; on mossy rocks in forest habitats; Appalachian Mountains .... *Physcia pseudospeciosa* J. W. Thomson

71. Lobes flat, finely divided and lacy, dissolving into granules (blastidia: fig. 19e) or soredia; East Temperate, rare on west coast; mostly on bark of various kinds, rarely on rock .... ***Physcia millegrana*** **Degel.**

71. Lobes convex, clearly downturned at the tips, with granular soredia forming on the lower surface of the lobe tips; western; on rock, especially limestone and sandstone .... ***Physcia tribacia*** **(Ach.) Nyl.**

[Note: Plate 664 in LNA as "*Physcia callosa,*" the name used for this species by N. Am. authors]

PHYSCONIA

1. Soredia and isidia absent; apothecia common 2
1. Soredia or isidia present; apothecia rare 5

2. On the ground or over rocks in arctic or alpine sites and in grasslands; thallus loosely attached over entire surface ....***Physconia muscigena*** **(Ach.) Poelt**
2. On bark or rocks at low elevations; thallus closely appressed to substrate 3

3. Isidia present, mostly cylindrical but sometimes flattened; southwestern ....***Physconia elegantula*** **Essl.**
3. Isidia absent, but often with lobules, especially on the apothecial margins 4

4. Lower surface white to tan throughout, corticate; lobules common on lobe surface and margins as well as on apothecial margins; eastern, rare ....*Physconia subpallida* Essl.
4. Lower surface brown to black in central, older parts of thallus, white to tan on lobe tips; lobules most conspicuous on apothecial margins although often present on lobe margins as well; western, especially California; common ....***Physconia americana*** **Essl.**

5.(1) Rhizines unbranched or dichotomously branched, pale tan to brown, mostly on or close to the margins; lobes wrinkled or bumpy (rugose) ....***Anaptychia elbursiana*** **(Szatala) Poelt** (syn. *Physconia thomsonii*) [Plate 683 in LNA]
5. Rhizines squarrose, black, more or less uniformly distributed; lobes smooth and even 6

6. Granular to elongate and somewhat branched isidia and lobules present on lobe margins; Alaska, Great Lakes to New England, and southern Rockies ....*Physconia grumosa* Kashiw. & Poelt
6. Isidia absent, soredia present 7

7. Soralia discrete 8
7. Soredia more or less continuous along the lobe margins; lobe margins turned up 9

8. Lower surface of lobe tips fibrous, lacking a cortex; soralia crescent-shaped or lip-shaped on the lobe margins, containing very coarse soredia; lobe margins turned down; common in west and frequent in northeast .... ***Physconia perisidiosa*** **(Erichsen) Moberg**

8. Lower surface of lobe tips smooth, with a cortex; soralia on short lateral lobes, not crescent-shaped, lacking downturned lobes; common temperate lichen, but absent in Pacific Northwest and southeastern coastal plain .... *Physconia leucoleiptes* (Tuck.) Essl.

9. Medulla pale yellow, K+ yellow, KC+ orange-yellow (secalonic acid) .... ***Physconia enteroxantha* (Nyl.) Poelt**

9. Medulla white, K–, KC– (secalonic acid absent) 10

10. Upper cortex with thick-walled cells (scleroplectenchyma); mostly eastern North America .... ***Physconia detersa* (Nyl.) Poelt**

10. Upper cortex with thin-walled cells (pseudoparenchyma: see fig. 5 d, f); western North America .... ***Physconia isidiigera* (Zahlbr.) Essl.**

PILOPHORUS

1. Apothecia sessile; thallus lacking soredia; spores 19-24 x 7-8 μm; Alaska and Yukon .... *Pilophorus dovrensis* (Nyl.) Timdal, Hertel & Rambold (syn. *Lecidea pallida*)

1. Apothecia at the summits of solid stalks, or apothecia lacking 2

2. Stalks sorediate at the tip, under 5 mm tall 3

2. Stalks not sorediate 4

3. Soredial mass KC+ pink to violet (lobaric acid); stalks smooth at base; primary thallus mostly granular .... ***Stereocaulon pileatum* Ach.**

3. Soredial mass KC–; stalks areolate at base; primary thallus areolate ..... ***Pilophorus cereolus* (Ach.) Th. Fr.**

4. Stalks under 5 mm tall; eastern, very rare .... *Pilophorus fibula* (Tuck.) Th. Fr.

4. Stalks becoming over 5 mm tall; western 5

5. Apothecia elongate, cylindrical; thallus olive to brown .... ***Pilophorus clavatus* Th. Fr.**

5. Apothecia spherical; thallus gray-green 6

6. Stalks with hemispherical areolae covering a black central support, usually under 10 mm tall; extreme west coast of Pacific Northwest .... *Pilophorus nigricaulis* Satô

6. Central stalk not pitch black; often taller than 10 mm 7

7. Stalks branched at tips, robust; rare; on soil and pebbles; Alaska and Yukon .... *Pilophorus robustus* Th. Fr.

7. Stalks unbranched or branched only once, slender; common, on rock; Pacific Northwest .... ***Pilophorus acicularis* (Ach.) Th. Fr.**

PLACIDIUM (includes *Catapyrenium, Clavascidium, Heteroplacidium*) See Breuss (2002b); Breuss & McCune (1994); Harris & Ladd (2005).

1. Growing on bark of hardwoods; squamules 2-5 mm wide, lower surface pale; mostly eastern Temperate and southwestern as well as California ....***Placidium arboreum* (Schweinitz *ex* Michener) Lendemer** (syn. *P. tuckermanii*)

1. Growing on rock or soil 2

2. Thallus gray to gray-brown when dry 3

2. Thallus brown to red-brown when dry 4

3. Thallus thick, blue-gray areolate, not truly squamulose, without a dark lower surface; on dry calcareous rocks Arizona and New Mexico north to Saskatchewan; initially parasitic on other lichens .... *Verrucaria inficiens* Breuss (= "*Catapyrenium plumbeum*" of many authors)

3. Thallus squamulose, squamules 1-3 mm wide, notched; lower surface black; cells of upper cortex small and poorly defined; Rocky Mountains, scattered in northeast ....*Catapyrenium cinereum* (Pers.) Körber

4. On rocks, southwestern; squamules convex, small (up to 1.5 mm diameter); spores broadly ellipsoid, 13-17 x 8-11 µm; attachment hyphae thin-walled ....*Placidium acarosporoides* (Zahlbr.) Breuss (syn. *Heteroplacidium acarosporoides*)

4. On soil 5

5. Squamules convex, small, 0.5-1.0(-2.0) mm in diameter, in small clusters; attachment hyphae thick-walled; arid western interior ....*Heteroplacidium congestum* (Breuss & McCune) Breuss

5. Squamules over 2 mm in diameter 6

6. Lower surface of squamules black, often smooth, with sparse attachment hyphae and no rhizines; cells of lower cortex spherical, thin-walled, sometimes arranged in vertical columns ....*Placidium lachneum* (Ach.) Breuss

6. Lower surface of squamules mostly pale brown, with or without rhizines; cells of lower cortex irregularly arranged 7

7. Arctic-alpine; spores 16-22 x 8-11 µm; squamules becoming elongate and forming rosettes .... *Placidium norvegicum* (Breuss) Breuss

7. Temperate to southern; spores 10-18 x 5-8 µm; squamules rarely forming rosettes 8

8. Spores in two rows in the ascus; brown rhizines sometimes seen among brown attachment hyphae; widespread temperate, but infrequent .... *Clavascidium umbrinum* (Breuss) Breuss

8. Spores in a single row in the ascus; rhizines present or absent 9

9. Lower surface attached with pale rhizines .... ***Clavascidium lacinulatum*** **(Ach.) M. Prieto** (syn. *Placidium lacinulatum*)

9. Lower surface attached only by wefts of pale hyphae, no rhizines 10

10. Squamules rimmed with black pycnidia that form knob-like projections; lower cortex poorly defined; mainly southwestern .... *Placidium pilosellum* (Breuss) Breuss

10. Pycnidia inconspicuous, buried in thallus, not marginal 11

11. Lower cortex barely discernible; medulla containing many spherical cells, 9-14 µm in diameter; squamules mostly appressed throughout; widely distributed in the west, rarer in the east .... *Placidium squamulosum* (Ach.) Breuss

11. Lower cortex well defined, 40-90 µm thick; medulla containing few spherical cells; squamules round, concave, slightly raised and free at margins, sometimes imbricate; mainly southeastern Arizona and Ozarks, rare elsewhere .... *Placidium chilense* (Räsänen) Breuss

PLACOPSIS Largely based on Brodo (1995) and Galloway (2005).

1. Thallus dispersed areolate, with abundant, often crowded apothecia, apothecia and areoles lying on a black bed of *Stigonema* cyanobacteria, appearing like a prothallus; ordinary flat to tuberculate cephalodia absent; coastal Pacific Northwest .... *Placopsis roseonigra* Brodo

1. Thallus with well-developed lobes along the margins; apothecia frequent or infrequent; cephalodia pinkish to brown, flat to lumpy, at least some of them central 2

2. Thallus granular isidiate, thin, usually rough, minutely pitted with pock-marks left by isidia that have broken loose; rare, west coast .... *Placopsis cribellans* (Nyl.) Räsänen

2. Thallus lacking isidia, usually sorediate; thallus thick, smooth 3

3. Thallus pale pinkish white when dry, sometimes becoming pinkish brown especially at the lobe tips; soralia in hemispherical mounds of greenish soredia; cephalodia with vague, shallow lobes only at the margins; containing 5-*O*-methylhiascic acid as well as gyrophoric acid ....***Placopsis lambii* Hertel & V. Wirth**

3. Thallus relatively dark, rosy-brown throughout; soralia usually excavate and irregular in shape, soredia dark gray when fresh; cephalodia deeply lobed, usually to the center 4

4. Thallus without pruina; containing only gyrophoric acid; along the west coast and into the arctic ....*Placopsis gelida* (L.) Lindsay *s. str.*

4. Thallus covered with a light or heavy pale gray-brown pruina; containing gyrophoric acid often accompanied by cryptostictic acid and some hiascic acids (minor); montane western ....*Placopsis fusciduloides* D.J. Galloway

PLACYNTHIELLA (including *Hertelidea*)

1. Thallus red-brown, isidiate, C+ pink when tested under a microscope (gyrophoric acid); rarely fertile; on wood or, less frequently, on peat or moss ....*Placynthiella icmalea* (Ach.) Coppins & P. James

1. Thallus dark brown to gray or olive-black, granular or sorediate, C– (lacking gyrophoric acid), often with apothecia 2

2. Thallus verrucose to areolate, pale brownish gray to dark gray, typically producing pale to dark, round soralia, UV+ white (perlatolic acid); spores 7-11(-16) x 3-5 µm; ascus tips K/I+ blue, uniform or as a tube; on wood 3

2. Thallus granular, dark brown to olive-black, UV– (lacking lichen substances); spores 10-15 x 4-7 µm; asci K/I– or uniformly pale blue (*Trapelia*-type); mainly on soil and peat, sometimes rotting wood 4

3. Exciple lacking granules or crystals; hypothecium dark brown; western and northeastern ....*Hertelidea botryosa* (Fr.) Printzen & Kantvilas (syn. *Biatora botryosa*)

3. Exciple containing small granules that shine in polarized light and dissolve in K; hypothecium greenish brown or greenish gray; mostly southeastern .... *Hertelidea pseudobotryosa* R.C. Harris, Ladd & Printzen

4. Granules large, 100-300 µm (0.1-0.3 mm) in diameter

....**Placynthiella oligotropha (J. R. Laundon) Coppins & P. James**

4. Granules less than 150 µm in diameter ....*Placynthiella uliginosa* (Schrader) Coppins & P. James

PLACYNTHIUM

1. Thallus typically with a conspicuous blue-black prothallus; thallus areolate, not lobed at the margins; spores 2- to 4-celled, 10-17 x 3.5-6 µm; on limestone; very widespread and common ....**Placynthium nigrum (Hudson) Gray**

1. Thallus lacking a well-developed prothallus; thallus distinctly lobed at the margins 2

2. Lower surface pale, without a black or blue-green hypothallus; marginal lobes only 0.1 mm wide; widespread but not common .... *Placynthium stenophyllum* (Tuck.) Fink

2. Lower surface consisting of a thin or thick mat of blue-black or blue-green hyphae (hypothallus); marginal lobes usually wider than 0.1 mm 3

3. Thallus lacking isidia or lobules, forming star-like clusters of short, narrow lobes that are flat and not grooved, 0.15-0.25 mm wide, up to 1 mm long; spores 2-celled; on dry limestone; eastern .... *Placynthium petersii* (Nyl.) Burnh.

3. Thallus isidiate and (or) lobulate; spores 2- to 4-celled; northern ....4.

4. Lobes 0.5-2.0 mm long and 0.1-0.2 mm wide; slightly convex and often grooved along their length; frequently producing slender, long, branched isidia on older portions; on calcareous rocks; boreal to arctic-alpine ....*Placynthium asperellum* (Ach.) Trevisan

4. Lobes often longer than 2 mm, 0.2-0.4 mm wide, flat; producing flattened isidia or lobules on the surface; on periodically submerged siliceous rocks; southern boreal-montane ....*Placynthium flabellosum* (Tuck.) Zahlbr.

PLATISMATIA (including *Esslingeriana*)

1. Lobes 0.5-5 mm wide, often elongate or finely divided and sometimes pendent 2

1. Lobes (3-)5-20 mm wide, rarely pendent 4

2. Pycnidia on the thallus surface, common; thallus with a yellowish tint; medulla yellowish and KC+ pink at least on lobe tips and apothecial margins .... **Esslingeriana idahoensis (Essl.) Hale & M. J. Lai**

2. Pycnidia, when present, along the lobe margins; thallus without a yellowish cast, KC– 3

3. Isidia absent .... ***Platismatia stenophylla*** **(Tuck.) W. L. Culb. & C. F. Culb.**

3. Isidia present along lobe margins .... ***Platismatia herrei*** **(Imshaug) W. L. Culb. & C. F. Culb.**

4.(1) Soredia and isidia absent 5

4. Soredia or isidia present 6

5. Northeastern U.S., and southeastern Canada; medulla PD–
....***Platismatia tuckermanii*** **(Oakes) W. L. Culb. & C. F. Culb.**

5. Western; medulla PD+ red (fumarprotocetraric acid)
.... ***Platismatia lacunosa*** **(Ach.) W. L. Culb. & C. F. Culb.**

6. Isidia laminal; surface with a network of sharp ridges and depressions, with white pseudocyphellae
.... ***Platismatia norvegica*** **(Lynge) W. L. Culb. & C. F. Culb.**

6. Isidia and (or) soredia present on lobe margins; lobe surface relatively smooth, without pseudocyphellae
....7

7. Isidia dominating; common and widespread in boreal regions
.... ***Platismatia glauca*** **(L.) W. L. Culb. & C. F. Culb.**

7. Soredia dominating; California to B.C. .... *Platismatia wheeleri* Goward et al.

PLATYGRAMME (including *Thecaria* and *Leiorreuma*; see also *Phaeographis* key)

1. Lirellae very prominent, oval to slightly elongate or Y-shaped, with thick pale green or gray walls; disk heavily white pruinose; exciple within the thin thalline coat, heavily carbonized including base; spores 55-110 x 15-23 µm; Florida .... ***Thecaria quassiicola*** **Fée** (syn. *Phaeographina quassiaecola* [sic])

1. Lirellae flat on the thallus or slightly immersed, branched; walls black; spores less than 100 µm 2

2. Spores (48-)69-80(-90) x 16-26 µm; lirellae slender, with a blue-gray pruinose disk
.... ***Platygramme caesiopruinosa*** **(Fée) Fée** (syn. *Phaeographina caesiopruinosa*)

2. Spores under 35 µm long 3

3. Spores 13-16 x 7-10 µm ....*Leucodecton glaucescens* (Nyl.) A. Frisch [see *Myriotrema* key, couplet 5]

3. Spores 21-35 x 8-13µm 4

4. Lirellae broad, flat, with very thin walls, not or very slightly pruinose; thallus K+ red or K– (sometimes containing norstictic acid); Louisiana to Virginia and Arkansas .... *Leiorreuma explicans* (Fink) Lendemer (syn. *Phaeographina explicans*)

4. Lirellae narrow, opening by a slit; lirellae heavily pruinose; containing no lichen substances; Florida .... *Phaeographis asteroides* (Fink) Lendemer & Knudsen

POLYBLASTIA (including *Agonimia*) Based on Thomson (1997) and Orange et al. (2009a).

1. Growing on soil, peat, or mosses; thallus dark greenish brown to brown, continuous 2
1. Growing on calcareous rocks 3

2. Perithecial wall entirely pale, lacking an involucrellum; spores narrowly ellipsoid, 28-75 x 12-30 µm .... *Agonimia gelatinosa* (Ach.) M. Brand & Diederich (syn. *Polyblastia gelatinosa*)
2. Perithecial wall black at summit (involucrellum present); spores broad, (17-)21-28 x 11-16 µm .... *Polyblastia sendtneri* Kremp.

3. Spores dark brown, broad, 50-70 x 26-40 µm; thallus thick, areolate .... *Polyblastia theleodes* (Sommerf.) Th. Fr
3. Spores colorless; spores mostly 25-45 x 14-23 µm; thallus thin, rimose-areolate, pale to dark gray 4

4. Perithecia 0.2-0.3 mm in diameter, superficial to half immersed in the thallus or rock, sometimes forming shallow pits .... *Polyblastia hyperborea* Th. Fr.
4. Perithecia 0.3-0.6 mm in diameter, prominent, not forming pits in rock .....*Polyblastia cupularis* A. Massal.

POLYCHIDIUM (including *Leptogidium*)

1. Branches dark brown, shiny, up to 200 µm (0.2 mm) in diameter, commonly producing dark red-brown apothecia up to 2 mm across; photobiont *Nostoc;* forming small cushions in moss over rocks; coastal and arctic-alpine in the west .... ***Polychidium muscicola* (Sw.) Gray**
1. Branches gray-olive to blue-green, 50-100 µm in diameter; apothecia not seen; photobiont *Scytonema*; forming wooly cushions on the undersides of conifer twigs and branches in very humid habitats, Oregon to Alaska 2

2. Longitudinal sections of branches reveal twisted cyanobacterial filaments and a loose medulla or irregular hyphae ....***Leptogidium contortum* (Henssen) T. Sprib. & Muggia** (syn. *Polychidium contortum*)

2. Sections of branches reveal straight cyanobacterial filaments and dense, straight hyphae in the medulla of the branches ....*Leptogidium dendriscum* (Nyl.) Nyl. (syn. *Polychidium dendriscum*)

PORINA (including *Pseudosagedia, Segestria, Strigula*) Based in part on Harris (1995), Orange et al. (2009b), and Clauzade and Roux (1985).

1. Growing directly on rock 2
1. Grown on bark or moss 7

2. Perithecia pale to dark brown; spores 4-celled 3
2. Perithecia black; spores 4- to 8-celled 5

3. Growing on maritime rocks exposed to salt spray; northwest coast; spores 17.5-22(-24) x (4.0-)4.5-6.5 µm .... *Porina pacifica* Brodo (see also couplet 6)
3. Growing on non-maritime rocks; eastern 4

4. Spores 15-20(-23) x (3.0-)3.5-5.0 µm; perithecia 0.13-0.3 mm in diameter ....*Segrestia leptalea* (Durieu & Mont.) R. C. Harris
4. Spores 20-30(-40) x (4.0-)5.0-6.0(-8.0) µm; perithecia 0.3-0.5 mm n diameter ....*Segrestia lectissima* Fr.

5.(2) Spores 8-celled, (28-)32-45(-50) x (4.5-)5.0-6.6 µm; perithecia 0.2-0.3 mm in diameter; eastern ....*Pseudosagedia guentheri* (Flotow) Hafellner & Kalb
5. Spores 4-celled, 13-32 x 3.5-6.5 µm; perithecia 0.15-0.35 mm in diameter 6

6. Outer wall of perithecium (involucrellum) K+ red-orange; conidia more than 5.0 µm long; on maritime rocks; northwest coast .... *Porina pacifica* Brodo (see also couplet 3)
6. Outer wall of perithecium K– or K+ purple; conidia under 4.0 µm long; on inland rocks; widespread .... *Pseudosagedia chlorotica* (Ach.) Hafellner & Kalb

7.(1) On moss or bark, especially tree bases; ascus tip clearly thickened, with a distinct ocular chamber; spores (3-)8-celled, 22-42 x 5.0-7.5 µm; northeast to Great Lakes region ...*Strigula stigmatella* (Ach.) R. C. Harris
7. On bark; ascus tip not or barely thickened, often lacking a distinct ocular chamber; spores 4- to many-celled 8

8. Thallus and sometimes perithecia with long, cylindrical isidia (sometimes sparse); spores 35-57 x 5.5-8 µm, 8(-9)-celled 9

8. Thallus lacking isidia 10

9. Perithecia 0.3-0.4 mm in diameter; thallus dark brown or olivaceous; central and eastern U.S. .... *Pseudosagedia isidiata* (R. C. Harris) R. C. Harris

9. Perithecia 0.5-0.7 mm in diameter; thallus pale brown; spores 8-celled; southeastern coastal plain .... *Porina scabrida* R. C. Harris

10. Spores 4-celled; spores 13-20 (-24) x 3.5-5.0 µm 11

10. Spores more than 4-celled 12

11. Perithecia black, less than 0.2 mm in diameter; thallus dark greenish; eastern and southwestern, on the bases of conifers and birch .... *Pseudosagedia aenea* (Wallr.) Hafeller & Kalb

11. Perithecia pale brown to pink; eastern and north central ..... *Segestria leptalea* (Durieu & Mont) R. C. Harris

12. Spores mostly 8-celled [but see note under *P. cestrensis*], 35-60 µm long 13

12. Spores 12- to 22-celled, 60-140 µm long 14

13. Perithecia pale to dark brown, covered with thalline tissue, 0.3-0.5 mm in diameter; thallus pale brown; spores 11-14 µm wide; southeastern coastal plain .... *Porina nucula* Ach.

13. Perithecia black, naked, less than 0.3 mm in diameter; thallus dark brown, membranous; spores under 7.5 µm wide; widespread, East Temperate, on hardwoods or bald cypress .... *Pseudosagedia cestrensis* (Tuck. *ex* Michener) R. C. Harris

[Note: Spores extremely variable, often more than 8-celled, some ± intermediate with *P. raphidosperma* (R. C. Harris, pers. comm.)]

14. Spores 9-15 µm wide, strongly tapered at one end; perithecia pale greenish gray, covered with thalline tissue, more than 0.3 mm in diameter; inner wall pale brown to yellowish brown; southeastern .... ***Porina heterospora* (Fink *ex* Hedrick) R. C. Harris**

14. Spores 2.5-5 µm wide, needle-shaped, equally tapered at both ends; perithecia naked, 0.2-0.3(-0.4) mm in diameter; inner wall dark; eastern .... *Pseudosagedia rhaphidosperma* (Müll. Arg.) R. C. Harris

PORPIDIA (including *Clauzadea, Farnoldia, Immersaria,* and *Pachyphysis*) See Gowan (1989) and Fryday (2005).

1. Thallus orange to yellow- or reddish-brown 2
1. Thallus gray or indistinct 6

2. Thallus sorediate; medulla IKI– 3
2. Thallus without soredia 4

3. Thallus dark rusty orange, rimose-areolate to areolate, producing dark gray coarse soredia in discrete, usually excavate soralia; containing confluentic acid .... *Porpidia melinodes* (Körber) Gowan & Ahti
3. Thallus pale orange often with gray patches, smooth to rimose or rimose-areolate, producing pale gray to creamy soralia either in mounds or depressed; containing stictic acid .... *Porpidia ochrolemma* (Vainio) Brodo & R. Sant.

4. Medulla IKI+ blue; thallus yellow-brown, areolate, areoles often having whitish margins; apothecia immersed (*Aspicilia*-like); spores 11-18(-26) x (6-)7-9(-14) µm; western Great Lakes and Arizona .... *Immersaria athroocarpa* (Ach.) Rambold & Pietschm.
4. Medulla IKI–; thallus orange, smooth to rimose-areolate; apothecia usually sessile 5

5. Thallus uniformly orange; containing confluentic acid; arctic-alpine .... ***Porpidia flavicunda (Ach.) Gowan*** *[called "Porpidia flavocaerulescens" in LNA; Plate 707]*
5. Thallus orange in patches, otherwise gray; containing stictic acid complex or no substances; widespread ..... .*Porpidia macrocarpa* (DC.) Hertel & Knopf ("oxydated" form)][Compare with *Lecidea lapicida;* see couplet 20 of *Lecidea* key.]

6.(1) On limestone; arctic-alpine, rather rare; epihymenium brown or green 7
6. On non-calcareous rock; temperate to boreal, common; epihymenium brown to olive-brown 11
[Note: *Farnoldia jurana* (couplet 10) is sometimes found on non-calcareous rocks.]

7. Thallus white and chalky, rather thick; medulla IKI+ blue-black; spores 11-15(-17) x 6-7 µm; containing confluentic acid .... *Porpidia speirea* (Ach.) Krempelh.
[Compare with *Porpidia grisea* Gowan: spores mostly 15-20 x 6-8 µm; on siliceous rock.]
7. Thallus gray to brownish, thin or indistinct; medulla IKI– or faint blue; lacking confluentic acid 8

8. Hymenium purplish; spores subglobular, 10-13 x 8-10(-12) μm, not halonate; exciple colorless externally, dark purplish black within; some paraphyses thick, moniliform, with a gelatinous sheath; central U.S., especially Ozarks .... *Pachyphysis ozarkana* R. C. Harris & Ladd

8. Hymenium not purplish; spores ellipsoid; exciple brown-black throughout; paraphyses not moniliform or with a gelatinous sheath; boreal to arctic-alpine 9

9. Apothecia red-brown when wet, very small (under 0.3 mm in diameter), flat, not pruinose; margins sometimes prominent; hypothecium and exciple dark brown; epihymenium brown; spores 6.5-12(-14) x 3.5-7 μm .... *Clauzadea monticola* (Ach. *ex* Schaerer) Hafellner & Bellem.

9. Apothecia black when wet, 0.4-3 mm in diameter, flat to convex, prominent margins; spores over 6.5 μm wide; epihymenium green to blue-green, at least in part; hypothecium and exciple black to brown-black, confluent 10

10. Exciple dark red-brown to almost black, becoming green toward the margin; apothecia flat when young, losing the margin and becoming convex, bumpy and pitted when mature, lacking pruina; spores 10.5-14 (-16) x (5.5-)6.5-8.5 μm .... *Farnoldia hypocrita* (A. Massal.) Fröberg

10. Exciple black or purple-black throughout; apothecia remaining flat, often pruinose, with a prominent margin; spores (13-)16-27 x 8-12 μm .... *Farnoldia jurana* (Schaerer) Hertel

[Note: *Farnoldia jurana* is not a synonym of *Melanolecia transitoria*, as is stated in LNA, p. 583.]

11.(6) Thallus sorediate in round, gray to creamy soralia 12

11. Thallus not sorediate 15

12. Medulla IKI+ blue, K–, PD– (containing confluentic acid) .... *Porpidia tuberculosa* (Sm.) Hertel & Knopf

12. Medulla IKI–, K+ yellow, PD+ orange (stictic acid) 13

13. Thallus rimose-areolate to areolate, usually fertile, with pruinose apothecia; frequent, Appalachian, typically on metal-rich rocks .... *Porpidia degelii* (H. Magn.) Lendemer & R. C. Harris, *ined.*

[Note: This is the lichen formerly considered to be a sorediate morphotype of *P. albocaerulescens*.]

13. Thallus very rarely fertile, apothecia not pruinose; rare 14

14. Thallus very thin, areolate to dispersed areolate .... *Porpidia soredizodes* (Lamy *ex* Nyl.) Laundon

14. Thallus well developed, smooth to rimose-areolate, often with orange patches; on metal-rich rocks .... pale morphotype of *Porpidia ochrolemma* (Vainio) Brodo & R. Sant.

15.(11) Apothecial disks gray due to a light or heavy pruina 16

15. Apothecial disks black, not pruinose; apothecial margin the same color as the disk; apothecia sessile, not immersed in thallus 19

16. Medulla IKI+ blue; apothecia sessile; west coast .... *Porpidia grisea* Gowan

16. Medulla IKI–; apothecia slightly to distinctly immersed in thallus 17

17. Exciple brown to brown-black within, darker at edge; apothecia with or without contrasting margins; containing confluentic acid; rather rare, boreal .... *Porpidia cinereoatra* (Ach.) Hertel & Knoph

[Note: Close to *P. contraponenda* (Arnold) Knoph & Hertel (syn. *P. diversa*), which has sessile apothecia with raised margins and contains 2'-*O*-methylmicrophyllinate.]

17. Exciple black and carbonaceous only at the edge, otherwise white; apothecial margins black, contrasting with the disk 18

18. East Temperate; thallus PD+ orange or yellow , K+ yellow or red (stictic or norstictic acid) ....***Porpidia albocaerulescens* (Wulfen) Hertel & Knoph**

18. Oregon to Alaska; thallus PD–, K– (containing 2'-*O*-methylsuperphyllinic acid) ....*Porpidia carlottiana* Gowan

19.(15) Apothecial margin brittle and radially cracked; thallus endolithic; spores 12-18 x 6-8 µm; Appalachian-Great Lakes-Ozarks distribution ....*Porpidia subsimplex* (H. Magn.) Fryday (syn. *Porpidia tahawasiana*)

19. Apothecial margin not brittle or radially cracked 20

20. Apothecia commonly 1-2.5 mm in diameter; spores mostly 13-23 x 7-10 µm; cells in the exciple mostly 3-6 µm in diameter; hymenium 80-120 µm high .... *Porpidia macrocarpa* (DC.) Hertel & Knoph

20. Apothecia usually less than 1.2 mm in diameter; spores 10-17 x 5-9 µm; cells in the exciple about (4.2-)5-8 µm in diameter (seen only in thin sections) 21

21. Apothecia 0.5-1.5 mm in diameter, disk black or, rarely, dark red-brown; exciple distinct from hypothecium, brown to red-brown within, black at edge; hymenium 60-75(-100) µm high; on dry, exposed rocks or pebbles ....***Porpidia crustulata* (Ach.) Hertel & Knoph**

[Note: *Porpidia thomsonii* Gowan is an arctic-alpine species with apothecia 0.7-2.0 mm in diameter and broad, melanized excipular hyphae, 5-8 µm wide. Its relationship with *P. macrocarpa* and *P. crustulata* is discussed by Gowan (1989), Gowan & Ahti (1993), and Fryday (2005).]

21. Apothecia 0.3-0.7(-0.9) mm in diameter, disk red-brown or less frequently, black; exciple merging into hypothecium, becoming paler at outer edge; hymenium 75-115 um high; Pacific Northwest and East Temperate; in shaded or aquatic habitats .... *Bryobilimbia ahlesii* (Körber) Fryday et al. (syn. *Lecidea ahlesii*)

PROTOBLASTENIA

1. Growing on soil, apothecia large, hemispherical up to 1.5 m in diameter; arctic tundra .... *Protoblastenia terricola* (Anzi) Lynge
1. On limestone 2

2. Thallus well developed, white to pale gray, rimose areolate; apothecia 0.5-1.2 mm in diameter; widespread and common .... ***Protoblastenia rupestris* (Scop.) J. Steiner**
2. Thallus very thin and continuous to endolithic 3

3. Apothecia hemispherical, 0.5-1.2 mm in diameter, sessile; infrequent, arctic .... *Protoblastenia calva* (Dickson) Zahlbr.
3. Apothecia flat to convex, 0.2-0.5 mm in diameter, often forming pits in the rock; common, temperate to arctic .... *Protoblastenia incrustans* (DC.) J. Steiner

PROTOPARMELIA Based on Brodo & Aptroot (2005), and Ryan, Nash, & Hafellner (2004).

1. Growing on bark or wood 2
1. Growing directly on rock 3

2. Granules or microsquamules round to hemispherical, (0.15-)0.2-0.5 mm in diameter, pale to, more frequently, dark reddish brown and shiny; convex apothecia frequently seen, apothecial margins thin and disappearing; lobaric acid absent; not parasitized by *Sphinctrina*; northwestern .... *Protoparmelia ochrococca* (Nyl.) P. M. Jørg. et al.
2. Granules or microsquamules round, flat, 0.08-0.2(-0.3) mm in diameter, pale grayish brown to olive-gray, dull; apothecia rare, with prominent apothecial margins; lobaric acid present; frequently parasitized by *Sphinctrina*; eastern .... *Protoparmelia hypotremella* Herk et al.

3. Spores with pointed tips, 8-13(-16) x 3-5(-8) µm; medulla usually K–, UV+ white, KC+ pink (lobaric acid); prothallus gray to whitish, rimose, thin and inconspicuous and often absent, black; common and widespread boreal to arctic .... ***Protoparmelia badia* (Hoffm.) Hafellner**

3. Spores with rounded tips; cortex usually K+ yellow or red; prothallus black to blue-black, usually conspicuous; uncommon 4

4. Prothallus producing tiny green-black granules consisting of 2-6 cells (thallospores); thallus sometimes sorediate, margin sometimes minutely lobed; apothecia uncommon, poorly developed; rare, montane; cortex K+ yellow or K– (stictic or lobaric acid)
.... *Protoparmelia nephaea* (Sommerf.) R. Sant. *ex* Poelt & Obermayer

4. Thallus lacking thallospores on the prothallus; thallus without soredia; apothecia usually abundant; cortex K+ red, PD+ yellow (norstictic acid) 5

5. Thallus lobed at the margins with marginal areoles larger than those in the center; western montane, rare in east ....*Protoparmelia cupreobadia* (Nyl.) Poelt in Poelt & Leukert

5. Thallus not distinctly lobed at margin, dispersed areolate, central areoles about the same size as those at the margin; southwestern mountains; reported from Great Lakes region
.... *Protoparmelia atriseda* (Fr.) R. Sant. & V. Wirth in V. Wirth

PROTOTHELENELLA Partially based on Orange et al. (2009c).

1. On rock in moist habitats; spores broad, length/width ratio about 2:1, 18-30 x 10-15 µm; thallus well developed, rimose-areolate, or barely perceptible; perithecia 0.2-0.5 mm in diameter, mostly buried in thallus; Ontario to Newfoundland, south to Pensylvania; rare in western arctic
.... *Protothelenella corrosa* (Körber) H. Mayerhofer & Poelt

1. On mosses, soil, and peat, arctic-alpine; spores narrow, length/width ratio about 3:1 2

2. Spores submuriform, 22-33 x 7-20 µm; perithecia 0.1-0.3 mm in diameter, prominent; Arizona, Colorado, Alberta, New York, Washington ....*Protothelenella sphinctrinoidella* (Nyl.) H. Mayerhofer & Poelt

2. Spores muriform, 38-50 x 10-15 µm; perithecia 0.2-0.5 mm in diameter, partly buried in thallus; B.C., Nunavut (E Pen Isl.) .... *Protothelenella sphinctrinoides* (Nyl.) H. Mayerhofer & Poelt

PSEUDEPHEBE

1. Branches very slender and entirely terete, 0.1-0.2 mm in diameter, not flattening at the tips; distance between axils mostly 1-3 mm ....***Pseudephebe pubescens* (L.) M. Choisy**

1. Branches 0.2-0.5 mm in diameter, usually becoming somewhat flattened at the tips; distance between axils 0.2-0.5(-1) mm .... ***Pseudephebe minuscula* (Nyl. *ex* Arnold) Brodo & D. Hawksw.**

PSEUDEVERNIA

1. Thallus abundantly isidiate on the upper surface; Appalachian-Great Lakes region .... ***Pseudevernia consocians*** **(Vainio) Hale & W. L. Culb.**
1. Thallus not isidiate 2

2. Lobes mostly under 1 mm wide; lower surface white except at the base; in the Appalachian mountains .... ***Pseudevernia cladonia*** **(Tuck.) Hale & W. L. Culb.**
2. Lobes 1-3 mm wide; lower surface dark gray to black except at the lobe tips; southwestern .... ***Pseudevernia intensa*** **(Nyl.) Hale & W. L. Culb.**

PSEUDOCYPHELLARIA

[Note: Based on new genetic research by Moncada et al. (2013), but this treatment assumes the eventual conservation of *Pseudocyphellaria* with a new type from the *P. crocata* group.]

1. Pseudocyphellae yellow; soralia yellow or gray (calycin present) 2
1. Pseudocyphellae and soralia (if present) white (calycin absent) 5

2. Photobiont green; medulla dark yellow throughout; lobes smooth and even; medulla PD–, K–; southeastern .... ***Crocodia aurata*** **(Ach.) Link** (syn. *Pseudocyphellaria aurata*)
2. Photobiont blue green; medulla white or yellow; lobes with a network of depressions and sharp ridges; medulla PD–, K– or PD+ orange, K+ yellow 3

3. Medulla yellow throughout; soralia entirely marginal in western populations, but laminal and marginal in the east .... *Pseudocyphellaria perpetua* McCune & Miadlikowska s. lat
   [Note: Moncada et al. (2014) place *P. perpetua* into synonymy with the older *P. hawaiiensis* H. Magn. but indicate that more than one species may be involved. More work is needed here.]
3. Medulla mostly white, except yellow near soralia and pseudocyphellae; soralia laminal and marginal 4

4. Soralia mostly gray; fine tomentum on upper lobe surface especially close to margins; Oregon to Alaska; rare .... *Pseudocyphellaria mallota* (Tuck.) H. Magn.
   [Note: This is classified in the genus *Parmostictina* by Moncada et al. 2013.]
4. Soralia yellow; lacking tomentum on upper surface of lobes; Appalachian-Great Lakes region and west coast .... ***Pseudocyphellaria crocata*** **(L.) Vainio**
   [Note: Moncada et al. (2014) suggest that North American material named as *P. crocata* is not the same species as material from SE Asia, which is the type locality.]

5.(1) Thallus greenish gray or pale brown, dull or scabrose; lobes smooth and even; algal layer grass-green; isidia and lobules along the lobe margins; medulla PD–, K–..... ***"Pseudocyphellaria" rainierensis* Imshaug**

5. Thallus dark reddish brown, rather shiny; lobes with a network of depressions and sharp ridges; algal layer dark blue green; isidia and lobules absent; medulla PD+ orange, K+ yellow (stictic acid) 6

6. Soredia absent; apothecia abundant
.... ***Anomalobaria anthraspis* (Ach.) B. Moncada & Lücking** (syn. *Pseudocyphellaria anthraspis*)

6. White to gray soredia present, mainly on ridges; apothecia rare..
.***Anomalobaria anomala* (Brodo & Ahti) B. Moncada & Lücking** (syn. *Pseudocyphellaria anomala*) (See also *P. mallota*, couplet 4)

PSORA (including *Miriquidica* [p.p.] and *Psorula*)

1. Apothecia mainly at the margins of the squamules; squamules pruinose, at least in part 2

1. Apothecia mainly on the surface of the squamules; squamules pruinose or not 4

2. Squamules strongly convex and fissured, yellow-brown to olive-brown
.... ***Psora cerebriformis* W. A. Weber**

2. Squamules more or less flat, smooth or slightly fissured, reddish or orange-brown 3

3. Squamules brick red to orange-brown, flat or slightly convex, with edges turned up
.... ***Psora decipiens* (Hedwig) Hoffm.**

3. Squamules pink to pinkish orange, depressed in the center, with edges turned down
.... ***Psora crenata* (Taylor) Reinke**

4.(1) Thallus bright yellow or yellow-green to yellowish brown, with yellow pigments in the cortex 5

4. Thallus brown, lacking yellow pigments in the cortex 6

5. Thallus bright yellow to yellow-green (rhizocarpic acid); squamules lying flat on the soil, with yellowish margins; medulla C–; mainly southwestern to midwestern arid areas .... ***Psora icterica* (Mont.) Müll. Arg.**

5. Thallus yellowish brown to olive (usnic acid); squamules ascending and on edge, with white pruinose margins; medulla frequently C+ pink (gyrophoric acid); arctic and western alpine, rare in eastern mountains
.... *Psora rubiformis* (Ach.) Hook.

6. Squamules 0.4-1.5(-3) mm wide 7

6. Squamules 2 mm wide or more 12

7. Parasitic on the cyanobacterial lichen, *Spilonema*; lower surface and margins of squamules blackish or blue-green .... ***Psorula rufonigra*** **(Tuck.) Gott. Schneider**

7. Not parasitic; lower surface and margins of squamules white or brown, not blackish green (except at edges in *Miriquidica scotopholis*) 8

8. Squamules not lobulate medulla C– 9

8. Squamule margins divided into tiny round lobules, sometimes lightly pruinose on the surface, but not along the margins; medulla C+ pink (gyrophoric acid) or C–... 10

9. Apothecia pale to dark red-brown, hemispherical; edges of squamules conspicuously white pruinose; containing no substances; Rocky Mountains to arctic .... ***Psora himalayana*** **(Church. Bab.) Timdal**

9. Apothecia dark brown to black, flat to convex; lecanorine when very young, lecideine when mature; edges of squamules not at all pruinose; containing miriquidic acid; southern California .... *Miriquidica scotopholis* (Tuck.) B. D. Ryan & Timdal

10. Medulla C– .... *Psora luridella* Tuck. (see also couplet 15)

10. Medulla C+ pink.. 11

11. Squamules loosely attached, often ascending; California .... ***Psora pacifica*** **Timdal**

11. Squamules closely attached and flat against the substrate; mostly high elevations in the Rocky Mountains .... *Psora montana* Timdal

12.(6) Squamules pruinose or whitened on the edges and sometimes the surface 13

12. Squamules almost entirely without pruina 16

13. Medulla K+ red (norstictic acid); squamules usually concave; southwestern U.S. east to the Ozarks, on soil .... *Psora russellii* (Tuck.) A. Schneider

13. Medulla K– 14

14. Found in central and eastern temperate North America on rock; apothecia rusty brown typically with persistent margins .... ***Psora pseudorussellii*** **Timdal**

14. Found in western or northern North America on rock or soil; apothecia red-brown to dark brown or almost black, soon becoming convex and marginless 15

15. Squamules pale yellowish brown to pale reddish brown, 2-5 mm wide, scattered or contiguous and crowded, occasionally overlapping, lightly pruinose on the surface and margins .... ***Psora tuckermanii*** **R. A. Anderson *ex* Timdal**

[Note: Compare with *P. rubiformis*, couplet 5.] *Psora luridella* can also key out here. It has a thinner upper cortex (40-70 vs. 60-120 µm thick) and lacks oxalate crystals in the medulla.]

15. Squamules dark reddish brown, 1-3 mm wide, overlapping like shingles, only the edges conspicuously white pruinose .... ***Psora himalayana*** **(Church. Bab.) Timdal**

16.(12) Medulla of thallus C+ pink or red 17

16. Medulla of thallus C– 19

17. Squamules pale gray, thick and somewhat convex .... ***Trapeliopsis glaucopholis*** **(Nyl. *ex* Hasse) Printzen & McCune** (syn. *T. californica*)

[Note: = most N. Am. specimens named as *T. wallrothii*; treated as *T. wallrothii* in LNA]

17. Squamules greenish brown or red-brown, thin or thick 18

18. Squamules olive to greenish brown, often white on the margin, thin, flat or curled inward and cup-like when dry, ascending or standing on edge, (2-)3-6(-10) mm wide .... ***Psora nipponica*** **(Zahlbr.) Gotth. Schneider**

18. Squamules (including the margin) red-brown, folded or flat, often shingled, adnate or ascending, 2-4(-6) mm wide; apothecia black or very dark brown; on soil or rock in California .... *Psora californica* Timdal

19.(16) Squamules uniformly reddish brown, shiny, smooth or fissured; apothecia very dark brown to black .... ***Psora globifera*** **(Ach.) A. Massal**

19. Squamules pale to medium brown usually with whitish margins, dull, somewhat fissured; apothecia typically red-brown 20

20. Western, on rock or soil; apothecia soon convex and marginless .... ***Psora tuckermanii*** **R. A. Anderson *ex* Timdal**

20. Central and eastern regions, on rock; apothecia mostly flat, with long persistent margins, finally convex and marginless .... ***Psora pseudorussellii*** **Timdal**

PUNCTELIA Based on Lendemer & Hodkinson (2010).

1. Lower surface and rhizines mostly black 2
1. Lower surface and rhizines pale to dark brown 6

2. Growing on rock; lobules absent; medulla KC+ red, C+ pink (gyrophoric acid) .... ***Punctelia stictica* (Duby) Krog**
2. Growing on bark or, rarely, rock 3

3. Lobes covered with lobules 4
3. Lobes with soredia 5

4. Medulla KC–, C– (fatty acids); Appalachians .... ***Punctelia appalachensis* (W. L. Culb.) Krog**
4. Medulla KC+ pink, C+ pink (gyrophoric acid); Texas .... *Punctelia subpraesignis* (Nyl.) Krog

5. Medulla KC+ pink, C+ pink (gyrophoric acid); Appalachians and west coast .... *Punctelia borreri* (Sm.) Krog
5. Medulla KC–, C– (fatty acids); rare, southeast and California .... *Punctelia reddenda* (Stirton) Krog

6. Soredia present; medulla C+ red (lecanoric acid) 7
6. Soredia absent 11

7. Saxicolous; soredia developing from breakdown of pustules; California .... *Punctelia punctilla* (Hale) Krog
7. Corticolous 8

8. Soredia coarsely granular, corticate to almost lobulate, on the upper surface of the lobes associated with pseudocyphellae and cracks; common in Ozarks .... *Punctelia missouriensis* G. Wilh. & Ladd
8. Soredia powdery, along the lobe margins and on the surface, lobules absent 9

9. Upper surface with ridges and depressions; soralia mostly marginal; lobes narrow (less than 2 mm) with pruinose tips; mainly Ozarks .... *Punctelia perreticulata* (Räsänen) G. Wilh. & Ladd
9. Upper surface smooth to slightly ridged; lobes pruinose or not 10

10. Lobe tips pruinose; west coast and southwest .... ***Punctelia jeckeri* (Roum.) Kalb** [Plate 739 in LNA, as "*Punctelia subrudecta*"]

10. Lobe tips without pruina; common in central and eastern North America ....*Punctelia caseana* Lendemer & Hodkinson (= most eastern N. Am. records of "*P. subrudecta*")

11. Isidia present, cylindrical, branched or unbranched; lobules absent; pycnidia and apothecia infrequent; medulla C+ red (lecanoric acid) .... ***Punctelia rudecta* (Ach.) Krog**

11. Isidia absent; lobules common; pycnidia and apothecia commonly seen; medulla C+ red or C– 12

12. Medulla C–, KC–, lichesterinic and protolichesterinic acids present; conidia rod-shaped; central and eastern U.S. and adjacent Canada .... ***Punctelia bolliana* (Müll. Arg.) Krog**

12. Medulla C+ red, medulla KC+ red (lecanoric acid); lichesterinic and protolichesterinic acids absent 13

13. Conidia long, thread-shaped; southwestern U.S. .... ***Punctelia hypoleucites* (Nyl.) Krog**

13. Conidia relatively short, bacilliform; central and eastern U.S. ....*Punctelia graminicola* (B. de Lesd.) Egan (syn. *P. semansiana*)

PYRENOPSIS (including *Cryptothele, Euopsis,* and *Phylliscum*)

[Note: The genus *Pyrenopsis* is in need of North American revision, and this key should be regarded as tentative at best. See also Schultz (2006).]

1. Thallus squamulose, clearly lifting from substrate and attached centrally (umbilicate); photobiont *Chroococcus*, cells 10-40 µm in diameter; spores 8-10 x 4-5 µm, 8-16 per ascus; paraphyses disappearing ....*Phylliscum demangeonii* (Moug. & Mont.) Nyl.

1. Thallus crustose, not lifting from substrate; photobiont *Gloeocapsa*, cells usually less than 10 µm in diameter 2

2. Paraphyses absent, replaced with well-developed periphyses; thallus thin, rimose-areolate; perithecia-like apothecia buried in thallus, opening by an ostiole; spores 7.5-10.7 x 4.5-7.5 µm; on moist rocks, Ontario, rare ....*Cryptothele granuliforme* (Nyl.) Henssen

2. Paraphyses persistent, branched or unbranched; possibly widespread 3

3. Disk broad, red-brown from beginning; green algae in margin (as well as the *Gloeocapsa* in the thallus) ....*Euopsis granatina* (Sommerf.) Nyl.

3. Disk poriform, sometimes broadening somewhat at it gets older; lacking any green algae 4

4. Spores (10-)11-13 x 7.5-8(-10) µm; northeastern .... ***Pyrenopsis polycocca* (Nyl.) Tuck.**

[Note: This should be compared with *P. haematina* P. M. Jørg. & Henss., described from Europe.]

4. Spores 12-20 x 7-10 µm (14-25 x 8-12 µm in Tuckerman's original description); eastern? .... *Pyrenopsis phaeococca* Tuck.

PYRENULA (including *Anthracothecium*) Based on Harris (1989, 1995). See also Aptroot (2012).

[Note: All species from the southeastern coastal plain, except as noted.]

1. Spores muriform 2
1. Spores mostly transversely septate (occasionally with a few cells divided longitudinally) 7

2. Spores 2-6 per ascus, over 100 µm long; perithecia with ostioles that are off-center (not at the summits) ...3
2. Spores 8 per ascus; ostioles at summit or off-center 4

3. Spores 2 per ascus, 130-180 x 35-50 µm, with squarish locules; perithecia solitary .... *Anthracothecium nanum* (Zahlbr.) R. C. Harris
3. Spores (2-)4(-6) per ascus, 100-150(-190) x 30-41 µm, with globular locules; perithecia solitary or in clusters of 2-3 joined at the ostioles .... *Pyrenula schiffneri* (Zahlbr.) Aptroot (syn. *Pyrenula falsaria*)

4. Thallus orange or yellow, K+ purple (anthraquinones); spores 23-35 x 11-17 µm .... ***Pyrenula ochraceoflavens* (Nyl.) R. C. Harris**
4. Thallus not orange or yellow; K– 5

5. Perithecia fused in groups by their off-center or lateral ostioles; spores 45-65 x 18-27 µm .... *Pyrenula ravenelii* (Tuck.) R. C. Harris
5. Perithecia solitary, ostioles at summit 6

6. Spores 45-60 x 16-22(-25) µm .... *Pyrenula leucostoma* Ach.
6. Spores 30-40(-53) x 11-15 µm .... *Pyrenula thelomorpha* Tuck.

7.(1) Spores 27-38 x 12-18 µm 8
7. Spores less than 25 µm long 10

8. Thallus and perithecial warts dark red (the pigment sometimes confined to the perithecial warts) .... ***Pyrenula cruenta* (Mont.) Vainio**
8. Thallus and perithecial warts greenish when fresh, whitish to brownish in herbarium 9

9. Overmature old spores containing a red oily substance; spores 4- to 6-celled .... *Pyrenula concatervans* (Nyl.) R. C. Harris

9. Overmature old spores simply empty, collapsing; spores 4-celled .... *Pyrenula punctella* (Nyl.) Trevisan

10.(7) Spore walls at the tips of the spores thin 11

10. Spore walls at the tips of the spores obviously thick 12

11. East Temperate; no orange pigment on summits of the perithecia; thallus UV+ yellow (lichexanthone) .... *Pyrenula pseudobufonia* (Rehm) R. C. Harris

11. Pacific Northwest near coast; scant or abundant orange pigment on summits of the perithecia; thallus UV– or dull whitish .... *Pyrenula occidentalis* (R. C. Harris) R. C. Harris

12. Perithecia fused into an extensive raised pseudostroma; spores 17-21 x 7-9 µm .... *Pyrenula anomala* (Ach.) Ach.

12. Perithecia solitary, not in a pseudostroma 13

13. Thallus scurfy, whitish, without a cortex; spores 19-25 x 8-12 µm; perithecia 0.3-0.4 mm in diameter; hemispherical, relatively thin walled .... *Pyrenula microcarpa* Müll Arg. (syn. *P. cinerea*)

13. Thallus smooth, with a cortex; spores 15-21 x 5.5-8; perithecia 1-1.5 mm in diameter, flattened-conical, thick walled .... *Pyrenula mamillana* (Ach.) Trevisan (syn. *P. marginata*)

PYRRHOSPORA (including *Ramboldia*) Based on revisions found in Kantvilas & Elix (1994, 2007); Kalb et al. (2008).

1. Thallus sorediate, often sterile 2

1. Thallus without soredia, fertile 3

2. Soredia white, produced in discrete round soralia; thallus C–, PD+ red (fumarprotocetraric acid); apothecia red-orange to blood red, with a thin margin visible in all but the oldest apothecia; spores narrow, 2.0-5.0 µm; boreal-montane

......*Ramboldia cinnabarina* (Sommerf.) Kalb et al. (syn. *Pyrrhospora cinnabarina*) [*not* Plate 745 in LNA; see below]

2. Soredia dull yellowish, diffuse, with thallus commonly becoming leprose; thallus C+ orange, PD– (xanthones); apothecia very dark brown to black, without margins when mature; spores broad, 6.0-7.0 µm; coastal, humid regions .... ***Pyrrhospora quernea* (Dickson) Körber**

3. Apothecia red-orange to blood red, K+ red-purple (russulone)....................4
3. Apothecia reddish brown or black, not red; K– 6

4. Thallus UV–; western boreal-montane; thallus PD+ red (fumarprotocetraric acid)....................
Ramboldia gowardiana **(T. Sprib. & Hauck) Kalb et al.** (syn. *Pyrrhospora gowardiana*) [Plate 745 in LNA as *Pyrrhospora cinnabarina*]
4. Thallus UV+ yellow (lichexanthone); southeastern, lowland 5

5. Thallus PD+ red, K– (fumarprotocetraric acid); common
.... ***Ramboldia russula*** **(Ach.) Kalb et al.** (syn. *Pyrrhospora russula*)
5. Thallus PD+ yellow, K+ red (norstictic acid); infrequent .... *Ramboldia haematites* (Fée) Kalb et al.

6.(3) Apothecia shiny black; thallus gray, well developed, often shiny, PD+ red (fumarprotocetraric acid); spores narrow, 2.8-4.5 µm .... *Ramboldia elabens* (Fr.) Kantvilias & Elix (syn. *Pyrrhospora elabens*)
6. Apothecia brown, pale to very dark, never shiny black; thallus yellowish to yellow-gray, PD–, C+ orange or C–; spores broad, (4.0-)5.2-6.8 µm; deciduous trees and shrubs; eastern North America and California
.... *"Pyrrhospora" varians* (Ach.) R. C. Harris
[Note: The generic placement of this species is still uncertain. There may be two entities within the broad "*P. varians*," one with a minutely verruculose to granulose, C– thallus; and one with a smooth to rimose, C+ orange thallus (including the type). They appear to be sympatric. More chemical and anatomical studies are needed on this species.]

RAMALINA (including *Niebla*)

1. Thallus pendent, 5-30 cm (2-20 inches) long, often with clustered or indistinct attachment points 2
1. Thallus growing in bushy or almost pendent tufts from a single point, generally less than 6 cm long 7

2. Branches round in cross section, very fine (mostly under 0.5 mm wide), the tips curled up often producing a few granules or soredia .... ***Ramalina thrausta*** **(Ach.) Nyl.**
2. Branches or stalks distinctly flattened, over 0.5 mm wide 3

3. Branches with expanded net-like tips .... ***Ramalina menziesii*** **Taylor**
3. Branches not producing net-like tips 4

4. Soredia abundant, in marginal and often laminal soralia .... ***Ramalina subleptocarpha*** **Rundel & Bowler**
4. Soredia absent 5

5. Branches 30-50 cm long, 0.2-1.3(-2) mm broad, twisted, with long white pseudocyphellae giving the branches a striped appearance; subtropical, Florida, Texas, and California ..... *Ramalina usnea* (L.) R. Howe

5. Branches 3-20 cm long, 2-6 mm broad, more or less flat; western 6

6. Branch surface with patches of fine fuzzy tomentum; thalli usually 3-5 cm long; infrequent, California .... *Ramalina puberulenta* Riefner & Bowler

6. Branches lacking hairs or tomentum; thalli (3-)5-10(-20) cm; very common, mainly California, north to B.C. .... ***Ramalina leptocarpha* Tuck.**

7.(1) Soredia present; apothecia rare 8

7. Soredia absent; apothecia often common 18

8. Soredia white to blue-gray; soralia in rounded mounds, becoming wooly in old specimens; branches mostly round in cross section, often spotted with black bands or dots; on coastal shrubs, trees, and rarely rocks along the west coast .... ***Niebla cephalota* (Tuck.) Rundel & Bowler**

8. Soredia white to yellowish green; soralia of various shapes, never becoming wooly; branches usually flattened, sometimes round in cross section, never spotted with black bands or dots 9

9. Branches at least partly hollow; medulla loose and webby 10

9. Branches solid throughout; medulla dense and compact 12

10. Branches with round to oval perforations, smooth; cortex thin, translucent; blastidiate soredia at branch tips; containing sekikaic acid complex; widespread in northern coastal or oceanic inland habitats .... ***Ramalina roesleri* (Hochst. *ex* Schaerer) Hue** (syn. *Fistulariella roesleri*)

10. Branches not perforate or lacerate, but with depressions or ridges; cortex thick, opaque; soredia powdery to granular, originating within the hollow branches and bursting out 11

11. Soredia forming in hood-like expansions of the lobe tips; containing evernic and obtusatic acids; boreal region, Appalachians and southwest .... ***Ramalina obtusata* (Arnold) Bitter**

11. Hoods not formed at lobe tips; soredia emerging from split lobe margins and openings in lobe surface; containing divaricatic acid; on coastal trees and sometimes rocks, southern to mid-California .... *Ramalina canariensis* J. Steiner

12.(9) Soralia very irregular in shape, at or close to the lobe tips, which often appear torn, frayed, or hood-shaped; evernic and obtusatic acids present (soralia PD–) 13

12. Soralia round, elliptical, or elongate; branch tips tapering, sometimes finely divided but not generally torn, frayed, or expanded; evernic and obtusatic acids absent (soralia PD+ or PD–) 14

13. Branches narrow or broad, 0.5-3 mm wide; soredia mainly on the lower surface of the branch tips, not developing within hood-shaped expansions, sometimes marginal in part; widespread .... ***Ramalina pollinaria* (Westr.) Ach.**

13. Branches broad, 1.5-4 mm wide; soredia developing in inflated, hood-shaped expansions of the branch tips; boreal region, Appalachians and southwest .... ***Ramalina obtusata* (Arnold) Bitter**

14. Soralia often producing tiny, isidia-like branchlets; branches angular or round in section, or flattened in part; subtropical to tropical regions; medulla PD–, containing sekikaic acid .... ***Ramalina peruviana* Ach.**

14. Soralia lacking isidia-like branchlets; branches mostly flattened; not southern 15

15. Soredia granular, usually concentrated at the branch tips; mainly on rocks in forest habitats 16

15. Soredia farinose, marginal or laminal, not concentrated at the branch tips; mostly on trees; sekikaic acid absent 17

16. Medulla PD–, KC– (sekikaic acid); common; Appalachian-boreal and southwest .... ***Ramalina intermedia* (Delise *ex* Nyl.) Nyl.**

16. Medulla PD+ red-orange, KC+ pink (protocetraric acid); Appalachian-Ozark .... *Ramalina petrina* Bowler & Rundel

17. Soralia mainly linear, along the branch margins, but also elliptical on the branch surface; branches (1-)2-4(-10) mm broad; soralia always PD–, K–; zeorin present; coastal California to Washington .... ***Ramalina subleptocarpha* Rundel & Bowler**

17. Soralia mainly elliptical, on the branch margins, and occasionally on the surface, rarely at the tips; branches mostly 0.5-3 mm broad; soralia PD– or PD+ yellow or red, K– or K+ red; zeorin absent; usnic acid alone, or with hypoprotocetraric, protocetraric, or salazinic acid; common, coastal and oceanic inland localities .... ***Ramalina farinacea* (L.) Ach.**

18.(7) On rocks or soil 19

18. On trees, or rarely rocks 23

19. Arctic, northwestern Alaska to Nunavut; on soil and rocks; branches hollow with frequent perforations, round in cross section except at base and axils, 0.3-1.0(1.3) mm wide; pycnidia absent or pale .... *Ramalina almquistii* Vainio (syn. *Fistulariella almquistii*)

19. On coastal rocks in California; pycnidia abundant and conspicuous, black 20

20. Branches basically round in cross section, but can have depressions and ridges; lacking cartilaginous cords in medulla 21

20. Branches distinctly flattened or angular in cross section; medullary cartilaginous cords present or absent 22

21. Usually fertile, or with blunt branch tips; branches sparsely divided, 1-2.5 mm in diameter .... ***Niebla combeoides* (Nyl.) Rundel & Bowler**

21. Usually sterile, with pointed tips having black pycnidia at the tips; branches much divided into regular dichotomies, under 0.5 mm in diameter .... *Niebla ceruchoides* Rundel & Bowler

22. Slender, cartilaginous cords present in medulla; basal tissue rarely blackened; common along the southern California coast .... ***Niebla homalea* (Ach.) Rundel & Bowler**

22. Lacking cartilaginous cords in the medulla; blackened basal plate often present; southern California, rare except on the foggy coastal islands .... ***Niebla laevigata* Bowler & Rundel**

23.(18) Branches at least partly hollow, with many perforations; medulla loose and webby .... ***Ramalina dilacerata* (Hoffm.) Hoffm.**

[Note: Two species may be included under this name: a very small (under 15 mm long) boreal taxon (*R. dilacerata s. str.*); and a much larger species (up to 50 mm long) confined to the Pacific Northwest coast. However, McCune & Geiser (2009) say that there are intermediates.]

23. Branches solid, without perforations; medulla dense and compact 24

24. Branches mottled with black spots and immersed black pycnidia; round to angular in cross section, often with depressions and ridges; apothecia at or close to branch tips; common on coastal shrubs and trees in southern California .... *Niebla ceruchis* Rundel & Bowler

24. Branches not mottled with black spots; pycnidia brown when present; branches flat or round in cross section; apothecia, when present, not at branch tips; mainly interior and southern regions 25

25. Branches smooth or with elongate striations 26

25. Branches with depressions and ridges or long grooves, or with verrucae, tubercles, papillae 27

26. Branches flattened throughout; spores fusiform, 16-25(-31) x 3-5 µm ...... ***Ramalina stenospora*** **Müll. Arg.**

26. Branches almost round in cross section except for the base; spores elongate ellipsoid to almost fusiform, 11.5-22 x 3-5 µm .... ***Ramalina montagnei*** **De Not.**

27. At least some apothecia on the branch surface or along the margins 28

27. Almost all apothecia on or close to the tips of the branches 30

28. Warts or tubercles abundant on the branches; divaricatic acid present .....***Ramalina complanata*** **(Sw.) Ach.**

28. Warts and tubercles absent; divaricatic acid absent 29

29. Western North America; branches more or less even in width, tapering only at the tip .... ***Ramalina leptocarpha*** **Tuck.**

29. Southcentral U.S.; branches tapering at the base and at the tips .... ***Ramalina celastri*** **(Sprengel) Krog & Swinscow**

30.(27) Branches with small white tubercles; medulla K+ red or purplish red 31

30. Branches with depressions and ridges, or with long grooves, but without tubercles; medulla K– 32

31. Spiny, perpendicular branches usually present, although sparse; spores ellipsoid, medulla PD+ red or yellow, C– (salazinic, norstictic, or protocetraric acids) ....***Ramalina willeyi*** **R. Howe**

31. Spiny, perpendicular branches absent; spores fusiform; medulla PD–, C+ pink to red or violet (rapidly disappearing) (cryptochlorophaeic and paludosic acids) .... ***Ramalina paludosa*** **B. Moore**

32. Branches broad and fan-shaped; pseudocyphellae depressed, between the vein-like ridges on the lower surface; southwestern or northcentral regions and prairie provinces ....***Ramalina sinensis*** **Jatta**

32. Branches divided and relatively narrow; pseudocyphellae raised or level with surface 33

33. Contains only usnic acid; southcentral or eastern U.S. and adjacent Canada .... ***Ramalina americana*** **Hale**

33. Contains divaricatic, norbarbatic and 4-O-demethylhypoprotocetraric acids, or stenosporic and glomelliferic acid; southeastern U.S. ....*Ramalina culbersoniorum* LaGreca

RHIZOCARPON Some measurements from Timdal & Holtan-Hartwig (1988) (marked with an asterisk:*).

1. Thallus brown, orange, gray, or white; medulla IKI– or rarely IKI+ blue 2
1. Thallus distinctly yellow (not orange); medulla IKI+ blue or rarely IKI– 19

2. Spores 2-celled 3
2. Spores 4-celled to muriform 7

3. Spores becoming dark brown or greenish; thallus thick, areolate, pale to dark brown or gray-brown; widespread boreal .... *Rhizocarpon badioatrum* (Flörke *ex* Sprengel) Th. Fr.
3. Spores colorless 4

4. Thallus brownish 5
4. Thallus white to creamy white 6

5. Spores (16.5-)17.5-20 (-25) x 8.5-11.0 µm; thallus usually thin, membranous; typically on shaded siliceous rocks in forests; Appalachian-Great Lakes distribution ....
.... ***Rhizocarpon infernulum* (Nyl.) Lynge** (syn. *R. infernulum* f. *sylvaticum*) [Plate 775 in LNA, as *R. hochstetteri*]
5. Spores 21-25 (-28) x (8.5-)10.0-12.0 µm; thallus rimose-areolate; typically on exposed siliceous rocks; British Columbia, Nova Scotia, Massachusetts, Missouri
.... *Rhizocarpon hochstetteri* (Körber) Vainio. *s. str.* [*not* Plate 775 in LNA ; see above]

6. Epihymenium reddish brown, often turning K+ reddish violet; spores 16-23 x 8-12 µm; medulla K– or K+ yellow (stictic acid); thallus thick, chalk-white, thick or thin. Rocky Mountains to arctic on calcareous rock
....***Rhizocarpon chioneum* (Norman) Th. Fr.**
6. Epihymenium olive green to bluish green, K+ greener; spores (11-)13.5-16(-18) x (5.5-)7.0-8.5 µm; thallus white to pale gray (rarely brownish), rimose-areolate to areolate; medulla K+ red or K+ yellow (norstictic or, rarely, stictic acid); Eastern Temperate on siliceous rock .... *Rhizocarpon cinereovirens* (Müll. Arg.)
[Note: The revision above is based, in part, on Fryday (2002).]

7.(2) Spores colorless 8
7. Spores dark brown or dark green when mature 15

8. Thallus pale orange to rusty orange 9
8. Thallus white, gray, or brownish, not orange; spores muriform 10

9. Apothecia under 0.6 mm in diameter, with a rough, uneven disk; spores 4-celled or with the occasional longitudinal septum, 15-20 x 5-8 µm; thallus dark rusty orange; mostly eastern montane west to the Great Lakes .... *Rhizocarpon oederi* (Weber) Körber

9. Apothecia 0.7-2.0 mm in diameter, disk smooth; spores muriform, many-celled, 22-37 x (10.5-)13-15(-17) µm; thallus pale orange .... *Rhizocarpon lavatum* (Fr.) Hazsl. [see couplet 12]

10. Thallus gray-brown, orange-brown, or brown to dark gray, thin, verruculose; no lichen substances or with stictic acid; common and widespread 11

10. Thallus white or pale gray; containing stictic or norstictic acid 13

11. Spores (17-)18-25(-30) x (8-)9.5-12(-13.5) µm, submuriform to few-celled muriform; thallus areolate-verruculose, usually in small patches; apothecia 0.4-0.8 mm in diameter .... ***Rhizocarpon reductum* Th. Fr.** (syn. "*R. obscuratum*")

[Note: According to Fryday (2000), *R. reductum* typically contains stictic acid, but eastern populations seem to lack it and may be a different taxon.]

11. Spores (22-)28-41(-50) x (10-)12-20 µm; few- to many-celled muriform 12

12. Thallus typically smooth, rimose-areolate, covering large areas, often pale orange; apothecia 0.7-2 mm in diameter; epihymenium brown to olive-brown; northeastern and northwestern North America .... *Rhizocarpon lavatum* (Fr.) Hazsl. [see also couplet 9]

12. Thallus consisting of small convex areoles, never orange; apothecia 0.3-0.6 mm in diameter; epihymenium blue-green; northeastern North America .... *Rhizocarpon timdalii* Ihlen & Fryday

13.(10) Thallus chalk white; apothecia scattered, not in concentric rows; on calcareous rock; mostly arctic, rare on northeast coast; contains stictic acid (K+ yellow) .... *Rhizocarpon umbilicatum* (Ram.) Flagey

13. Thallus gray; apothecia in concentric rows or scattered; on siliceous or somewhat calcareous rocks; Great Lakes to east coast 14

14. Thallus continuous, rimose to chinky areolate; apothecia typically arranged in concentric rings; medulla K+ yellow or K–, PD+ orange or PD– (stictic acid, sometimes in traces) .... *Rhizocarpon petraeum* (Wulfen) A. Massal. (syn. "*R. concentricum*" of many authors)

14. Thallus contiguous or dispersed areolate; apothecia scattered; medulla K+ red, PD+ yellow (norstictic acid) .... *Rhizocarpon rubescens* Th. Fr. [see note on next page]

[Note: This is the lichen previously called *R. plicatile*. The latter name is a synonym of *Stereocaulon plicatile* (Leight.) Fryday & Coppins (Fryday and Coppins 1996).]

15.(7) Spores 1-2 per ascus; thallus thick, areolate; medulla IKI– 16

15. Spores 8 per ascus; thallus areolate, often scattered over a black prothallus; medulla IKI+ blue 18

16. Areoles peltate (free at the margins), brown; mainly western coastal mountains .... *Rhizocarpon bolanderi* (Tuck.) Herre

16. Areoles convex, contiguous or dispersed, pinkish brown to gray 17

17. Spores mostly 1 per ascus; temperate to boreal .... ***Rhizocarpon disporum* (Nägeli *ex* Hepp) Müll. Arg**

17. Spores mostly 2 per ascus; widespread, temperate to arctic .... *Rhizocarpon geminatum* Körber

18.(15) Thallus brown to gray-brown; medulla or cortex C+ pink (sometimes faint), KC+ pink, K+ yellow or K– (gyrophoric acid and sometimes stictic acid); very common and widespread .... *Rhizocarpon grande* (Flörke *ex* Flotow) Arnold

18. Thallus pale brownish gray; medulla and cortex C–, KC–, medulla K+ red (norstictic acid); boreal-montane and arctic .... *Rhizocarpon eupetraeum* (Nyl.) Arnold

19(1) Spores 2-celled 20

19. Spores muriform, dark; epihymenium brown to olive-brown 22

20. Spores dark, 18-32 x 10-15 µm; medulla IKI+ blue; epihymenium intense blue-green; Arctic to eastern montane .... *Rhizocarpon eupetraeoides* (Nyl.) Blomb & Forssell

20. Spores colorless; medulla IKI– 21

21. Spores over 18 µm long .... *Rhizocarpon inarense* (Vainio) Vainio

21. Spores under 18 µm long .... *Rhizocarpon superficiale* (Schaerer) Vainio

22.(19) Thallus greenish-yellow, with crescent-shaped areoles surrounding some apothecia; epihymenium unchanged or becoming more intensely green with K .... ***Rhizocarpon lecanorinum* Anders**

22. Thallus lemon yellow, with apothecia sunken between the areoles; epihymenium K+ violet or purple 23

23. Parasitic on other crustose lichens; apothecia often convex; spores 17-32 x 10-16 µm (Fletcher et al. 2009b); western, Arizona and California to B.C and Saskatchewan ....*Rhizocarpon viridatrum* (Wulfen) Körber

23. Growing directly on rock, not parasitic; apothecia usually flat 24

24. Spores 28-60(-70) x 15-25 µm; hymenium greenish; western montane, Great Lakes, and northeast coast ....***Rhizocarpon macrosporum* Räsänen**

24. Spores 24-40 x 11-16 µm; hymenium colorless; widespread boreal to arctic-alpine ....***Rhizocarpon geographicum* (L.) DC.**

RHIZOPLACA Based in part on Ryan (2002) and McCune and Rosentreter (2007).

1. Thallus attached directly to rock 2

1. Thallus growing on soil, unattached (vagrant) 7

2. Thallus crustose, dispersed areolate with very convex areoles; apothecia orange, pruinose ....***Rhizoplaca subdiscrepans* (Nyl.) R. Sant.**

2. Thallus attached by a central holdfast (umbilicate) 3

3. Thallus whitish to gray, margins curling inward, lacking usnic acid; apothecia marginal or submarginal; cortex C+ yellowish, medulla PD–, C– (pseudoplacodialic acid in cortex); California 4

3. Thallus yellow-green, containing usnic acid 5

4. Thalli under 8 mm across; apothecia brown, without pruina, marginless when mature; spores elongate, somewhat curved, (12-)15-20(-23) x (3-)4-5 µm; southern California .... *Rhizoplaca glaucophana* (Nyl. *ex* Hasse) W. A. Weber

4. Thallus 7-15 mm across; apothecia black, white pruinose, with persistent lecanorine margins; spores broadly ellipsoid or almost spherical, 9-14 x 6.5-8.5 µm; central California .... *Rhizoplaca marginalis* (Hasse) W. A. Weber

5. Apothecial disks orange, pruinose ....***Rhizoplaca chrysoleuca* (Sm.) Zopf**

5. Apothecial disks yellow, brown, olive, or black, with or without pruina 6

6. Lower surface rough, broken into areoles with white cracks; apothecial disks yellowish brown, not pruinose; medulla PD+ orange (pannarin) ....***Rhizoplaca peltata* (Ramond) Leuckert & Poelt**

6. Lower surface smooth; apothecial disks yellow-brown to greenish or black, pruinose; medulla PD+ bright yellow, or less frequently PD– (with or without psoromic acid) .... ***Rhizoplaca melanophthalma* (DC.) Leuckert & Poelt *s. lat.***

7.(1) Thallus lobes clearly flattened throughout although strongly curled and contorted; apothecia usually abundant, yellow-brown to greenish black; lobes lacking whitish warts; medulla usually PD+ yellow (psoromic acid) .... ***Rhizoplaca melanophthalma* (DC.) Leuckert & Poelt *s. lat.***

[Note: Recent genetic work has shown that *R. melanophthalma* in its broad sense is composed of at least five cryptic species with few reliable chemical or morphological distinctions. See Leavitt et al. (2013).]

7. Thallus lobes becoming finely divided, appearing subfruticose in part with narrowly channeled lobe tips; apothecia lacking; small white warts produced especially on the margins or tips of the lobes; medulla PD– ....***Rhizoplaca haydenii* (Tuck.) W. A. Weber**

RINODINA (including *Amandinea* [p.p.]) Based largely on Sheard (2010).

[Note: Spore measurements from Sheard (2010) have an asterisk (*). For an illustration of spore types, see Plate 925.]

1. Spores 4-celled to submuriform, on moss, dead vegetation, soil, and wood 2
1. Spores 2-celled 3

2. Spores consistently 4-celled; Pacific Northwest and western montane, rare in Hudson Bay lowlands .... *Rinodina conradii* Körber
2. Spores 4-celled to submuriform; mainly southern California to southern Arizona and New Mexico .... *Rinodina intermedia* Bagl.

3. On soil, mosses, or dead vegetation; mainly arctic-alpine 4
3. On bark, wood, or rock; mostly temperate 9

4. Spores with uniformly thin walls at maturity ....*Phaeorrhiza*
4. Spores with unevenly thickened walls (*Physcia*- or *Physconia*-type) 5

5. Apothecia becoming convex to hemispherical, often pruinose when young, margins often excluded, 0.7-1.1(-1.5) mm in diameter, cortex thin; spores (20.5-)22-27(-30) x (9.5-)11-12(-14.5) µm, often tapered; usually containing atranorin, sometimes variolaric acid or the orange pigment skyrin (K+ purple), lacking sphaerophorin or zeorin .... *Rinodina mniaraea* (Ach.) Körber
5. Apothecia remaining flat when mature, 0.35-1.5 mm in diameter, pruinose or not, margins typically persistent; zeorin present or absent 6

6. Spores averaging 25-36 µm long, *Physcia*-type (Plate 925A); apothecia up to 1.5 mm in diameter, with an expanded lower cortex; zeorin present or absent 7

6. Spores averaging 20-24 µm long; apothecia usually under 0.5 mm in diameter, lower cortex expanded or not; containing zeorin 8

7 Apothecia without pruina; thallus brownish; lower cortex of apothecia often IKI+ pale blue, with clear columnar cells; thallus granular to verrucose, UV± dull gray (sphaerophorin), zeorin absent .... ***Rinodina turfacea* (Wahlenb.) Körber**

7. Apothecia typically pruinose; thallus white; lower cortex IKI–, opaque; ± zeorin, lacking sphaerophorin .... *Rinodina roscida* (Sommerf.) Arnold

8. Spores *Physcia*- or *Physconia*-type, often having pale, nipple-like tips, (18-)22-24(-29) µm long*; lacking pannarin (apothecial sections PD–); apothecia 0.3-0.5 mm, sometimes pruinose .... *Rinodina terrestris* Tomin

8. Spores almost all *Physcia*-type, tips not especially pale, but sometimes pointed at one or both ends, (17-)20-22(-26) µm long*; reported to contain pannarin, but apothecial sections usually PD–; apothecia 0.4-0.8 mm, never pruinose .... *Rinodina olivaceobrunnea* Dodge & Baker

9.(3) On rock 10

9. On bark or wood 21

10. Maritime, mostly on shoreline rocks in salt-spray zone, especially calcareous or bird-frequented rocks 11

10. On non-maritime rocks 12

11. Spores (10-)11-15 x (6-)7-8.5 µm, like *Physcia*-type, but swelling at the septum in K; on both east and west coasts .... *Rinodina gennarii* Bagl.

11. Spores (16-)20-23(-25) x (8-)9-11(-13) µm; *Physconia*- or *Physcia*-type, not swelling at septum in K; on west coast .... *Rinodina pacifica* Sheard

12. On calcareous rock; mature spores with a pigmented band around the middle near septum, spore often bulging at septum showing two layers, walls becoming thin at tips (*Bischoffii*-type; see Plate 925C), 16-20 x 10-12.5 µm; thallus typically scanty, dispersed areolate 13

12. On siliceous rock; spores lacking a pigmented band, not double-walled at septum 14

13. Hymenium usually clear; apothecial margins gray to pale brown, thick or thin, usually persistent, but can be lacking if the apothecia are immersed; widespread temperate to montane, rare in the arctic .... *Rinodina bischoffii* (Hepp) A. Massal.

13. Hymenium always containing granules or oil; apothecial margins dark brown to black or often excluded; mainly in the arctic .... *Rinodina calcigena* (Th. Fr.) Lynge

14. Thallus brownish to olive, cortex K– (atranorin absent) 15

14. Thallus creamy white to pale brownish gray, dispersed areolate, rimose-areolate to verrucose, cortex K+ yellow (atranorin) 18

15. Western montane; spores (15.5-)18-19(-22) x (7.5-)9.5-10.5(-12) µm*; thallus dull brown, thick, verrucose to areolate; apothecia at first sunken, later broadly attached; containing no substances .... *Rinodina milvina* (Wahlenb.) Th. Fr.

15. Central to eastern regions; apothecia broad to narrowly attached, never sunken into thallus 16

16. Spores 12-16 x 7.5-9.0 µm, *Pachysporaria*-type; thallus thin, rimose; cortex and apothecial sections C–; East Temperate .... *Rinodina siouxiana* Sheard

16. Spores much larger 17

17. Spores 18-22(-24) x (8-)10-12(-14) µm, locules becoming round or triangular with thick walls; thallus dull, verruculose to verrucose or rimose-areolate, sometimes dispersed areolate, cortex often KC+ pink, C+ pink (± 5-*O*-methylhiasic acid), zeorin present; common, central plains to eastern seaboard, in moist habitats .... *Rinodina tephraspis* (Tuck.) Herre

17. Spores 23-41 x 11-17 µm, locules angular, walls remaining thickened only at cell tips; thallus more or less smooth, shiny, olive-brown, C– (no substances); rare on rock, northeastern .... ***Rinodina ascociscana* (Tuck.) Tuck.** [see also couplet 48]

18.(14) Spores ellipsoid to narrowly ellipsoid, length/width ratio more than 1.9, locules angular or rounded at the ends (*Physcia*- or *Physconia*-type) 19

18. Spores ellipsoid to broadly ellipsoid, length/width ratio less than 1.9, locules hourglass-shaped (*Mischoblastia*-type, see Plate 925D) 20

19. Spores 17-22 x 8-11 µm, *Physcia*-type, retaining thick angular end walls with age; western mountains, Arizona and California to Alberta .... *Rinodina confragosa* (Ach.) Körber

19. Spores (19-)21-25(-31) x (9-)10-12(-16) µm, *Mischoblastia*- to *Pachysporaria*-type in older spores; west coast, California to Vancouver Island .... ***Rinodina bolanderi* H. Magn.**

20. Thallus continuous, rimose-areolate; spores (15.5-)17-20(-22) x (8.5-)10-12(-13.5) µm; apothecia frequently have black margins and resemble *Buellia*; East Temperate and southwestern .... *Rinodina oxydata* (A. Massal.) A. Massal.

20. Thallus dispersed areolate to areolate; spores (19-)20.5-24.5(-26) x (9.5-)11-13(-15) µm; apothecia at first immersed, later sessile, rarely with a pigmented margin; eastern ....*Rinodina destituta* (Nyl.) A. Zahlbr.

21.(9) Thallus edge distinctly lobed with a lower cortex .... ***Hyperphyscia syncolla* (Tuck. *ex* Nyl.) Kalb**

21. Thallus edge indefinite or definite, not lobed, lacking a lower cortex 22

22. Soredia or blastidia present 23

22. Soredia and blastidia absent 31

23. Thallus PD+ orange to red-orange (pannarin or stictic acid present) 24

23. Thallus PD– (pannarin absent) 27

24. Soralia round and discrete, gray, on a very thin, sometimes inconspicuous thallus; spores finally *Pachysporaria*-type, 19-29 x 11-16 µm*; containing stictic acid, zeorin, atranorin, and chloroatranorin; coastal Pacific Northwest .... *Rinodina stictica* Sheard & Tønsberg

24. Soralia irregular, greenish, yellowish or brownish; containing pannarin 25

25 Thallus thick, areolate with convex to flat areoles or granular; spores *Physcia*-type, 15-20 x (7-)8-10(-11) µm; lacking zeorin or secalonic acid; eastern North Temperate .... *Rinodina excrescens* Vainio

25. Thallus thin, or with scattered areoles; containing zeorin 26

26. Thallus brownish gray to yellowish brown, consisting of very small scattered areoles bursting into yellowish granular soredia (often containing secalonic acid); spores with angular locules (*Physcia*-type), less than 20 x 12 µm; Appalachian-Great Lakes and Pacific Northwest .... *Rinodina efflorescens* Malme

26. Thallus gray, thin and membranous, breaking into greenish gray soralia (not yellowish; lacking secalonic acid); spores with round locules (*Pachysporaria*-type), more than 20 x 12 µm; Appalachian-Great Lakes .... *Rinodina willeyi* Sheard & Giralt

27.(23) Thallus composed of small, dispersed, gray to brownish areoles, with patches of coarse blastidia ranging into soredia, UV+ dull or bright blue-white (sphaerophorin); spores *Physcia*- or *Physconia*-type), 22-34 x 10-20 µm*; distribution California to Alaska ....*Rinodina disjuncta* Sheard & Tønsberg

27. Thallus continuous to dispersed areolate, soredia or blastidia and medulla UV– (lacking sphaerophorin); spores mostly under 20 µm long 28

28. Apothecia extremely rare; spores with thin end walls, not at all thickened except at septum (*Physconia*-type); thallus composed of flat to subsquamulose areoles, producing fine, greenish white soredia; thallus and soredia K+ yellow, UV– (atranorin and zeorin); Pacific Northwest and Great Lakes region .... *Rinodina degeliana* Coppins

28. Apothecia usually present, sometimes common; spores mostly with thickened end walls; lacking any lichen substances 29

29. Spores *Pachysporaria*- or *Mischoblastia*-type; central to northeastern and southeastern Arizona .... *Rinodina papillata* H. Magn. (see also couplet 53)

29. Spores mainly *Physcia*-type; western 30

30. On junipers in dry open habitats of the western interior; apothecia sunken in thallus at first, becoming erumpent; apothecial margins often becoming thin and disappearing ....*Rinodina juniperina* Sheard

30. On deciduous trees in moist to mesic climates, west coast to Idaho; apothecia superficial; apothecia margins persistent ....*Rinodina santae-monicae* H. Magn.

31.(22) Spores 12-32 per ascus, 11-19 x 6-9 µm 32

31. Spores 8 per ascus 33

32. Spores mostly 16-32 per ascus, walls more or less uniform except for septum (*Physconia*-type); apothecia crowded, 0.4-0.7 mm in diameter, margins thick; thallus areolate, brown to gray-brown; central and northeastern regions, deciduous trees, especially poplars .... ***Rinodina populicola* H. Magn.**

32 Spores mostly 12-16 per ascus, walls remaining thick at tips (*Physcia*-type); apothecia scattered, 0.2-0.4 mm in diameter, margins thin; thallus pale, very thin and membranous; west coast and northeast, smooth-barked deciduous trees, rarely poplars .... *Rinodina polyspora* Th. Fr.

33. Spores thin-walled and even in thickness when mature, 10-15 x 5-7µm 34

33. Spores conspicuously thick-walled, usually more than 15 µm long, walls unevenly thickened when mature 36

34. Immature spores clearly with a thickened mid-septum (*Physcia*- or *Physconia*-type); central plains to west coast, rare in northeast .... *Rinodina pyrina* (Ach.) Arnold

34. Immature spores lacking a thickened mid-septum; north-central to east coast 35

35. Spores even in outline, not constricted; thallus rugose, rimose, or more or less smooth; coastal plain, Massachusetts to Gulf of Mexico .... *Amandinea milliaria* (Tuck.) E. Lay & P. May (syn. *Rinodina milliaria*)

35. Spores becoming slightly constricted at septa; thallus verruculose or areolate; northcentral to northeastern regions, not coastal plain .... *Amandinea dakotensis* (H. Magn.) P. May & Sheard (syn. *Rinodina dakotensis*)

36.(33) Thallus distinctly yellowish, K+ yellow (xanthones and zeorin); spores *Pachysporaria*-type, mostly 14-17 x 7-8 µm; Florida to Louisiana .... *Rinodina lepida* (Nyl.) Müll. Arg.

36. Thallus not distinctly yellowish, lacking xanthones 37

37. Thallus gray, white, or creamy, K+ yellow (atranorin); western 38

37. Thallus gray, brown, or olive, thin or thick, K– (lacking atranorin) 43

38. Apothecia 0.6-1.5 mm in diameter; thallus well developed; spores (19)21-25(-31) x (9-)10-12(-16) µm; thallus thick, verrucose to areolate; maritime California to southern British Columbia .... ***Rinodina bolanderi* H. Magn.**

38. Apothecia 0.4-0.7(-1.0) mm in diameter; spores 15-20(-23) x 8-11 µm 39

39. Pannarin present in epihymenium (PD+ orange crystals); apothecial cortex expanded below with columnar cells; apothecial disks sometimes pruinose 40

39. Apothecia without pannarin or pruina; cortex expanded or not 41

40. Epihymenium granular on the surface; spores with a dark septum, swelling in K (Plate 925E), averaging 17.5-18.5 µm long; zeorin absent; California to southern British Columbia, mainly coastal .... *Rinodina marysvillensis* H. Magn.

40. Epihymenium granular between tips of paraphyses; spores with a pale septum, not swelling in K, averaging 19-20 µm long; sometimes containing zeorin; western montane .... *Rinodina aurantiaca* Sheard

41. Apothecial cortex expanded to 70 µm at base, with columnar cells, IKI+ blue; containing zeorin .... *Rinodina capensis* Hampe in A. Massal.

41. Apothecial cortex not expanded at base with columnar cells, IKI–; lacking zeorin 42

42. Spores swollen at the septa, especially in K (*Dirinaria*-type); California north to B.C .... *Rinodina californiensis* Sheard

42. Spores not swollen at septa *Physcia*- to almost *Physconia*-type; mainly Rocky Mountains .... *Rinodina boulderensis* Sheard

43.(37) Spores (22-)25-36 µm long 44

43. Spores 12-25 µm long 49

44 Apothecial cortex expanded and cellular at base, usually IKI+ blue 45

44. Apothecial cortex thin, eastern 48

45. Northern, extending southward in mountains 46

45. Mainly California to Pacific Northwest 47

46. Thallus reddish brown to gray-brown, granular to verrucose; sphaerophorin present, zeorin absent; on wood and bone, rarely on bark; widespread arctic-alpine west of Hudson Bay .... ***Rinodina turfacea* (Wahlenb.) Körber** (see also couplet 7)

46. Thallus pale to dark gray or gray-brown, thin, areolate; no lichen substances; corticolous, especially on poplar and aspen; mostly southern boreal .... *Rinodina austroborealis* Sheard

47. Medulla UV+ white (sphaerophorin); lower cortex of apothecia columnar, IKI+ blue; spores with dark central septum; California to southern Oregon .... *Rinodina badiexcipula* Sheard

47. Medulla UV– (no substances); lower cortex cellular, IKI–; spore septum not darkened; California to southern B.C. and Idaho .... *Rinodina oregana* H. Magn.

48.(44) Spores *Physcia*-type, 23-41 x 11-17 µm; thallus shiny, olive to brownish; apothecial sections and thallus PD– (lacking lichen substances) ....***Rinodina ascociscana* (Tuck.) Tuck.**

48. Spores *Pachysporaria*-type, (21-)28-30(-36) x (9-)14-16(-20) µm*; thallus continuous and often shiny, pale gray; apothecial sections and thallus PD+ orange (pannarin) ....*Rinodina adirondackii* H. Magn.

49.(43) Apothecia soon becoming biatorine, disk-colored, disks often pruinose; spores (17-)20-23(-25) x 8-10(-12) µm, *Physcia*- or *Physconia*-type; containing zeorin and variolaric acid; California to southern B.C. ....*Rinodina hallii* Tuck.

49. Apothecia with a persistent, thallus-colored, thalline margin, never biatorine (but see *R. maculans* below); disks usually not pruinose (except in *R. papillata*) 50

50. Spores very thick-walled, with round to almost triangular lumina creating an hourglass shape (*Pachyporaria* –type, see Plate 925F) ; no substances 51

50. Spores with angular lumina, at least when mature, *Physcia*-, *Physconia*- or *Dirinaria*-type 55

51. Eastern to mid-western (but see *R. papillata*, couplet 53) 52

51. Southwestern 54

52. Spore length to width ratio more than 2:1, (12-)13-17(-23) x 6.5-8.5(-10) µm; thalline margin often poorly developed; coastal plain, Massachusetts to Texas ....*Rinodina maculans* Müll. Arg. (syn. *R. applanata*)

52. Spores length to width ratio less than 2:1 53

53. Thallus continuous, rugose, sometimes granular, not shiny, rarely sorediate with soredia 10-20 µm in diameter; spores (13-)15-17.5(-19.5) x (6.5-)8.0-10(-11) µm; northeast and mid-west ....*Rinodina pachysperma* H. Magn.

53. Thallus smooth, shiny, areolate, frequently producing granules or granular blastidia at the areole margins, ca. 150-300 µm in diameter; spores (14-)17.5-18.5(-21.5) x (7.5-)9.0-10(-11.5) µm*; northeast to Ozarks and southeastern Arizona ....*Rinodina papillata* H. Magn.

54. Apothecia with persistent margins; spores (12-)13-16(-18) x (6.5-)7.5-9.0(-10) µm; southwestern montane ....*Rinodina coloradiana* H. Magn.

54. Apothecial margin becoming excluded in old convex apothecia; spores (13-)14.5-18(-23) x (7.0-)8-10(-13) µm; mostly southern California .... *Rinodina herrei* H. Magn.

55.(50) From Great Lakes and central plains eastward 56

55. West of Mississippi to the arctic 59

56. Thallus PD+ orange (pannarin); thallus fairly thick, with convex to flat gray areoles .... *Rinodina excrescens* Vainio [see also couplet 25]

56. Thallus PD– (lacking pannarin); thallus thin, often inconspicuous 57

57. Spores *Dirinaria*-type, swelling at septum in K; thallus thin and inconspicuous; southern boreal to arctic-alpine .... *Rinodina metaboliza* Vainio

57. Spores *Physcia*-type, not swelling at septum in K 58

58. Apothecia at first sunken in thin, membranous, brownish gray thallus, gradually emerging and marginate, scattered; spores (15-)16-20(-23) x 7.5-10(-12) µm; zeorin present; Appalachian-Great Lakes distribution; on deciduous trees, especially sugar maples .... *Rinodina subminuta* H. Magn.

58. Apothecia sessile, clustered in crowded clumps; thallus minutely areolate, greenish gray to brown or gray-brown; spores (13.5-)15-18(-21) x (6.0-)6.5-8.0(-9.0) µm, septum very dark in all spores; zeorin absent; branches and twigs of deciduous and coniferous trees and on wood; widespread North Temperate to southern boreal and montane .... *Rinodina freyi* H. Magn. (syn. *R. glauca*)

59.(55) Spores *Physconia*-type (Plate 925B), but immature spores can be *Physcia*-like; mainly western montane 60

59. Spores *Physcia*- or *Dirinaria*-type sometimes resembling *Physconia*-type when overmature 61

60. Mostly on wood, less often on conifer bark; apothecia often crowded, margins persistent; spores (17-)18.5-22(-24) x (8-)9.0-11 µm; thallus thin to areolate, brownish .... *Rinodina archaea* (Ach.) Arnold

60. Mostly on deciduous trees and shrubs, less frequently on conifers; apothecia usually scattered, not contiguous, margins thin and sometimes disappearing; spores 15-18.5(-20) x 8.0-9.7 µm .... *Rinodina orculata* Poelt & M. Steiner

61. Spores (17.5-)22-23(-27) x (9-)10.5-11(-13) µm*, swelling at septum with K (*Dirinaria*-type); California, on deciduous shrubs and trees .... *Rinodina endospora* Sheard

61. Spores mostly 14-19 x 7-9 µm 62

62. Spores clearly *Dirinaria*-type, swelling at septum in K; mainly central plains and southern boreal areas, scattered in the arctic .... *Rinodina metaboliza* Vainio [see also couplet 57]

62. Spores *Physcia*-type, not swelling at septum in K 63

63. Thallus usually obvious, greenish gray to brownish, areolate or rimose; amphithecial cortex indistinct, poorly developed; apothecia usually crowded in patches, margins well developed and persistent; widespread North Temperate to southern boreal and western montane, typically on twigs .... *Rinodina freyi* H. Magn. (syn. *R. glauca*) [see also couplet 58]

63. Thallus extremely thin or with indistinct flat areoles, greenish gray to brown; amphithecial cortex usually distinct, thick or thin; apothecia scattered; margins thin, brown, persistent or disappearing; typically on smooth-barked trees and shrubs 64

64. Amphithecial cortex distinct, expanded at the base; apothecia mostly 0.4-0.8 mm in diameter; California to B.C. and northern Rockies .... *Rinodina laevigata* (Ach.) Malme

64. Amphithecial cortex thin, distinct or indistinct, not expanded; apothecia mostly 0.25-0.4 mm in diameter; arctic-alpine .... *Rinodina septentrionalis* Malme

ROCCELLA

1. Thallus with many soralia containing powdery, white soredia; apothecia absent .... ***Roccella gracilis* Bory** (syn. *R. peruensis*) [Plate 791 in LNA as "*R. babingtonii*"]

1. Thallus lacking soralia, but with numerous apothecia .... *Roccella decipiens* Darb.(syn. *R. fimbriata, R. babingtonii*)

SARCOGRAPHA Based on Harris (1995).

1. Spores 6-celled, 22-24 x 6-7 µm; lirellae pruinose, in clusters on raised, white verrucae; Florida .... *Sarcographa medusulina* (Nyl.) Müll. Arg.

1. Spores 4-celled 2

2. Hymenium clear, without oil droplets; spores 15-20 x 5-6 µm; sections K+ red (norstictic acid); lirellae sinuous and branched, lightly pruinose, in clusters on barely raised white patches; Florida .... *Phaeographis intricans* (Nyl.) Staiger (syn. *Sarcographa intricans*)

2. Hymenium containing tiny oil droplets 3

3. Sections K– (no substances); spores 17-21 x 7-8.5 µm; clusters of lirellae on verrucae that are not white; coastal plain, Florida to North Carolina .... *Sarcographa tricosa* (Ach.) Müll. Arg.

3. Sections K+ yellow (stictic acid); spores 16-21 x 5.5-6 µm; clusters of lirellae on white, well-defined verrucae; Florida .... ***Sarcographa labyrinthica* (Ach.) Müll. Arg.**

SARCOGYNE (including *Polysporina*) Based in part on Knudsen and Standley (2007) and Knudsen (2007b).

1. Apothecial disks with lumps of carbonized tissue either as spots or forming sterile ridges; apothecial margins heavily carbonized 2

1. Apothecial disks smooth, without carbonized lumps; apothecial margins carbonized or biatorine 4

2. Apothecia convex, 0.3-0.6 mm in diameter; growing on other saxicolous crustose lichens or independently and endolithic or forming a thin, white areolate thallus

   .... *Polysporina subfuscescens* (Nyl.) K. Knudsen & Kocourk. (= *P. lapponica* of North American authors)

2. Apothecia flat to concave; on rocks; thallus endolithic 3

3. Apothecia mostly 0.2-0.4 mm in diameter, deeply concave, with thick, cracked margins, often partially embedded in rock; on calcareous rocks .... *Polysporina urceolata* (Anzi) Brodo

3. Apothecia (0.2-)0.3-1.0 mm in diameter, flat or rough; superficial; usually on siliceous rocks, rarely calcareous .... ***Polysporina simplex* (Taylor) Vězda**

4. Apothecia pruinose; exciple almost biatorine with a black, not carbonized, outer edge and colorless within; on calcareous rocks .... ***Sarcogyne regularis* Körber**

4. Apothecia not pruinose; exciple carbonized or not; on siliceous (non-calcareous) rocks 5

5. Exciple not carbonized, only black at outer edge, yellowish within; apothecial margins thin, sometimes disappearing; apothecia often dividing and proliferating; widespread .... *Sarcogyne similis* H. Magn.

5. Exciple carbonized and brittle, colorless to yellowish within; apothecial margins persistent and prominent; apothecia clustered or single but generally not splitting and proliferating; widespread in northern and southern regions 6

6. Apothecia 1-3(-6) mm in diameter, narrowly attached, usually round; hypothecium brown to black

   .... *Sarcogyne clavus* (DC.) Kremp.

6. Apothecia 0.3-0.7 mm in diameter, broadly attached, usually irregular or angular in shape; hypothecium colorless .... *Sarcogyne hypophaea* (Nyl.) Arnold (= *S. privigna* of most authors)

SCOLICIOSPORUM See also *Bacidia*.

1. Mainly on rock, rarely wood; thallus thin, without soredia; spores 19-30(-32) x 2-3 µm, 4(-8)-celled. apothecia red-brown to black; widespread temperate, especially in the east .... *Scoliciosporum umbrinum* (Ach.) Arnold

1. On bark, especially conifers 2

2. Thallus with yellowish soralia; apothecia pale to very dark brown; spores 3- to 8-celled, 22-40 x 1.5-3.5 µm, twisted and S-shaped; Pacific Northwest in humid forests .... *Scoliciosporum sarothamni* (Vainio) Vězda

2. Thallus dark green granular, not sorediate; apothecia dark brown to black; spores 5- to 8-celled, 18-35(-40) x 3-5 µm, curved and tapered (comet-shaped); eastern temperate to southern boreal .... ***Scoliciosporum chlorococcum* (Stenh.) Vězda**

SOLORINA

1. Thallus consisting of two parts: a brownish to gelatinous crustose layer containing cyanobacteria (blue-green in section), and scattered red-brown apothecia with squamulose margins containing green algae; spores 4 per ascus; arctic-alpine .... ***Solorina spongiosa* (Ach.) Anzi**

1. Thallus foliose, with cephalodia in the form of a discontinuous blue-green layer of cyanobacteria below the green algal layer, or forming warts on the lower surface 2

2. Lower surface and medulla bright orange; upper surface not pruinose; arctic-alpine .... ***Solorina crocea* (L.) Ach.**

2. Lower surface pale brown; medulla white; upper surface pruinose or not 3

3. Spores 8 per ascus; thallus well developed or only surrounding apothecium, not pruinose; medulla C+ pink (methyl gyrophorate and tenuorin); southern Rockies to mostly western arctic .... *Solorina octospora* (Arnold) Arnold

3. Spores 2 or 4 per ascus; thallus often white pruinose or scabrose, at least at the lobe margins; medulla C– (lacking methyl gyrophorate) 4

4. Spores 4 per ascus, up to 42 µm long; widespread and common, especially on calcareous soils and rock .... ***Solorina saccata* (L.) Ach.**

4. Spores 2 per ascus, up to 100 µm long; southern Rockies to arctic .... *Solorina bispora* Nyl.

SPHAEROPHORUS (including *Acrocyphus* and *Bunodophoron*) Largely based on M. Wedin et al. (2009).

1. Medulla orange, K+ red-purple; rare in Pacific Northwest .... *Acrocyphus sphaerophorioides* Lév.
1. Medulla white, K– or K+ yellow 2

2. Thallus branches strongly flattened; ascomata flattened and hood-like; spores more or less globose, colorless to gray; often containing stictic acid (K+ yellow, PD+ orange); on mossy tree trunks and rocks in humid forests of Pacific Northwest .... *Bunodophoron melanocarpum* (Sw.) Wedin
2. Thallus branches round in cross section except sometimes at the axils; ascomata spherical; spores broadly ellipsoid, violet to black; lacking stictic acid (K–, PD–), but sometimes with thamnolic acid (K+ yellow, PD+ orange)or squamatic acid (K–, PD–) 3

3. Thallus branches all approximately the same length, usually under 20 mm long, dichotomously branched; fruiting bodies rare; medulla IKI– .... ***Sphaerophorus fragilis* (L.) Pers.**
3. Thallus with a stout main stem 30-80 mm long, and finer side branches; fruiting bodies common or rare; medulla IKI+ blue 4

4. On rocks and soil in arctic and alpine habitats, and in humid forests in northeastern North America; few clusters of coralloid branchlets ......*Sphaerophorus globosus* (Hudson) Vainio [*not* Plate 805 in LNA; see below]
4. On trees and mossy rocks in humid forests of the Pacific Northwest 5

5. Thallus with many clusters of slender coralloid branchlets; branches relatively smooth; medulla dense; ascomata uncommon .... ***Sphaerophorus tuckermanii* Räsänen** [Plate 805 in LNA, as *S. globosus*]
5. Thallus with almost no clusters of coralloid branchlets; main branches with conspicuous dents and depressions (scrobiculate) especially close to the base; medulla rather loose and cottony; ascomata common .... *Sphaerophorus venerabilis* Wedin, Högnabba & Goward

SPORASTATIA

1. Thallus copper brown, with distinctly radiating lobes at the margins; western arctic to the alpine southern Rockies .... ***Sporastatia testudinea* (Ach.) A. Massal.**
1. Thallus gray, not lobed at the margins; rather rare in Rockies, infrequent in arctic .... *Sporastatia polyspora* (Nyl.) Grummann

STAUROTHELE Based in part on Thomson (2002).

1. Spores 8 per ascus, colorless; thallus pale olive, rimose, continuous; northeast to central regions, on calcareous rock .... ***Willeya diffractella*** **(Nyl.) Müll. Arg.** (syn. *Endocarpon diffractellum, Staurothele diffractella*) [Plate 808 in LNA]
1. Spores 2 per ascus, brown; thallus shades of brown, or absent 2

2. Thallus mostly endolithic, scanty, with black perithecia scattered over the rock; western, frequent in the dry interior, on calcareous rocks .... *Staurothele elenkenii* Oxner
2. Thallus well developed; perithecia buried in thallus to various degrees 3

3. Thallus continuous or rimose to rimose-areolate, without any lobes; perithecia buried in thallus to various degrees, not in separate areoles; prothallus absent 4
3. Thallus areolate, the perithecia buried in separate areoles 5

4. Perithecia becoming prominent, producing low or tall bumps; involucrellum black at summit; thallus dark brown, rarely pale; semiaquatic on siliceous rocks; widespread temperate to boreal .... ***Staurothele fissa*** **(Taylor) Zwackh**
4. Perithecia not, or barely, forming bumps, entirely buried in thallus showing only brown to blackish ostiole or brown disk (involucrellum) at summit of perithecia on thallus surface; involucrellum usually brown; thallus reddish brown; on wet or dry mainly siliceous rocks in Rocky Mountains, rare in Great Lakes region .... *Staurothele clopimoides* (Anzi *ex* Arnold) Stein

5. On wet or, less frequently, dry rock; thallus usually with a lobed or fibrous prothallus; areoles containing perithecia are larger than areoles that are sterile; widespread in northern, central, and western North America .... ***Staurothele drummondii*** **(Tuck.) Tuck.**
5. On dry rock; thallus lacking a fibrous or lobed prothallus; fertile and sterile areoles about the same size; mainly central and southwestern U.S. 6

6. With black involucrellum around ostiole; areoles constricted at the base; thallus dark brown to black; on non-calcareous or calcareous rock .... *Staurothele areolata* (Ach.) Lettau
6. Ostiole not surrounded by an involucrellum; areoles not constricted at the base; thallus yellowish brown to olive-brown; on calcareous rocks, western to central regions, also reported from northeast .... *Staurothele monicae* (Zahlbr.) Wetm.

STEREOCAULON

[Note: In this genus, "PD–" can mean no color change or only pale yellow (due to atranorin).]

1. Stalks sorediate 2
1. Stalks without soredia 5

2. Stalks 10-40 mm tall; conspicuous cephalodia on the stalks; Pacific Northwest 3
2. Stalks under 5 mm tall; soredia at stalk summit; cephalodia on crustose primary thallus 4

3. Phyllocladia, in part, flattening, with lower surface and margins sorediate, side branches also sorediate; medulla PD– (lobaric acid) .... ***Stereocaulon coniophyllum* I. M. Lamb**
3. Phyllocladia all coarsely granular, grading into granular soredia, none flattening; medulla PD+ orange (stictic and norstictic acids) .... *Stereocaulon spathuliferum* Vainio

4. Sorediate mass KC+ pink to violet (lobaric acid); primary thallus mostly granular .... ***Stereocaulon pileatum* Ach.**
4. Sorediate mass KC–; primary thallus areolate .... ***Pilophorus cereolus* (Ach.) Th. Fr.**

5.(1) Growing attached directly to rock 6
5. Growing on soil (sometimes soil over rock) 15

6. Phyllocladia consisting of convex to flat, more or less round squamules, with dark green centers and pale margins; medulla PD+ orange (stictic acid often with norstictic acid) .... ***Stereocaulon vesuvianum* Pers.**
6. Phyllocladia uniform in color 7

7. Tomentum on stalks thick or thin, usually pinkish; cephalodia abundant, consisting of large spherical granules more or less buried in the tomentum, or lumpy galls; phyllocladia squamulose, often deeply lobed .... rare, saxicolous forms of *Stereocaulon grande* (H. Magn.) H. Magn. [see also couplet 22]
7. Tomentum usually thin or absent, or if thick, then gray, not pinkish; cephalodia common, sparse, or absent 8

8. Phyllocladia flattened and lobed, rarely coralloid; containing lobaric acid; mainly Great Lakes region east to New England 9
8. Phyllocladia mostly cylindrical, coralloid or granular, not flattened (but see *S. sterile* below, couplet 12) 10

9. Thallus prostrate and clearly dorsiventral, at least at the margins of the colony; stalks covered with a thin or thick gray tomentum; cephalodia absent ....***Stereocaulon saxatile* H. Magn.**

9. Thallus forming tight cushions of erect, "woody" stalks that are entirely devoid of tomentum; cephalodia frequently present, lumpy, brown to olive .... *Stereocaulon glaucescens* Tuck.

10. Phyllocladia mostly in warty or granular clusters at the stalk tips like cauliflower; cephalodia rare; containing porphyrilic acid; arctic-alpine, especially in Alaska .... *Stereocaulon botryosum* Ach.

10. Phyllocladia mostly cylindrical or coralloid, not especially concentrated at stalk tips; cephalodia frequent or rare, forming grape-like clusters when present; lacking porphyrilic acid 11

11. Thallus PD+ orange (stictic acid), dorsiventral in part; apothecia commonly present, terminal; cephalodia infrequent and inconspicuous; Appalachian-Great Lakes region north to Labrador
....***Stereocaulon dactylophyllum* Flörke**

11. Thallus PD– (lobaric acid); apothecia rarely seen 12

12. Thallus forming low prostrate cushions 1-2 cm high; phyllocladia cylindrical, thick or thin, some flattened somewhat; cephalodia common, bulbous; California to Alaska
....*Stereocaulon sterile* (Savicz) I. M. Lamb *ex* Krog

12. Thallus more or less erect, not dorsiventral; phyllocladia mostly slender, never flattened 13

13. Phyllocladia long, like "bony" fingers; stems slender and not sinewy; cephalodia common or infrequent, in grape-like clusters; Appalachian ....***Stereocaulon tennesseense* H. Magn. *ex* Degel.**

13. Phyllocladia short or long; cephalodia common, large and bulbous, brown to purple 14

14. Stalks under 3 cm tall; stems entirely free of tomentum, quite thick and sinewy at the base like a tree trunk, branching into finer branches at the top; coralloid phyllocladia sometimes grading to squamulose; humid boreal regions in Pacific Northwest, Great Lakes region, and northeast
....*Stereocaulon subcoralloides* (Nyl.) Nyl.

14. Stalks 2-5(-8) cm tall; thin tomentum often present; phyllocladia never squamulose, only coralloid to verrucose; Pacific Northwest and northeastern boreal area ..... *Stereocaulon intermedium* (Savicz) H. Magn.

15.(5) Stalks usually under 2.5 cm high; phyllocladia verrucose, sometimes lobed 16

15. Stalks mostly 2-8 cm long, erect or mat-forming, with distinct main stems; stems not woody or brittle; phyllocladia granular or squamulose, rarely coralloid 19

16. Phyllocladia warty, with dark centers and pale margins; containing porphyrillic acid; mainly eastern arctic and Alaska ....*Stereocaulon arenarium* (Savicz) I. M. Lamb

16. Phyllocladia uniform in color; lacking porphyrillic acid 17

17. Primary thallus disappearing; stalks mat-forming, without main stems; stems woody, brittle; arctic-alpine ....***Stereocaulon rivulorum* H. Magn.**

[Note: *Stereocaulon incrustatum* Flörke is very similar but has knobby, globose phyllocladia 0.2-0.5 mm in diameter that remain separate, not partly fusing as in *S. rivulorum*, the stalks are more erect and separate often with a dark, patchy, thin tomentum, and the cephalodia are globose and usually conspicuous. It is rare in North America in arctic Canada, Alaska and Colorado.]

17. Primary thallus persistent, granular to squamulose; stalks usually erect and distinct 18

18. Mostly northeastern, Great Lakes to east coast, with scattered boreal localities farther north and west; cephalodia black, with a fuzzy or grainy surface, containing *Stigonema* (fig. 1a) ....*Stereocaulon condensatum* Hoffm.

18. Western montane and arctic; cephalodia brown, smooth and often fissured, containing *Nostoc* (fig. 1d) ....*Stereocaulon glareosum* (Savicz) H. Magn.

19.(15) Phyllocladia mostly granular, clustered; cephalodia in fibrous tufts, dark brown or olive-black, abundant; boreal to arctic ....***Stereocaulon paschale* (L.) Hoffm.**

19. Phyllocladia flat, deeply or shallowly lobed, or warty to granular; cephalodia granular to bulbous, not fibrous, abundant or sparse 20

20. Tomentum thin or scattered in patches, not pinkish; stalks typically long, sparsely branched, 3-5(-6) cm long, covered with minutely squamulose to almost granular phyllocladia and granular cephalodia; apothecia mostly lateral; medulla PD+ orange (stictic acid); western, New Mexico to B.C. ....*Stereocaulon myriocarpum* Th. Fr.

20. Tomentum thin to thick, white to pinkish; stalks well branched, with squamulose to warty phyllocladia; medulla PD– or PD+ orange 21

21. Tomentum thick or sometimes thin, often pinkish; cephalodia common and conspicuous, coarsely granular and covered with tomentum or lumpy and bare; medulla PD– (lobaric acid) 22

21. Tomentum thick, puffy, creamy white, rarely pinkish; cephalodia tiny, dark blue-green granules buried in the tomentum 23

22. Boreal; phyllocladia typically squamulose, not warty-granular; stalks usually erect, 4-8 cm tall; apothecia occasional, terminal .... ***Stereocaulon grande* (H. Magn.) H. Magn.**

22. Arctic-alpine; phyllocladia typically warty or verrucose to granular, some becoming squamulose; stalks prostrate or erect, 1-5 cm tall; apothecia rare, attached laterally ....*Stereocaulon alpinum* Laurer *ex* Funck

23. Medulla PD– (lobaric acid); thallus usually prostrate, forming dorsiventral mats; common in Pacific Northwest, uncommon elsewhere .... ***Stereocaulon sasakii* Zahlbr.**

23. Medulla PD+ orange (stictic acid); thallus erect, not typically forming dorsiventral mats; common in the northeast and into the northern and western boreal regions .... ***Stereocaulon tomentosum* Fr.**

[Note: Different lichenologists differ on the correlating growth form of the two chemotypes. Some say that *S. tomentosum*, which contains stictic acid, is usually prostrate, not erect, and the opposite is true of *S. sasakii*, which containing lobaric acid. More field observations are needed here.]

STICTA See also MacDonald et al. (2003), Galloway and Thomas (2004).

[Note: New genetic research (Moncada et al. 2013) has shown that the classification of species of *Lobaria, Sticta,* and *Pseudocyphellaria* needs major realignment. The traditional treatment below is therefore provisional.]

1. Lacking soredia, isidia, or lobules; arctic-alpine on the ground .... *Sticta arctica* Degel.

1. With soredia, isidia, or lobules; temperate to tropical 2

2. Soredia produced on or close to the lobe margins; west coast and southern Appalachians .... ***Sticta limbata* (Sm.) Ach.**

2. Soredia absent, isidia or tiny lobules present 3

3. With cylindrical or branched isidia; widespread 4

3. With flattened isidia-like lobules (phyllidia) 6

4. Isidia along the lobe margins or cracks in the thallus, coralloid, often sparse; surface of the lobes smooth and often shiny .... ***Sticta beauvoisii* Delise**

[Note: Northern and western specimens that key to *S. beauvoisii* may represent a distinct species.]

4. Isidia covering the thallus surface, which is dull and granular in appearance 5

5. Lobes broad and round; usually on bark, rarely on rock; common .... ***Sticta fuliginosa* (Hoffm.) Ach.**

5. Lobes divided into multiple smaller lobes; usually on rock and humus; rare .... *Sticta sylvatica* (Hudson) Ach.

6.(3) Southeastern 7

6. Southwestern (Arizona) 8

7. Medulla mustard yellow to orange, K+ purple; slightly to elaborately lobulate on thallus margins ....*Sticta fragilinata* T. MacDonald

7. Medulla white, K–; thallus with small round lobes ....*Sticta carolinensis* T. Macdonald

8. Lobes often strongly wrinkled and with small depressions ....*Sticta leucoblephera* (Müll. Arg.) D. J. Galloway

8. Lobes lacking depressions and wrinkles ....*Sticta xanthotropa* (Kremp.) D. J. Galloway

STRANGOSPORA (including *Piccolia, Biatorella, Steinia,* and *Sarcosagium*) Based in part on Knudsen & Lendemer (2007), James et al. (2009), and Knudsen & Ryan (2007).

1. Growing on soil, peat, or over moss 2

1. Growing on wood or bark 4

2. Spores spherical, ca. 5-6 µm in diameter, 12-16 per ascus; apothecia brown; hypothecium brown; widespread temperate, but inconspicuous .... *Steinia geophana* (Nyl.) Stein.

2. Spores ellipsoid, 5-8 x 2-2.5 µm, numerous within the ascus 3

3. Apothecia yellow to reddish yellow, up to 1.4 mm in diameter; arctic-alpine ....*Biatorella hemisphaerica* Anzi

3. Apothecia red-brown, translucent when wet, up to 0.5 mm in diameter; widespread, mainly temperate, but recorded from Alaska .... *Sarcosagium campestre* (Fr.) Poetsch & Scheidem.

4.(1) Apothecia yellow, orange, or red 5

4. Apothecia reddish-brown to black 7

5. Thallus yellow or yellow-green; apothecia yellow-orange, K+ deep purple (anthraquinones); spores 2.0-2.5 µm in diameter; southeastern coastal plain .... *Piccolia nannaria* (Tuck.) Lendemer & Beeching

5. Thallus gray or indistinct; apothecia K– or K+ purple; spores 3-4.5 µm in diameter 6

6. Apothecia scarlet red, 0.1-0.3 mm, K–; widespread in central regions and California, rare in northeast

....*Strangospora microhaema* (Norman) R.A. Anderson

6. Apothecia yellow to orange, 0.1-0.5 mm in diameter, pruina K+ purple; widespread except in northeast ....*Piccolia ochrophora* (Nyl.) Hafellner

7.(4) Apothecia dark brown to black, 0.3-0.5 mm in diameter; epihymenium bluish or greenish-gray, usually HNO_3+ purple; rare and widely distributed B.C. to New Brunswick, but frequent in southern California ....*Strangospora moriformis* (Ach.) Stein.

7. Apothecia reddish brown, 0.2-0.5 mm in diameter; epihymenium yellow-brown to red-brown, HNO_3–; rare, mostly North Temperate .... *Strangospora pinicola* (A. Massal.) Körber

TELOSCHISTES (including *Seirophora*) Based in part on Frödén, Ryan & Kärnefelt (2004).

1. Branches producing granular soredia; apothecia rare 2

1. Branches without soredia; apothecia common 3

2. Thallus consisting of tangled tufts of slender branches, 2-7(-10) cm tall; widespread subtropical ....***Teloschistes flavicans* (Sw.) Norman**

2. Thallus consisting of masses of foliose to subfruticose lobes, often fringed or covered with fibrils, 0.5-1 cm tall; western intermontane region ....*Seirophora contortuplicata* (Ach.) Fröden (syn. *Teloschistes contortuplicatus*)

3. Thallus in tufts up to 2 cm high; apothecial margins abundantly ciliate; widespread, but scattered, especially in prairies .... ***Teloschistes chrysophthalmus* (L.) Th. Fr.**

3. Thallus in tufts 2-7(-10) cm high; apothecial margins with or without cilia 4

4. Apothecia yellow to orange; branches with a tendency to become strap-shaped; Texas ....*Teloschistes exilis* (Michaux) Vainio (*not* LNA, Plate 827)

4. Apothecia red-orange to dark orange; branches somewhat flattened to angular or terete (especially near tips); widespread in subtropical regions, especially coastal southern California ....non-sorediate morph of ***Teloschistes flavicans* (Sw.) Norman** [Plate 827 in LNA as "*Teloschistes exilis*"]

THELENELLA Based almost entirely on Harris (1995); also Orange et al. (2009d).

1. Growing on mosses and plant remains; mainly western and northeastern mountains; spores 60-110 x 20-27 µm, 2-4 per ascus (var. *muscorum*), or 50-65 x 14-19 µm, 8 per ascus (var. *octospora*) ....*Thelenella muscorum* (Fr.) Vainio

1. Growing on bark or rock 2

2. Spores becoming brown, 25-40 x 10-17 µm, muriform; on bark, California
.... *Thelenella hassei* (Zahlbr.) H. Mayerh.

2. Spores remaining colorless 3

3. Growing on bark; spores muriform, 8 per ascus 4

3. Growing on rock 6

4. Spores 24-34 x 7.5-10 µm; central U.S. .... *Thelenella pertusariella* (Nyl.) Vainio

4. Spores 11-17 µm wide 5

5. Southeastern; spores broadest toward the base, 30-38 x 13-15 µm .... *Thelenella rappii* R. C. Harris

5. California to coastal B.C., Minnesota; spores cylindrical or broadest in the middle, 25-42 x 11-17 µm
.... *Thelenella modesta* (Nyl.) Nyl.

6.(3) Spores 30-45 x 12-19 µm, 8-10 x 4-celled; Ozarks to southeast .... *Thelenella luridella* (Nyl.) H. Mayrh.

6. Spores 24-36 x 9-13 µm, 7-9 x (2-)3-4-celled 7

7. California and Arizona; thallus thick, rimose-areolate........................ *Thelenella inductula* (Nyl.) H. Mayrh.

7. Ozarks and New Jersey; thallus thin, continuous to rimose .... *Thelenella brasiliensis* (Müll. Arg.) Vainio

THELIDIUM Based in part on Clauzade & Roux (1985); Harris & Ladd (2005); Fink (1935); Lendemer & Harris (2008).

[Note: Distributions given in the couplets are partly based on unconfirmed reports in CNALH and should be considered as very approximate; all are poorly known.]

1. Spores mostly 4-celled 2

1. Spores 2-celled 4

2. Perithecia immersed in rocks, forming pits, 0.1-0.25(-0.5) mm in diameter; spores 30-55 x 11-17 µm; Missouri, Maine .... *Thelidium incavatum* Nyl *ex* Mudd

2. Perithecia ± superficial; not or infrequently forming pits in rock 3

3. Perithecia 0.4-0.8 mm in diameter; spores 29-52 x 12-18 µm; Pennsylvania to Ontario, scattered in the arctic, on calcareous rock .... *Thelidium papulare* (Fr.) Arnold

3. Perithecia 0.1-0.3 mm in diameter; spores 23-36 x 10-15 µm; central to eastern U.S. .... *Thelidium zwackhii* (Hepp) A. Massal.

[Note: The distinctions between *Thelidium microbolum* (Tuck.) Hasse, reported from California to central U.S., and *T. zwackhii*, are unclear at present. Hasse's material of *T. microbolum* represents a species of *Verrucaria* (Nash 2002: 510).]

4. Perithecia immersed in rock, 0.15-0.4 mm in diameter; spores 20-42 x (8-)10-18 µm; widespread (especially Ozarks) .... *Thelidium decipiens* (Nyl.) Krempelh.

4. Perithecia superficial 5

5. Spores 15-21 x 6-9 µm; perithecia 0.1-0.2 mm in diameter; B.C., central U.S., and arctic .... *Thelidium minutulum* Körber

5. Spores 19-32(-41) x 9-14(-16) µm; perithecia 0.25-0.4 (-1.0) mm in diameter, usually superficial but can form pits; Eastern, Ontario, Alberta/B.C. mountains, arctic .... ***Thelidium pyrenophorum* (Ach.) Mudd**

THELOCARPON Based largely on Orange et al. (2009d).

1. Spores (6-)8-12 x 3-5(-6) µm; on bare or alga-covered soil or stones; eastern .... *Thelocarpon superellum* Nyl.

1. Spores smaller than 6.5 x 3 µm 2

2. Spores subglobose, 1.5-5 x 1.5-2 µm; paraphyses persistent; on rock, wood, or burnt soil; East Temperate, common, and rare in California .... *Thelocarpon laureri* (Flotow) Nyl.

2. Spores ellipsoid or bacilliform, 4-6.5 x 1.0-2.0 µm 3

3. Paraphyses disappearing, but periphyses (around the ostiole) are present; on rotting wood and moist rocks; eastern and California .... *Thelocarpon intermediellum* Nyl.

3. Paraphyses persistent; periphyses absent; typically on other lichens, but also on soil or wood; Great Lakes region westward to B.C. .... *Thelocarpon epibolum* Nyl.

THELOMMA

1. Thallus distinctly lobed at the margin, verrucose-areolate in the center, dull, often pruinose; spores 2-celled, 15-20 x 10-12 µm; on wood and sometimes rocks .... ***Thelomma californicum* (Tuck.) Tibell**

1. Thallus areolate to verrucose, not lobed 2

2. Thallus sterile, producing dark brown soredia-like granules or isidia on the surface of the verrucae

.... *Thelomma ocellatum* (Körber) Tibell [See couplet 5]

2. Thallus fertile, producing mazaedia 3

3. Spores 2-celled; on wood 4

3. Spores 1-celled, spherical, 13-16 µm in diameter; on rocks 6

4. Spores (12-)13-16(-17) x 7-9 µm; medulla IKI–; Florida to Carolinas .... *Thelomma carolinianum* (Tuck.) Tibell

4. Spores 22-28 x 14-15 µm; on wood 5

5. Medulla IKI–; thallus areolate, brownish gray, not granular-isidiate, with regular round mazaedia; mainly coastal California to Alaska .... *Thelomma occidentale* (Herre) Tibell

5. Medulla IKI+ blue; thallus verrucose, gray, forming dark brown soredia-like corticate granules on the surface of the verrucae; mazaedia breaking through the surface in irregular or round patterns; B.C. and Montana to California .... *Thelomma ocellatum* (Körber) Tibell

6.(3) Thallus dusky to dark yellow, verrucae up to 2.5 mm in diameter; cortex KC–, UV+ blue-white (divaricatic acid); California .... *Thelomma santessonii* Tibell

6. Thallus gray to yellowish gray, verrucae smaller; cortex KC+ pink, UV+ white (3-chlorodivaricatic acid); coastal and montane California to B.C. .... ***Thelomma mammosum* (Hepp) A. Massal.**

THELOPSIS Based on Harris (1979), Lendemer et al. (2013).

1. Spores 1-celled, subglobose, 5-7.5 µm in diameter; mainly Ozarks, on hardwoods .... *Thelopsis flaveola* Arnold

1. Spores 2- to 6-celled or submuriform 2

2. Spores 2-celled, 12-15 x 5-8 µm; southern California and Colorado; ascomata subglobose, thallus-colored, with white ostioles, up to 1.0 mm in diameter; on bark and rock .... *Thelopsis isiaca* Stizenb.

2. More than 2-celled; southeastern to central regions; ascomata pink to pale brown 3

3. Spores 4(-6)-celled, 12-18(-21) x 5-6 µm; ascomata up to 0.3 mm in diameter; on bark of deciduous trees or sandstone .... *Thelopsis rubella* Nyl.

3. Spores submuriform, 9-14 x 5-7 µm; ascomata tiny, ca. 0.1 mm in diameter; on bark of deciduous trees

.... *Thelopsis inordinata* Nyl.

THELOTREMA (including *Reimnitzia*) Based largely on Harris (1995).

1. Thallus isidiate; spores brown 2
1. Thallus not isidiate; spores colorless 3

2. Isidia sparse, slender or thick, unbranched; spores submuriform, 16-28 x 7-12 µm, apothecia frequent, immersed, disk 0.5-1.6 mm across, pruinose, with ragged margins; thallus PD– (no substances); coastal plain .... *Reimnitzia santensis* (Tuck.) Kalb (syn. *Thelotrema santense*)
2. Isidia abundant, long and branched; spores 2-celled, 8-11 x 5-6 µm; apothecia rarely produced; thallus PD+ yellow (psoromic acid); Florida .... *Thelotrema eximium* R. C. Harris

3. Spores muriform; contain no substances 4
3. Spores only transversely septate, (4-)6- to 15-celled, 25-45 x 6-10 µm 7

4. Spores 25-35(-45) x 12-15(-18) µm, (2-)6-8 per ascus; apothecia less than 0.5 mm in diameter, immersed; ostiole black, toothed; coastal plain .... *Thelotrema defectum* Hale
4. Spores over 50 µm long 5

5. Spores 1 per ascus, 105-140 x 24-30 µm; apothecia with a wide open ostiole up to 0.5 mm wide; coastal plain .... *Thelotrema monospermum* R. C. Harris
5. Spores 2-4 per ascus, 65-90 x 15-20 µm 6

6. In section, exciple seen as a thin, torn membrane within the apothecial cavity and free from the walls; mainly along the northeastern and northwestern coasts, Appalachians, rare in Florida .... ***Thelotrema lepadinum (Ach.) Ach.***
6. In section, exciple thick, filling the apothecial cavity and often covering the disk; coastal plain to N.C. .... *Thelotrema adjectum* Nyl.

7.(3) Thallus K+ yellow, PD+ orange (stictic acid); ostioles broad, up to 1.5 mm across, pruinose; Florida and Louisiana to N.C. .... *Thelotrema dilatatum* (Müll. Arg.) Hale
7. Thallus K–, PD– (no substances) 8

8. Apothecia largely buried in the thallus or bark, 0.3-0.6 mm across; thallus thin, with an imperceptible medulla or within the bark 9

8. Apothecia prominent in small warts or hills; thallus well developed or thin 10

9. Spores 20-45 x 6-8 µm, (6-)8- to 12-celled; exciple even at opening; disk not or thinly pruinose; mainly southeastern U.S. .... *Thelotrema subtile* Tuck.

9. Spores wider, 30-40(-55) x 7.5-10 µm, 8- to 10-celled; exciple ragged and toothed at opening; disk heavily pruinose; Pacific Northwest and eastern U.S. .... *Thelotrema petractoides* P. M. Jørg. & Brodo

10. Spores (4-)6- to 10-celled, 20-38(-42) x 8-10 µm; apothecia 0.3-0.6 mm across, in warts; thallus thin, within bark; northern (B.C., New England, Newfoundland), rare .... *Thelotrema suecicum* (H. Magn.) P. James

10. Spores 8- to 15-celled, 25-45 x 6-8 µm; apothecia 0.2-0.35(-0.5) mm across, forming smooth hills; thallus well developed, often areolate, with a thick medulla; coastal plain, common .... *Thelotrema lathraeum* Tuck.

TONINIA

1. Spores 4- to 8-celled, 23-55 µm long 2

1. Spores 1- to 4-celled, 12-24 µm long 5

2. Thallus heavily and coarsely pruinose; epihymenium gray, K+ violet; Rocky Mountains, Yukon to New Mexico .... ***Toninia alutacea*** **(Anzi) Jatta**

2. Thallus without pruina; epihymenium greenish to brown or red-brown, K– or K+ red 3

3. Squamules very convex, becoming columnar, shiny or dull reddish brown, with pseudocyphellae that develop into small pits; epihymenium brown to greenish; montane Colorado and coastal California .... *Toninia bullata* (G. Meyen & Flotow) Zahlbr.

3. Squamules somewhat convex or forming a continuous verrucose thallus; lacking pseudocyphellae and pits; widespread in western U.S. and adjacent southern Canada 4

4. Epihymenium reddish brown, K+ red; thallus dark olive-brown, rarely reddish brown .... *Toninia ruginosa* (Tuck.) Herre

4. Epihymenium green, K–; thallus pale to dark brown, not olive .... *Toninia squalida* (Ach.) A. Massal

5.(1) Epihymenium brown to green, K– 6

5. Epihymenium gray to violet, K+ violet 9

6. Thallus crustose, areolate to rimose-areolate, without pruina; spores ellipsoid, 2-celled, 10-15.5 x 4.5-6 µm; Colorado and northern B.C. .... *Toninia phillippea* (Mont.) Timdal

6. Thallus squamulose, pruinose or not; spores cylindrical to fusiform, 1- to 4-celled, 12-25 x 3-7 µm 7

7. Spores (2-)4-celled, cylindrical, 3.0-5.5 µm wide; widespread; squamules ± flat, lacking pores, pruinose or not, scattered or contiguous .... *Toninia aromatica* (Sm.) A. Massal.

7. Spores 1- to 2(-4)-celled, fusiform; pruina entirely absent from the thallus and apothecia; areoles and squamules flat to convex 8

8. Thallus smooth, uniform brown, usually shiny, with deep depressions and pores; spores 1- to 2-celled, 3.5-5 µm wide ....***Toninia tristis* (Th. Fr.) Th. Fr. subsp. *asiae-centralis* (H. Magn.) Timdal**

8. Thallus mottled gray-brown, dull, rough, lacking pits and depressions; spores 2- to 4-celled, 5-7 µm wide .... *Bilimbia lobulata* (Sommerf.) Hafellner & Coppins

9.(5) Thallus usually, and apothecia always, without pruina; spores 10-16.5 x 3.5-4.5 µm; southern Rockies and southern California .... *Toninia massata* (Tuck.) Herre

9. Pruina present on the thallus and often the apothecia, sometimes sparse or patchy 10

10. Squamules very small, 0.4-1(-2) mm across, almost granular in appearance, often scattered, covered with coarse pruina; arctic and south into the Canadian Rockies .... *Toninia arctica* Timdal

10. Squamules medium to large, 1-4(-5) mm across 11

11. Pruina coarse and granular; southern Rockies; shaded soil and rock .... *Toninia subdiffracta* Timdal

11. Pruina fine and powdery 12

12. Squamules strongly convex and folded throughout; thallus not lobed at the margins or forming rosettes; pruina sometimes confined to most exposed surfaces of squamules and often lacking on the apothecia; on soil; very widespread and common .... ***Toninia sedifolia* (Scop.) Timdal**

12. Squamules flattened and somewhat lobed at the margins, convex in the center; thallus often forming rosettes; pruina uniform and dense on the thallus as well as the apothecia; on calcareous rock; southern Rockies .... *Toninia candida* (Weber) Th. Fr.

TRAPELIA

[Note: The genus badly needs a North American revision. Many if not most North American collections are unreliably identified with inconsistently used names.]

1. Growing on bark or wood; thallus consisting of scattered, thin, dark green to brownish areoles with small, round, bright or pale green soralia; apothecia rarely present; Pacific Northwest and eastern U.S. .... *Trapelia corticola* Coppins & P. W. James

1. Growing on siliceous rock 2

2. Thallus sorediate 3
2. Thallus without soredia 4

3. Thallus more or less continuous, rimose to areolate with flat areoles, pale greenish gray; greenish or slightly brownish soredia developing along thallus cracks; apothecia rare; East Temperate, infrequently California to B.C., common on shaded rocks .... *Trapelia placodioides* Coppins & P. James
3. Thallus with scattered to contiguous, dull green to brownish, flat to convex areoles breaking into excavate or convex brown or greenish brown soralia; red-brown, biatorine apothecia common; distribution uncertain but reported from California, rare .... *Trapelia obtegens* (Th. Fr.) Hertel

4. Thallus areolate, the areoles usually scattered, sometimes coalescing and appearing continuous, but some are always separate and sometimes lobed; apothecia very dark red-brown, 0.2-0.4(-0.7) mm in diameter, biatorine, or thallus-colored margins very thin; apothecial base (in section) lacking a prominent, colorless stipe; most specimens containing significant amounts of 5-*O*-methylhiascic acid; common and widespread especially on exposed pebbles and small stones ..... ***Trapelia glebulosa*** **(Sm.) J. R. Laundon** (syn. *T. involuta*)
4. Thallus continuous and smooth or becoming rimose to areolate, but areoles never lobed; 5-*O*-methylhiascic acid absent or present in traces 5

5. Thallus typically thick, continuous, becoming rimose-areolate to verruculose, rarely areolate; apothecia prominent and thick, 0.3-0.7(-1.2) mm in diameter, at least some with 2-3(-5) concentric brown excipular layers; apothecial base (medial section) showing an elongated, colorless, stem-like stipe; on boulders in shaded woodlands in northeastern North America .... *Trapelia stipitata* Brodo & Lendemer, ined.
5. Thallus thin and smooth, often membranous, sometimes rimose in places; apothecia flat, thin, (0.2-)0.3-0.5(-0.7) mm in diameter, usually with thin, single, intermittent, thallus-colored margins, with 1(-2) excipular layers (not visible from above); apothecial base (in section) often with a short, broad, colorless stipe; in exposed or shaded localities; widespread ...... *Trapelia coarctata* (Turner *ex* Sm.) M. Choisy *s. str.*

TRAPELIOPSIS

1. Thallus squamulose, thick, not producing soredia, pale gray; on rocks and soil, California to Oregon .... ***Trapeliopsis glaucopholis* (Nyl. *ex* Hasse) Printzen & McCune** (syn. *T. californica*)
   [Note: = most N. Am. specimens named as *T. wallrothii*; treated as *T. wallrothii* in LNA]
1. Thallus areolate to verrucose, not squamulose; usually sorediate in part; widespread 2

2. On soil, peat, or wood 3
2. Almost exclusively on wood; thallus pale to dark green or greenish gray to greenish black; apothecia almost always dark lead gray to black 5

3. Thallus dark green or greenish gray, membranous to gelatinous, breaking into large or small, irregular soralia; rarely fertile, but apothecia black; Appalachian-Great Lakes .... *Trapeliopsis gelatinosa* (Flörke) Coppins & P. James
3. Thallus pale gray to greenish white or pinkish, well developed; some apothecia pink to brown, others almost black 4

4. Soralia in limited patches, rarely covering large areas, lacking orange spots; apothecia common, pink, brown, or dark gray, sometimes with all colors represented on the same thallus or even the same apothecium; common and widespread .... ***Trapeliopsis granulosa* (Hoffm.) Lumbsch**
4. Soredia sometimes comprising almost the entire thallus, orange in spots, reacting K+ purple; apothecia rare, pink to brown; infrequent in Pacific Northwest, rare in northeast .... *Trapeliopsis pseudogranulosa* Coppins & P. James

5. Thallus areolate or granular, bursting into patches of dark soredia; apothecia scattered, flat, smooth, with persistent margins; spores 7-10 x 3-5 µm; on hard, weathered wood; widespread .... ***Trapeliopsis flexuosa* (Fr.) Coppins & P. James**
5. Thallus granulose, usually dissolving into fine soredia (becoming leprose); apothecia usually clustered and fusing, convex, without margins but often with a vaguely paler edge; spores 10-15 x 4-5 µm; on soft, well rotted wood; northeastern .... *Trapeliopsis viridescens* (Schrader) Coppins & P. James

TRYPETHELIUM (including *Astrothelium, Bathelium, Laurera,* and *Pseudopyrenula*) Based largely on Harris (1995).

[Note: With the exception of *T. virens* (East Temperate), all species are found in tropical or subtropical areas, especially southeastern coastal plain.]

1. Perithecia buried in a wart-like pseudostroma or in thallus 2
1. Perithecia discrete, not buried in a pseudostroma or in thallus 12

2. Pseudostroma brown, often shiny, with an outer layer formed of jigsaw puzzle-like cells, medulla generally pigmented; 8 spores per ascus; Florida 3
2. Pseudostroma not brown, outer cells not jigsaw puzzle-like 4

3. Spores 4-celled, 18-28 x 6-9 µm; medullary pigment K+ yellow .... *Bathelium carolinianum* (Tuck.) R. C. Harris
3. Spores muriform, 40-50 x 12-17 µm; medullary pigment K+ red-purple .... *Bathelium madreporiforme* (Eschw.) Trevisan

4. Spores thin-walled, muriform 5
4. Spores thick-walled, unevenly thickened, muriform or only transversely septate, mostly less than 75 x 30 µm; spores 8 per ascus 6

5. Spores 200-270 x 30-40 µm, 4 per ascus; thallus yellow or yellowish green .... ***Laurera megasperma* (Mont.) Riddle**
5. Spores 90-130 x 20-30 µm, 8 per ascus; thallus olive-brown .... *Laurera subdisjuncta* (Müll. Arg.) R. C. Harris

6. Spores 6- to 16-celled 7
6. Spores 4-celled 9

7. Medulla of pseudostromata yellow, orange or tan, K+ purple (anthraquinones) 8
7. Medulla of pseudostromata not pigmented, K–; spores 6- to 12-celled, 38-52 x 7-10 µm; East Temperate .... ***Trypethelium virens* Tuck. *ex* E. Michener**

8. Spores 10- to 14-celled, 40-50 x 9-12 µm; medulla yellow to tan; southeastern coastal plain .... *Trypethelium eluteriae* Sprengel

8. Spores 13- to 16-celled, 60-75(-85) x 11-12 µm; medulla orange; Florida .... *Trypethelium subeluteriae* Makhija & Patwardhan

9.(6) Perithecia grouped two or three together sharing a single ostiole, pruinose at least around the ostiole, dusted with yellow or orange anthraquinones (K+ purple); spores 28-35 x 10-13 µm .... ***Astrothelium versicolor* Müll. Arg.**

9. Perithecia not sharing an ostiole and not pruinose; anthraquinones absent or present; spores 18-28 x 6-10 µm 10

10. Perithecia superficial; thallus whitish, without a cortex; hymenium containing many oil drops; spore locules with angular walls .... *Pseudopyrenula diluta* (Fée) Müll. Arg. (syn. *P. subnudata*)

10. Perithecia entirely immersed in pseudostromata with only the ostioles visible; thallus dark brownish to olive, with a cortex or below the bark cells; spore locules lens-shaped 11

11. Thallus orange, K+ purple .... ***Trypethelium aeneum* (Eschw.) Zahlbr.**

11. Thallus olive to pale yellowish olive, K– .... *Trypethelium variolosum* Ach. (syn. *T. ochroleucum*)

12.(1) Perithecia conical; thallus light brown to pale orange; excipulum not distinguishable from involucrellum; perithecial cavity colorless; spore locules lens-shaped; spores with a gelatinous halo .... ***Trypethelium tropicum* (Ach.) Müll. Arg.**

12. Perithecia hemispherical; thallus whitish; excipulum distinct from involucrellum at least at base, perithecial cavity yellowish; spore locules with angular walls, spores without a halo .... *Pseudopyrenula diluta* (Fée) Müll. Arg.

TUCKERMANOPSIS (including *Tuckermanella*)

[Note: The spelling change from "*Tuckermannopsis*" to "*Tuckermanopsis*" recommended and accepted by Index Fungorum is adopted here.]

1. Soredia produced along the lobe margins .... ***Tuckermanopsis chlorophylla* (Willd.) Hale**

1. Soredia absent 2

2. Lobes long and strap-shaped, flat; dichotomously branched, forming shrubby clumps; linear-elongate pseudocyphellae sometimes present along the margins ... ***Tuckermanopsis subalpina* (Imshaug) Kärnefelt**

2. Lobes rounded or somewhat elongated, flat to crinkled and crisped; branching usually irregular; pseudocyphellae present or absent, never linear-elongate 3

3. Isidia present on lobe margins and sometimes upper surface, cylindrical to spherical; apothecia absent; pycnidia sparse or very inconspicuous; lobes elongated .... ***Tuckermanella coralligera* (W. A. Weber) Essl.** (syn. *Tuckermanopsis coralligera*)

3. Isidia absent, although marginal lobules or warty tubercles sometimes produced; apothecia abundant; pycnidia abundant and conspicuous; lobes rounded 4

4. Lobes flat 5

4. Lobes undulating or crisped at the margins 7

5. Medulla C+ red (olivetoric acid); southwestern .... *Tuckermanella weberi* (Essl.) Essl.

5. Medulla C– 6

6. Thallus brown, at least when dry, brown to olive when wet; pseudocyphellae sparse; apothecia on or close to the lobe margins; boreal to North Temperate .... ***Tuckermanopsis sepincola* (Ehrh.) Hale**

6. Thallus olive when dry, grass green when wet; pseudocyphellae abundant and conspicuous; apothecia on the lobe surface; Appalachian-Great Lakes and southwestern to prairies .... ***Tuckermanella fendleri* (Nyl.) Essl.** (syn. *Tuckermanopsis fendleri*)

7. Warty tubercles and (or) lobules frequent on the lobe surface and margins; pseudocyphellae abundant and conspicuous, especially on the tubercles; medulla (especially in the apothecial margin) usually yellow or orange in spots .... ***Tuckermanopsis platyphylla* (Tuck.) Hale**

7. Warty tubercles absent; lobules present or absent, pseudocyphellae absent; medulla white throughout 8

8. Medulla C–, KC–, UV– (protolichesterinic acid present); lobules often present on the lobe margins .... ***Tuckermanopsis orbata* (Nyl.) M. J. Lai**

8. Medulla C+ pink or KC+ pink to red (protolichesterinic acid absent); lobules absent 9

9. Medulla C–, KC+ pink, UV+ blue-white (alectoronic acid) ...***Tuckermanopsis americana* (Sprengel) Hale**

9. Medulla C+ pink, KC+ red, UV– (olivetoric acid) .... *Tuckermanopsis ciliaris* (Ach.) Gyelnik

UMBILICARIA (including *Lasallia*)

1. Lobes pustulate or blistered, with depressions on the lower surface corresponding to pustules on the upper surface 2

1. Lobes not pustulate: smooth and even, wrinkled, bumpy, or with a network of depressions and sharp ridges; if warty, the warts do not have corresponding depressions on the lower surface 4

2. Rhizines sparse; lobes usually crowded and overlapping, convex or undulating and crisped at the margins .... ***Lasallia caroliniana* (Tuck.) Davydov et al.** (syn. *Umbilicaria caroliniana*) (see also couplet 6)

2. Rhizines absent; lobes usually solitary and flat 3

3. Lower surface black and coarsely papillate like very rough sandpaper; upper surface rather smooth, with low pustules having sloping sides .... ***Lasallia pensylvanica* (Hoffm.) Llano**

3. Lower surface pale to dark brown, rather smooth or slightly roughened; upper surface very rough, with abundant pustules having almost vertical sides .... ***Lasallia papulosa* (Ach.) Llano**

4.(1) Rhizines present, sparse or abundant 5

4. Rhizines absent 15

5. Lower surface black; upper surface smooth to scabrose 6

5. Lower surface pale to dark brown, pink, or gray; upper surface rough 12

6. Rhizines sparse, thick and irregular, lacking plates on lower surface; thallus large, smooth and shiny, up to 10 cm across, with crowded, strongly folded, overlapping lobes; southern Appalachians and Alaska .... ***Lasallia caroliniana* (Tuck.) Davydov et al.** (syn. *Umbilicaria caroliniana*) [Plate 859 in LNA as *Umbilicaria*]

6. Rhizines abundant, with or without plates; thallus small or large 7

7. Thallus relatively round and flat, upper surface dull or mat, never shiny; rhizines developing out of a granular lower surface and forming a velvety mat, plates absent 8

7. Thallus upper surface shiny or dull; rhizines not forming a short velvety mat; lower surface not granular, but plates of tissue usually obvious 10

8. Thallus usually brown, but sometimes with a grayish "bloom"; thin and membranous or moderately thick, rather fragile, 4-15(-30) cm in diameter .... ***Umbilicaria mammulata* (Ach.) Tuck.**

8. Thallus usually pale gray or almost white, thick and stiff like cardboard, rather tough, 2-8(-12) cm in diameter 9

9. Rhizines all covered with a coating of black granules (fig. 35b); temperate to boreal and western montane

.... ***Umbilicaria americana*** **Poelt & T. H. Nash**

9. Rhizines having black granules only at the base, the remainder naked (fig. 35a), pale tan to black; arctic-alpine .... *Umbilicaria vellea* (L.) Ach.

10.(7) Upper surface dull, often pruinose, scabrose or slightly verruculose; thallus small, usually under 3 cm in diameter; rhizines tend to be flattened, knobby, or ball-tipped; mainly western arctic and alpine .... *Umbilicaria cinereorufescens* (Schaerer) Frey

10. Upper surface smooth and shiny, never pruinose or scabrose; thallus up to 6 cm in diameter; rhizines ropy or branched, not flattened except where they emerge from plates 11

11. Lower surface covered with irregular plates of tissue with short, highly branched rhizines developing from the plate margins; rhizines confined to lower surface; apothecia abundant, flat, usually angular, with concentric ridges; western montane .... *Umbilicaria angulata* Tuck.

11. Lower surface rough, but plates confined to area immediately around attachment; many rhizines unbranched or forked, ropy, developing from the lower cortex (not plates), some branched, often forming clumps on the upper surface; apothecia rare, convex, round, with radiating ridges; western North America and Newfoundland .... *Umbilicaria polyrrhiza* (L.) Fr.

12.(5) Thallus dull gray to gray-brown, 2-8 cm across, with coarse granular soredia developing from the breakdown of the upper cortex close to the downturned margins; lower surface rough, with abundant, tapered, unbranched or forked rhizines; apothecia very rare; mainly northeastern, scattered elsewhere .... *Umbilicaria hirsuta* (Sw. *ex* Westr.) Hoffm.

12. Thallus gray or brown, surface without soredia, but covered with crystal-like deposits at least close to the umbilicus; apothecia often present 13

13. Lower surface gray; rhizines unbranched, smooth; apothecial disks convex with thick concentric ridges .... ***Umbilicaria proboscidea*** **(L.) Schrader**

13. Lower surface pink to pale brown; rhizines dichotomously branched; apothecial disks more or less flat, with or without concentric ridges 14

14. Marginal cilia or cilia-like rhizines absent, but rhizines abundant on lower surface; apothecia broadly attached and adnate; apothecial disks with central buttons of sterile tissue or smooth and even; cortex KC+ red, C+ pink, although the reactions are sometimes faint and difficult to see (gyrophoric acid) .... ***Umbilicaria virginis*** **Schaerer**

14. Marginal cilia or cilia-like rhizines common and abundant; nonmarginal rhizines sparse; apothecia constricted at the base and raised; apothecial disks with concentric ridges of sterile tissue; cortex KC–, C– (gyrophoric acid absent) .... ***Umbilicaria cylindrica*** **(L.) Delise *ex* Duby**

15.(4) Isidia present on thallus surface, mostly granular .... ***Umbilicaria deusta*** **(L.) Baumg.**

15. Isidia absent 16

16. Upper surface of thallus pruinose, or with coarse, crystal-like deposits 17

16. Upper surface of thallus entirely without pruina or deposits 21

17. Lower surface pruinose at least close to margins 18

17. Lower surface not pruinose 19

18. Apothecia adnate; apothecial disks with concentric ridges of sterile tissue; thallus thin to moderately thick .... ***Umbilicaria proboscidea*** **(L.) Schrader**

18. Apothecia raised; apothecial disks with central buttons of sterile tissue; thallus thick and stiff .... ***Umbilicaria polaris*** **(Schol.) Zahlbr.** (= *U. kraschenınnikovii* of most authors) [Plate 864 in LNA]

19. Upper surface bumpy but without a reticulate pattern of ridges; apothecia abundant, raised, with smooth and even disks; mainly Alaska to alpine Washington and western Northwest Territories .... ***Umbilicaria rigida*** **(Du Rietz) Frey**

19. Upper surface with a network of depressions and sharp ridges, or wrinkled; western montane and arctic 20

20. Reticulate sharp ridges extending to the thallus margins; lower surface of well-developed specimens typically showing deep folds around the attachment point; apothecia not common, adnate, with disks having central buttons of sterile tissue .... ***Umbilicaria decussata*** **(Vill.) Zahlbr.**

20. Reticulate ridges becoming shallow and often fading entirely at the thallus margins; lower surface of well-developed specimens smooth, without deep folds around attachment .... *Umbilicaria lyngei* Schol.

21.(16) Lower surface with a network of rough and papillate membranes 22

21. Lower surface smooth, papillate, or tuberculate, without membranes 23

22. Margins of the thallus perforated with irregular holes and forming finely divided lobes; apothecial disks with concentric ridges; medulla PD+ orange, K+ yellow (stictic acid) .... ***Umbilicaria torrefacta*** **(Lightf.) Schrader**

22. Margins of the thallus not perforated and lobed; apothecial disks with radiating ridges; medulla PD–, K– .... ***Umbilicaria muehlenbergii*** **(Ach.) Tuck.**

23. Thallus round and flat; upper surface verruculose to ridged 24

23. Thallus forming cushions of crowded, overlapping lobes; upper surface smooth, not covered with warts or areoles 26

24. Upper surface obscurely verruculose at the margins, becoming more strongly warty and even reticulate-ridged toward the center; lower surface uniformly black, sooty, with very fine black granules; apothecia rarely seen; rare .... *Umbilicaria nylanderiana* Zahlbr.

24. Upper surface partly or entirely covered with convex warts or areoles, lacking reticulate ridges; lower surface gray to dark brown, sometimes almost black; apothecia commonly produced 25

25. Upper surface areolate with warts over a smooth, black, basal layer that is visible between the areoles on at least part of the thallus; thallus 1-5 cm across, up to 0.3 mm thick; lower surface more or less uniform brown to brownish black; common .... ***Umbilicaria hyperborea*** **(Ach.) Hoffm.**

25. Upper surface entirely covered with rounded warts or worm-like ridges, lacking patches of smooth black areas; thallus 5-10(-20) cm across, up to 1.1 mm thick; lower surface often grayish with a black area around the umbilicus; arctic to western montane .... *Umbilicaria arctica* (Ach.) Nyl.

26. Lower surface black; lobes margins turned up; apothecia rare .... ***Umbilicaria polyphylla*** **(L.) Fr.**

26. Lower surface pale or dark brown, lobe margins turned down; apothecia abundant .... ***Umbilicaria phaea*** **Tuck.**

USNEA Based in part on Lendemer et al. (2013), Clerc (2011, and unpublished notes on northeastern *Usneae*), Halonen et al. (1998), and McCune & Geiser (2009).

[Note: See figs. 36 and 37 for *Usnea* terminology.]

1. Thallus pendent 2

1. Thallus growing in bushy tufts from a single point, or almost pendent 16

2. Surface of main branches rough and eroded, with the cortex thin and crumbling; main branches almost undivided, but with short to moderately long perpendicular side branches abundant all along the filaments .... ***Usnea longissima*** **Ach.**

2. Surface of main branches with a continuous cortex, with depressions and ridges, or with tubercles or papillae, not at all eroded 3

3. Axis pink to brown; medulla white or reddish, usually CK+ orange (diffractaic acid; but see *U. trichodea*) ... 4

3. Axis and medulla white 5

4. Medulla usually pink to red, dense or lax, 20-30%; older branches with abundant whitish warts; cortex very thick and hard .... ***Usnea ceratina*** **Ach.**

4. Medulla white, dense, 16-21%; whitish warts absent; cortex thin .... ***Usnea trichodea*** **Ach.**

[Note: Some specimens of *U. trichodea* contain constictic acid either with or instead of diffractaic acid.]

5. Isidia present 6

5. Isidia absent 13

6. In Florida; isidia mostly developing within soralia on branch tips; medulla K+ red (norstictic and galbinic acids) .... ***Usnea dimorpha*** **(Müll. Arg.) Motyka**

6. Boreal to temperate; isidia not developing within soralia, although sometimes in small clusters; medulla K– or K+ red (but norstictic acid absent) 7

7. Branches smooth, without papillae, with circular, thickened cracks (fig. 37e) at least at the base 8

7. Branches with papillae, with or without circular, thickened cracks; medulla KC–; PD– or PD+ yellow, K– or K+ red (lacking protocetraric acid) 9

8. Medulla KC+ pink, PD+ red, K– (protocetraric acid); Appalachians and west coast .... ***Usnea subgracilis*** **Vainio** (syn. *U. hesperina*)

8. Medulla KC–, PD+ yellow, K+ red (salazinic acid); Appalachians to Maritimes .... *Usnea merrillii* Motyka

9. Surface often dented or pitted; thallus more or less uniform in color; cortex very thin and fragile (cortex 7-10%, medulla 22-26%, axis 28-38%); medulla PD–, K–, or PD + yellow, K+ red (salazinic acid) .... ***Usnea scabrata*** **Nyl.**

9. Surface more or less even, not pitted or dented; cortex moderately thick and tough 10

10. Axis extremely broad, almost filling the branch; medulla very thin and dense, about the same width as cortex (7-13%); containing salazinic acid; widespread boreal and west and east coasts .... *Usnea silesiaca* Motyka (syn. *U. madierensis*)

10. Axis usually under 50%, medulla medium, dense or loose (10-20%) 11

11. Thickened, bone-like joints abundant all along the branches; contains salazinic acid .... *Usnea chaetophora* Stirton

11. Thickened, bone-like joints sparse or present only near the base 12

12 Main branches dark, especially at the base; cortex 7-10%, medulla 14-21%, axis 36-43%; medulla PD+ yellow, K+ red (salazinic acid); boreal and western temperate .... ***Usnea dasopoga* (Ach.) Nyl.** (syn. *Usnea filipendula, U. dasypoga*)

12. Main branches not darker than other branches; cortex about as thick as medulla (10-20%); medulla PD+ yellow, K– (baeomycesic and squamatic acids); Pacific Northwest .... *Usnea pacificana* P. Halonen

13.(5) Medulla loose and webby; branches dented and pitted, not divided into segments by circular cracks .... ***Usnea cavernosa* Tuck.**

13. Medulla thin and compact 14

14. Branches strongly ridged, often broken into small segments, but not bone-like (with swollen circular cracks); medulla PD+ yellow, K+ red, KC–, CK– (norstictic acid); rare, Appalachian-Great Lakes distribution .... *Usnea angulata* Ach.

14. Branches terete, not dented, pitted, or ridged, but with swollen circular cracks and bone-like segments (fig. 37e) at least at the base 15

15. Medulla PD+ red, K–, KC+ pink, CK + orange (protocetraric acid) .... ***Usnea subgracilis* Vainio** (syn. *U. hesperina*)

15. Medulla PD+ yellow, K+ red (salazinic acid) .... *Usnea silesiaca* Motyka (see couplet 10)

[Note: Isidia can be very sparse or even absent in *Usnea merrillii* and *U. chaetophora* (see couplets 8 and 11).]

16.(1) Thallus with apothecia 17

16. Thallus without apothecia 20

17. In central to eastern North America; medulla rather dense 18

17. In southwestern U.S.; medulla loose, with white cottony hyphae 19

18. Medulla usually red, at least in part, or sometimes white; base of thallus not blackened; medulla K– or K+ red (norstictic acid); papillae, if present, wide at the base; very common .... ***Usnea strigosa* (Ach.) Eaton**

18. Medulla white; base of thallus blackened; medulla K+ red (salazinic acid); papillae usually present, cylindrical; northeastern, uncommon .... *Usnea subfusca* Stirton

19. Spiny fibrils abundant and crowded on the back of the apothecia; base of thallus narrowly or irregularly blackened or not blackened at all; medulla K–, PD– (salazinic acid absent) .... ***Usnea parvula* Motyka** (syn. *U. cirrosa*)

19. Spiny fibrils very sparse on the back of the apothecia; base of thallus extensively blackened; medulla usually K+ red, PD+ orange or yellow (salazinic acid) .... ***Usnea intermedia* (A. Massal.) Jatta** (syn. *U. arizonica*)

[Note: *Usnea quasirigida* Lendemer & Tavares (syn. *U. rigida* Mot. *non* Vainio) is a similar western species with protocetraric acid (K–, PD+ red-orange).]

20.(16) On rock 21

20. On trees or wood 23

21. Branches spotted with black bands, coalescing into more or less blackened tips; round soralia at the branch tips, lacking isidia; northern arctic with rare occurrences in the mountains of Washington and Oregon .... *Usnea sphacelata* R. Br. (syn. *Neuropogon sulphureus*)

21. Branches uniform in color or becoming brownish, without black bands; isidia developing in soralia; temperate; containing norstictic acid; mainly Appalachian 22

22. Isidia with black tips; also containing galbinic acid .... ***Usnea amblyoclada* (Müll. Arg.) Zahlbr.**

22. Isidia not black at tips; also containing salazinic acid .... *Usnea halei* P. Clerc

23.(20) Branches clearly constricted at base and usually the axils, giving the branches an inflated appearance (fig. 37d) 24

23. Branches not constricted at base, more or less uniform in diameter 28

24. Isidia absent or rare, soredia present 25

24. Both isidia and soredia developing within soralia, concentrated at the branch tips or not 27

25. Axis and medulla usually pale yellow; cortex often red-spotted; containing norstictic acid, psoromic acid, or fatty acids; Pacific Northwest and northeast .... *Usnea flavocardia* Räsänen (syn. *U. wirthii*)

25. Axis and medulla entirely white; cortex never red-spotted 26

26. Soredia in excavate soralia mainly at the branch tips; thallus spinulose; medulla K–, PD+ orange-red (protocetraric acid); western and Great Lakes region .... ***Usnea glabrata* (Ach.) Vainio**

26. Soralia convex, not confined to branch tips; thallus often spinulose; medulla K+ yellow to red (galbinic, norstictic, and salazinic acids); Appalachian-Ozark and southern California ....*Usnea dasaea* Stirton

27.(24) Soralia scattered over branches, remaining discrete; thallus black at the base for 1-2 mm .... ***Usnea fragilescens* Hav. *ex* Lynge var. *mollis* (Vainio) P. Clerc**

27. Soralia concentrated at branch tips, becoming confluent; thallus usually more or less uniform in color, but sometimes darkening at the base .... ***Usnea cornuta* Körber**

28.(23) Thallus cortex distinctly reddish to red-brown or orange, especially at the base, sometimes mottled 29

[Note: If spotted with red, see couplet 25.]

28. Thallus cortex shades of yellowish green (usnic yellow), not reddish 30

29. Red pigment typically covering branches and extending deep into cortex; medulla PD+ yellow to orange (norstictic and salazinic acids or stictic acid group); west coast and East Temperate .... ***Usnea rubicunda* Stirton**

29. Orange pigment only mottling branches, only superficial on cortex; medulla PD+ red (fumarprotocetraric and protocetraric acids); locally common Maine to Nova Scotia .... *Usnea subrubicunda* P. Clerc

30. Axis and medulla pink or red (rarely white in *U. ceratina*) 31

30. Axis and medulla white 34

31. Axis broad and hollow; mainly Florida and Georgia 32

31. Axis solid throughout; medulla PD–, K– (norstictic acid absent) 33

32. Branches round in cross section, not ridged; isidia in clusters; medulla PD+ yellow, K+ red (norstictic acid) .... ***Usnea baileyi* (Stirton) Zahlbr.**

32. Branches somewhat ridged with isidia along the ridges; medulla PD–, K– (diffractaic acid) .... *Usnea perplectata* Motyka

33. Conspicuous whitish warts or tubercles present on branch surface; medulla thick, loose or dense; containing diffractaic acid .... ***Usnea ceratina* Ach.**

33. Whitish warts absent; medulla thin and dense; lacking diffractaic acid .... ***Usnea mutabilis* Stirton**

34.(30) Isidia absent, soralia abundant 35

34. Isidia present, either directly on the branches or in soralia 38

35. Medulla usually K+ yellow (stictic acid complex) or K+ red (norstictic acid); soralia level with branch surface or mounded, round and relatively small; base of thallus blackened .... *Usnea glabrescens* (Nyl. *ex* Vainio) Vainio

35. Medulla K– or K+ red (salazinic acid); base of thallus pale or blackened 36

36. Axis almost filling branch, medulla very thin and dense .... *Usnea silesiaca* Motyka

36. Axis narrower than 50%; medulla thick and typically loose 37

37. Soralia excavate, wider than half the branch diameter; soredia fine to coarse .... ***Usnea lapponica* Vainio**

37. Soralia mounded to excavate, often coalescing; soredia white, coarsely granular .... *Usnea substerilis* Motyka

38.(34) Branches ridged, at least partly angular in cross section; isidia present, not arising from soralia; cortex thin and fragile; lacking papillae .... ***Usnea hirta* (L.) F. H. Wigg.**

38. Branches not ridged, uniformly round in cross section; isidiate soralia present or absent; cortex thick or moderate, not fragile; papillae present, at least on basal or older branches 39

39. Thallus not blackened at base; branches rather smooth, with few or inconspicuous papillae; cortex much thicker than medulla and very hard (cortex 18-22%, medulla 6-10%, axis 40-58%); medulla PD+ red, KC+ pink to red (protocetraric acid) .... ***Usnea subscabrosa* Nyl. *ex* Motyka**

[Note: Sparsely pigmented specimens of *U. subrubicunda,* also PD+ red, may key out here. It usually has papillae. See couplet 29.]

39. Thallus blackened at the base; main branches bumpy with abundant papillae; cortex thinner than, or equal to, medulla; medulla PD– or PD+ yellow or orange (protocetraric acid absent) 40

40. Axis almost filling branch, medulla very thin and dense, cortex thick and hard; containing salazinic acid .... *Usnea silesiaca* Motyka

40. Axis narrower than 50%, not filling branch 41

41. Cortex about as thick as the dense medulla (10-20%); medulla PD+ yellow (baeomycesic and squamatic acids); Pacific Northwest .... *Usnea pacificana* P. Halonen

41. Cortex thinner than medulla (cortex 7-14%, medulla 13-25%, axis 33-45%); medulla PD–, or PD+ orange or yellow, KC– 42

42. Isidia scattered or clustered, abundant or sparse, not normally arising from soralia; branches relatively soft and pliable; medulla K+ red, PD+ yellow to orange (salazinic acid) .... ***Usnea dasopoga* (Ach.) Nyl.** (syn. *U. filipendula, U. dasypoga*)

42. Isidia mostly clustered in round soralia (together with soredia); branches relatively stiff 43

43. Isidia very sparse, only on young soralia; large coalescing soralia dominating the thallus; medulla lax; containing salazinic acid often with barbatic acid, rarely usnic acid alone .... *Usnea substerilis* Motyka (see also couplet 35)

43. Isidia abundant, arising in clusters from small scattered soralia; medulla dense or lax 44

44. Main branches more or less equally dichotomous, tips not twisted or contorted; medulla typically dense, usually K–, PD–, UV+ (squamatic acid), sometimes K+ dark yellow, PD+ orange, UV– (thamnolic acid), salazinic acid absent .... *Usnea subfloridana* Stirton

44. Main branches in unequal dichotomies, tips twisted or contorted; medulla lax, K+ red, PD+ yellow to orange (salazinic acid) .... ***Usnea diplotypus* Vainio** *s. lat.*

VERRUCARIA (including *Bagliettoa, Hydropunctaria, Wahlenbergiella, Placopyrenium*) Based in part on Brodo and Santesson (1997) [salt water species], Thüs and Schultz (2009) [freshwater species], Lendemer and Harris (2008) [calcareous species]; Krzewicka (2012).

[Note: Distribution notes are not included in this key because the frequency of misidentified specimens in North American herbaria makes range maps unreliable.]

1. On dry rocks 2

1. On at least periodically submerged rocks 11

2. Thallus pale to dark brown gray-brown or black; medulla sometimes black in part 3
2. Thallis white, pale gray, or barely perceptible; medulla, if discernable, entirely white 8

3. Medulla black at least in basal part; thallus well developed, forming delimited patches; on calcareous rocks 4
3. Medulla with no black layer; thallus thin or thick, pale to dark brown 5

4. Thallus dark reddish brown, areolate to rimose-areolate; perithecia protruding above thallus surface, with a thick black wall that continues to the base of the medulla, 1-3 per areole; sterile black columns not present (and therefore, jugae absent); spores 17-24(-27) x (7.5-)8.5-11 µm; perithecia typically protruding, but sometimes immersed in thallus .... *Verrucaria nigrescens* Pers.
4. Thallus gray to grayish brown, smooth, rimose-areolate, with black lines between the areoles, many perithecia per areole, marginal areoles sometimes slightly elongate and subsquamulose; sterile black columns frequently extending from black lower medulla to surface resulting in black dots (jugae) at surface; perithecia entirely buried in thallus, with a black ostiole at the surface; walls colorless or black; spores 12.5-16 x 5.5-7.0(-8.0) µm .... *Placopyrenium fuscellum* (Turner) Gueidan & Cl. Roux (syn. *Verrucaria fuscella*)

5. Thallus dispersed to contiguous areolate to subsquamulose, smooth, dark brown; perithecia immersed, only black at the summit; spores broad, 13-16 x 7-10 µm; initially parasitic on other lichens ....*Heteroplacidium compactum* (A. Massal.) Gueidan & Cl. Roux (syn. *Verrucaria compacta*)
5. Thallus areolate to rimose-areolate; spores ellipsoid to narrowly ellipsoid; never parasitic 6

6. Spores 22-30(-35) x (9-)10-13(-15) µm; perithecia prominent, one per areole; recorded from the southwest .... *Verrucaria fuscoatroides* Servit
6. Spores less than 13-21 x 6-10 µm; perithecia partly or entirely immersed 7

7. Thallus thick, forming large areoles bumpy (verruculose) on the surface; perithecia entirely immersed in thallus, with only a black "cap" at the surface; spores (13-)17-21 x (6.0-)7.5-9.0(-10) µm; on calcareous rocks ....*Verrucaria glaucovirens* Grummann
7. Thallus thin, rarely thick, rough and irregular surface, rimose areolate or dispersed to contiguous areoles, sometimes almost granular; perithecia partly buried in thallus; spores 14-20(-25) x (4.5-)6.0-7.5 µm; usually on siliceous rocks .... *Verrucaria nigrescentoidea* Fink

8.(2) Thallus very thick, composed of continguous or scattered areoles, pale gray to bluish gray, surface rough and scabrose or pruinose; initially parasitic on other lichens; in the arid southwest; spores broad, 12-16 x 7-11 µm .... *Verrucaria inficiens* Breuss [See comments in *Placidium* key, couplet 3.]

8. Thallus very thin, never areolate; widespread 9

9. Thallus largely endolithic, forming only a whitish stain; perithecia tiny, 0.15-0.3 mm in diameter, immersed in tiny pits in the rock; spores (18-)23-31 x 9-14 µm
.... ***Bagliettoa calciseda* (DC.) Guedian & Cl. Roux** (syn. *Verrucaria calciseda*)

9. Thallus usually visible on the rock surface; perithecia 0.25-0.4 mm in diameter 10

10. Excipulum pale at base; thallus white; perithecia not more than half immersed in thallus or rock; spores ellipsoid, 18-28 x 9-14 µm .... *Verrucaria muralis* Ach.

10. Excipulum black (carbonaceous) at base; thallus greenish gray; perithecia 2/3 immersed; spores broadly or narrowly ellipsoid, 16-24 x (8-)10-13 µm .... *Verrucaria calkinsiana* Servít

11.(1) In freshwater habitats, streams or on lake shores 12

11. In salt water habitats in the intertidal zone 17

12. On calcareous rocks; thallus membranous to thick, smooth and continuous, pale brown to greenish; perithecia covered with thallus at first, then ± exposed at tip; involucrellum spreading, distinct from excipulum; spores 18-30 x (10-)11-14(-16) µm .... *Verrucaria elaeomelaena* (A. Massal.) Arnold

12. On siliceous rocks; thallus smooth, continuous to rimose 13

13. Spores mostly over 25 µm long; thallus thick; black medullary basal layer absent 14

13. Spores mostly under 25 µm long; spores usually not halonate 15

14. Involucrellum 25% to 75% to base of perithecium, fused to excipulum; perithecia mostly immersed; thallus pale; spores 25-30 x 10-12 µm, often halonate; widespread .... *Verrucaria aethiobola* Wahlenb. [Note: Small-spored specimens sometimes included within *V. aethiobola* can be placed in *V. cernaensis* Zschacke, not yet recorded from North America, but which should be sought (Krzewicka 2012).]

14. Involucrellum extending to base of perithecium, spreading; perithecia prominent but partly covered with thallus tissue; thallus typically dark; spores (21-)25-36(-41) x (8-)11-15(-18) µm, never halonate
.... *Verrucaria margacea* (Wahlenb.) Wahlenb.

15. Medulla with a continuous black basal layer; perithecia largely immersed with only the summit showing; involucrellum fused with black basal layer of medulla; thallus pale greenish to gray, smooth to rimose .... *Verrucaria praetermissa* (Trevisan) Anzi

15. Medulla lacking a continuous black basal layer 16

16. Thallus pale, grayish green to brownish; perithecia prominent, up to 1.3 mm in diameter, with a spreading involucrellum forming a colorless triangle between the excipulum and involucrellum; lacking any black basal tissue .... *Verrucaria hydrela* Ach.

16. Thallus dark brown to almost black; perithecia mostly immersed; involucrellum thick, fused to excipulum to the base of the perithecium and sometimes spreading at the base to form a partial black basal layer .... *Verrucaria funkii* (Sprengel) Zahlbr.

17.(11) Thallus brown, very thin and membranous, smooth, without any black bumps or ridges (jugae); spores 8-11(-12) x 4.5-6.5 µm; on both coasts .... *Verrucaria halizoa* Leighton (syn. *V. microspora*)

17. Thallus black to dark olive-brown when dry, thick, smooth or rough, or, if thin, then with conspicuous jugae 18

18. Thallus very thick, distinctly lobed at the margins, with radiating black ridges; perithecia entirely immersed and showing at the surface only as pale ostioles; spores almost spherical, 7-10 x 6.3-8 µm; frequent, from Vancouver Island to southeastern Alaska .... "*Verrucaria*" *epimaura* Brodo

18. Thallus thick or thin, not at all lobed at the margins; black ridges and bumps present or absent; perithecia immersed or somewhat prominent; spores ellipsoid 19

19. Thallus smooth or with inconspicuous cracks when dry, never areolate; greenish when wet, lacking any jugae; perithecia entirely immersed, not creating bumps on the surface; spores 8-10(-12) x 4-8 µm; fairly common on both coasts .... *Wahlenbergiella mucosa* (Wahlenb.) Gueidan & Thüs (syn. *Verrucaria mucosa*)

19. Thallus with abundant or sparse jugae appearing as lines or dots; thallus green, brown, or black................20

20. Thallus thick, rimose-areolate when dry, mostly black when wet, normally with a rough upper surface caused by abundant jugae; medulla with a black basal layer; perithecia immersed but creating bumps on the thallus surface; spores 10-20 x 6-10 µm; extremely common on both coasts .... ***Hydropunctaria maura* (Wahleb.) Keller, Gueidan & Thüs** (syn. *Verrucaria maura*)

20. Thallus thick or thin, lacking a black basal layer in the medulla; not common species 21

21. Spores 11-16 x 6-10 µm; jugae extremely abundant .... *Verrucaria amphibia* Clemente

21. Spores 7-12 x 4-7 µm; jugae sparse to abundant 22

22. Thallus rimose-areolate, at least in part; black jugae on surface and sometimes edging areoles; perithecia more or less immersed in thallus, but involucrellum showing at the surface ....*Verrucaria degelii* R. Sant.

22. Thallus very thin to membranous, continuous to rimose with only scattered cracks; jugae never edging the thallus or areoles; perithecia prominent...23

23. Thallus green to olive; jugae thick (50-)100-200 µm wide and high; perithecia up to 0.35 mm in diameter, frequently very coarsely ridged with jugae ....*Wahlenbergiella striatula* (Wahlenb.) Gueidan & Thüs (syn. *Verrucaria striatula*)

23. Thallus brownish, with scattered abundant or sparse jugae 20-50 µm wide; perithecia smooth or ridged, mostly under 0.25 mm in diameter 24

24. East and west coasts; excipulum dark brown to black at base; spores 8-9 x 4.5-7 µm; common .... *Verrucaria erichsenii* Zschacke

24. East coast only; exciple pale below; spores 6-11 x 3-5 µm; infrequent .... *Verrucaria ditmarsica* Erichsen

VULPICIDA

1. Soredia present on the lobe margins; pycnidia sparse or very inconspicuous .... ***Vulpicida pinastri*** **(Scop.) J.-E. Mattsson & M. J. Lai**

1. Soredia absent; pycnidia abundant and conspicuous 2

2. Growing on soil or moss; thallus lacking rhizines; apothecia rare .... ***Vulpicida tilesii*** **(Ach.) J.-E. Mattsson & M. J. Lai**

2. Growing on bark or wood; thallus with sparse rhizines; apothecia abundant 3

3. Lobes smooth, with rounded depressions, or bumpy; pycnidia prominent; eastern U.S. .... ***Vulpicida viridis*** **(Schwein.) J.-E. Mattsson & M. J. Lai**

3. Lobes with a network of depressions and sharp ridges or wrinkled; pycnidia entirely buried in thallus with just the ostiole showing; western montane ... ***Vulpicida canadensis*** **(Räsänen) J.-E. Mattsson & M. J. Lai**

XANTHOPARMELIA (including *Arctoparmelia*)

1. Thallus dark brown, cortical pigment usually turning HNO_3+ blue-green .... "*Neofuscelia*" group (see *Melanelia* key)
1. Thallus yellowish green or greenish yellow, containing usnic acid in cortex 2

2. Soredia present on upper surface of thallus in hemispherical mounds 3
2. Soredia absent 4

3. Upper surface dull, grayish toward center of thallus; lower surface dull, dark gray to black; medulla KC+ red, UV+ white (alectoronic acid); arctic and northern boreal .... *Arctoparmelia incurva* (Pers.) Hale
3. Upper surface shiny at lobe tips, brownish toward center of thallus; lower surface shiny black; medulla K+ yellow to red, PD+ yellow-orange (stictic and norstictic acids); southwestern montane habitats and coastal Nova Scotia; rare .... *Xanthoparmelia mougeotii* (Schaerer) Hale

4. Isidia present 5
4. Isidia absent 15

5. Lower surface and rhizines black 6
5. Lower surface and rhizines pale to dark brown 8

6. Medulla PD+ red, K+ yellow (fumaraprotocetraric acid); southeastern .... *Xanthoparmelia piedmontensis* (Hale) Hale
6. Medulla PD+ orange, K+ yellow or red (norstictic and [or] stictic acids) 7

7. Thallus closely or loosely attached to rock; containing stictic and cryptostictic acids; widespread in east and west .... ***Xanthoparmelia conspersa* (Ehrh. *ex* Ach.) Hale**
7. Thallus loosely attached to rock; containing stictic acid alone; southeastern .... *Xanthoparmelia isidiascens* Hale

8.(5) Medulla K+ yellow, orange, or red 9
8. Medulla K– 12

9. Isidia mainly cylindrical, often branched, not globular; widespread, especially in northeast, also southwest; medulla PD+ orange, K+ dark yellow (stictic acid) .... ***Xanthoparmelia plittii* (Gyelnik) Hale**

9. Isidia globular at least at first, later cylindrical; western; medulla PD+ yellow-orange, K+ red (salazinic or norstictic acid) 10

10. Containing norstictic and connorstictic acids alone; lacking salazinic acid .... *Xanthoparmelia maricopensis* T. H. Nash & Elix

10. Containing salazinic acid 11

11. Also containing barbatic and (or) diffractaic acid and sometimes norstictic acid; Arizona to California, uncommon .... *Xanthoparmelia schmidtii* Hale

11. Containing salazinic acid alone; west temperate, common .... ***Xanthoparmelia mexicana* (Gyelnik) Hale**

12.(8) Medulla PD– or PD± pale orange; southwestern 13

12. Medulla PD+ distinct yellow or red 14

13. Medulla CK–, UV– (hypoprotocetraric acid); common .... ***Xanthoparmelia weberi* (Hale) Hale**

13. Medulla CK+ orange, UV+ white (diffractaic and squamatic acids); infrequent .... *Xanthoparmelia ajoensis* (T. H. Nash) Egan

14. Medulla PD+ bright yellow (psoromic acid); western .... ***Xanthoparmelia lavicola* (Gyelnik) Hale**

14. Medulla PD+ red (fumarprotocetraric acid); southeastern .... *Xanthoparmelia subramigera* (Gyelnik) Hale

15.(4) Lower surface dark gray to white or rarely pale brownish, dull and velvety; old thalli forming full or partial concentric rings; medulla KC+ red, PD–, UV+ white (alectoronic acid); mainly boreal to arctic 16

15. Lower surface brown or black, shiny, not velvety; old thalli not normally forming concentric rings; medulla KC–, PD+ yellow, orange, or red, UV– (alectoronic acid absent) 17

16. Lower surface gray .... ***Arctoparmelia separata* (Th. Fr.) Hale**

16. Lower surface white or pale brown .... ***Arctoparmelia centrifuga* (L.) Hale**

17. Growing mainly on soil; western 18

17. Growing on rock 22

18. Contains salazinic acid (with or without traces of norstictic acid); lower surface pale to dark brown (rarely black in spots), with abundant rhizines 19

18. Contains norstictic acid (lacking salazinic acid); lower surface brown or black, with or without rhizines 21

19. Upper surface mottled with a network of pale blotches (maculae); lobes 1.5-3 mm wide, strongly inrolled but not forming closed tubes; interior semi-arid grasslands, Colorado to Alberta and Saskatchewan .... *Xanthoparmelia camtschadalis* (Ach.) Hale

19. Upper surface uniform in color, lacking maculae 20

20. Lobes 1-2(-4) mm wide, convex, but not strongly inrolled; thallus attached to soil and pebbles .... ***Xanthoparmelia wyomingica* (Gyelnik) Hale**

20. Lobes 1.5-5 mm wide, strongly inrolled forming tubes; thallus not attached (vagrant), on arid soils ....***Xanthoparmelia chlorochroa* (Tuck.) Hale**

21. Lower surface mostly brown, darkening toward the center, relatively smooth, with rhizines; lobes inrolled but not forming tubes ....*Xanthoparmelia neochlorochroa* Hale

21. Lower surface entirely black, typically with reticulate ridges; rhizines absent; lobes strongly inrolled forming tubes .... *Xanthoparmelia norchlorochroa* Hale

22.(17) Lobes convex; thallus very loosely attached, ascending; pycnidia absent and apothecia rare .... ***Xanthoparmelia wyomingica* (Gyelnik) Hale**

22. Lobes flat; thallus closely or somewhat loosely attached to substrate; pycnidia and apothecia abundant 23

23. Lower surface and rhizines black, sometimes brown at thallus edge 24

23. Lower surface and rhizines pale tan or brown throughout 27

24. Medulla K– or brownish, PD+ red or yellow 25

24. Medulla K+ yellow or red, PD+ yellow-orange; East Temperate, and southwest 26

25. Medulla PD+ red (fumarprotocetraric acid); lobes 1-2(-3) mm wide; upper surface smooth, not wrinkled; southern U.S., especially Ozarks ....***Xanthoparmelia hypomelaena* (Hale) Hale**

25. Medulla PD+ yellow (psoromic acid); lobes 2.5-5 mm wide, upper surface wrinkled; southwestern ....*Xanthoparmelia nigropsoromifera* (T. H. Nash) Egan

26. Thallus closely attached to rock and difficult to remove intact; lobes short and crowded, often overlapping and finely divided; containing stictic and norstictic acids ....*Xanthoparmelia angustiphylla* (Gyelnik) Hale (= most North American records of *X. hypopsila*)

26. Thallus loosely attached and easily removed; lobes strap-shaped; containing salazinic acid ....*Xanthoparmelia hypofusca* (Gyeln.) Hodkinson & Lendemer (= North American records of *X. tasmanica*)

27.(23) Medulla K–, PD+ red or yellow 28

27. Medulla K+ yellow or red, PD+ distinct yellow or orange (fumarprotocetraric acid absent) 31

28. Medulla PD+ yellow (psoromic acid); southwestern ....*Xanthoparmelia psoromifera* (Kurok.) Hale

28. Medulla PD+ red (fumarprotocetraric acid) 29

29. Appalachians, rare in southern Arizona; lobes loosely attached and overlapping, 1-2.5 mm wide; contains physodalic and constipatic acids as minor products ....*Xanthoparmelia monticola* (J. P. Dey) Hale

29. Southwestern; lobes mostly tightly attached, thallus becoming verrucose in older central parts; lacking physodalic or constipatic acids 30

30. Lobes broadening at periphery, 1-3(-4) mm wide; common ....***Xanthoparmelia novomexicana* (Gyelnik) Hale**

30. Lobes finely divided, 0.5-1 mm wide; frequent .....*Xanthoparmelia tuberculata* (Gyeln.) T. H. Nash & Elix

31.(27) Thallus loosely attached to substrate over entire surface or shingled; containing salazinic acid 32

31. Thallus closely appressed to substrate at least over most of the surface; lobes rounded, not constricted behind the tips; maculae absent 34

32. Spotted with conspicuous maculae; rare, if present at all, in western North America .... *Xanthoparmelia stenophylla* (Ach.) Ahti & D. Hawksw. (syn. *X. somloënsis s. str.*)[*not* Plate 912 in LNA; see below]

32. Lacking maculae; very common 33

33. East Temperate; lobes 2-4(-5) mm wide, typically flat or with weak depressons, but sometimes convex, long, strap-shaped, dichotomously branched, sometimes developing narrow lobules, square or truncated at the tips ....***Xanthoparmelia viriduloumbrina* (Gyelnik) Lendemer**
[Note: = eastern North American records of *X. somloënsis*; Plate 912 in LNA.]

33. Western, from southwest to British Columbia; lobes 1-3 mm wide, typically somewhat convex but sometimes flat, typically short and divergent, not regularly dichotomous, often developing abundant, narrow lobules in the thallus center; tips rounded to square .... *Xanthoparmelia coloradoensis* (Gyelnik) Hale [= almost all western records of *X. somloënsis*]

[Note: The distinctions given above for separating *X. viriduloumbrina* and *X. coloradoensis* are largely tendencies; the morphological overlaps are great as they are with other salazinic acid-containing, loosely adnate western species with pale brown lower surfaces. The problem should be taken up with molecular techniques.]

34.(31) Containing stictic and norstictic acids; thallus typically shiny, closely to moderately attached to substrate; lobes often divided and crenulate; widespread in northeast and parts of the west .... ***Xanthoparmelia cumberlandia* (Gyelnik) Hale**

34. Containing salazinic or norstictic acid; thallus very closely attached to substrate; western 35

35. Contaning salazinic acid; thallus dull, rounded lobes 1-5(-8) mm wide .... ***Xanthoparmelia lineola* (E. C. Berry) Hale**

35. Containing norstictic acid; thallus shiny, lobes 0.8-1.5 mm wide .... *Xanthoparmelia californica* Hale

XANTHORIA (including *Rusavskia, Seirophora,* and *Xanthomendoza*)

1. Soredia absent; apothecia abundant 2

1. Soredia present; apothecia absent or occasional 8

2. Thallus loosely attached over entire surface, forming round cushions of overlapping lobes 3

2. Thallus closely appressed to substrate, forming round, flat rosettes 5

3. Rhizines absent (thallus sometimes attached by stout holdfasts); conidia ellipsoid; on rocks, bark, and wood ....***Xanthoria polycarpa* (Hoffm.) Th. Fr.** (syn. *Polycauliona polycarpa* (Hoffm.) Frödén et al.)

3. Rhizines abundant, both attached to the substrate and unattached; conidia rod-shaped; on trees, rarely on rock 4

4. Spores mostly (11.5-)12.5-15.5 x 5-7.5 μm, septum about 1/3 the length of the spore or less, 1.5-4 μm wide; arid western interior ....*Xanthomendoza montana* (L. Lindblom) Søchting et al.

4. Spores mostly 15.5-18 x 7.0-8.0(-9.5), septum more than 1/3 the length of the spore, 4-8.5 μm wide; widespread, but not in western interior region ....***Xanthomendoza hasseana* (Räsänen) Søchting et al.** (syn. *Xanthoria hasseana*)

5.(2) Lobes mostly wider than 1 mm, rounded, flat or concave, clearly foliose; lower surface smooth, with scattered long rhizines; common on northeast coast, sporadic on west coast and southern Ontario .... ***Xanthoria parietina* (L.) Th. Fr.**

5. Lobes mostly narrower than 1 mm, short or elongated to linear, appearing crustose at least in center; rhizines absent or very rare 6

6. Lower surface with a cortex, wrinkled; lobes convex; on rock .... ***Rusavskia elegans* (Link) S. Y. Kond. & Kärnefelt** (syn. *Xanthoria elegans*)

6. Lower surface mostly without a cortex and appearing crustose; lobes convex or flat 7

7. On bark or wood; lobes often pruinose, rather thick; apothecia mostly under 1.5 mm in diameter, concave to finally flattened; common in California .... *Xanthoria tenax* L. Lindblom

7. On rock .... *Caloplaca* key, couplet 21

8.(1) Thallus very closely appressed to substrate, almost crustose, without rhizines; coarse soredia (schizidia) laminal, originating from the disintegration of pustules; on rock .... ***Rusavskia sorediata* (Vainio) S. Y. Kondr. & Kärnefelt** (syn. *Xanthoria sorediata*)

8. Thallus loosely attached to ascending, clearly foliose or squamulose; rhizines present or absent; soredia rarely laminal; on bark, wood, or rock 9

9. Rhizines abundant; pycnidia buried in the thallus, not conspicuous 10

9. Rhizines absent or very sparse; pycnidia conspicuous or not 13

10. Soredia mostly greenish yellow, within crescent-shaped hood- or lip-like expansions of the lobe tips between the upper and lower cortices ......***Xanthomendoza fallax* (Hepp) Søchting et al.** (syn. *Xanthoria fallax*)

10. Soredia yellow to orange, on the lobe margins or sometimes laminal, or on the lower surface of the lobe tips, not between the upper and lower cortices 11

11. Mostly northcentral to northeastern North America; soredia marginal at first, not produced in hood-shaped expansions; conidia always rod-shaped .... ***Xanthomendoza ulophyllodes* (Räsänen) Søchting et al.** (syn. *Xanthoria ulophyllodes*)

11. Western montane and California; soredia produced on lower surface of lobes 12

12. Soredia frequently produced in hood-shaped expansions; conidia both ellipsoid and rod-shaped .... *Xanthomendoza oregana* (Gyelnik) Søchting et al.

[Note: Rare specimens of *X. galericulata* with abundant rhizines will key out here. They have only rod-shaped conidia. See couplet 16.]

12. Soredia on the lower surface of the edges of the lobe tips, which are not expanded or hood-shaped; conidia all rod-shaped .... ***Xanthomendoza fulva* (Hoffm.) Søchting et al.** (syn. *Xanthoria fulva*) [northern morphotype with rhizines]

13.(9) Lobes flattened to almost cylindrical, branched, erect, sometimes almost fruticose in habit; soredia produced on or just under the lobe tips 14

13. Lobes flat to strongly convex, rounded or divided; soredia on the lower surface or margins of the lobe tips 15

14. Pycnidia immersed and inconspicuous; conidia ellipsoid, 2-3.5 µm long; on bark or branches, or on rock .... ***Xanthoria candelaria* (L.) Th. Fr.** (syn. *Polycauliona candelaria.*)

14. Pycnidia up to 0.3 mm wide, protruding and very conspicuous; conidia 3-4 x 1.5-2 µm; on calcareous rock or soil.. *Seirophora contortuplicata* (Ach.) Fröden (syn. *Teloschistes contortuplicatus*)

15. On calcareous rock; lobes mostly 0.4-2 mm broad, very convex, almost hood-like; soredia spreading over lower surface of lobes; pycnidia conspicuous or not; conidia rod-shaped, 4-5(-6) µm long; rare, in the western mountains...................... *Xanthomendoza mendozae* (Räsänen) Søchting et al. (syn. *Xanthoria mendozae*)

15. On bark; conidia rod-shaped, 3-5 µm long 16

16. Lobes variable in width, 0.4-1.4 mm broad, margins involute or flat; soredia produced under helmet-shaped, incurved lobe tips; pycnidia rare, but prominent when produced; dry areas of western U.S. .... *Xanthomendoza galericulata* Lindblom

16. Lobes finely divided, 0.2-0.5 mm broad, often in little rosettes; soredia on lower surface of lobe tips close to the edge on recurved lobe tips exposing the soredia; pycnidia prominent, resembling orange pimples 17

17. Lobes regularly dichotomously branched, always narrow; thallus yellow-orange; eastern to South Dakota and Nebraska ....*Xanthomendoza weberi* (S.Y. Kondr. & Kärnefelt) L. Lindblom (syn. *Xanthoria weberi*)

17. Lobes irregular, not dichotomous, narrow to rounded; thallus orange; mostly western (scattered in the east) .... ***Xanthomendoza fulva* (Hoffm.) Søchting et al.** (syn. *Xanthoria fulva*)

XYLOGRAPHA (includes *Lignoscripta*) See Spribille et al. (2014) and Ryan (2004a,b).

1. Epihymenium blue-green (HNO_3+ red); spores 12.5-17(-22) x 9-12 µm; ascomata pure black, linear to fusiform, sometimes with persistent margins that can be white pruinose; on juniper lignum in arid habitats; Arizona, Colorado, Utah, North Dakota .... *Lignoscripta atroalba* B. D. Ryan & T. H. Nash

1. Epihymenium colorless to dark brown, never greenish (HNO_3–); spores less than 9.5 µm wide; ascomata very pale to very dark brown, sometimes appearing black; margins present or absent, never pruinose 2

2. Thallus sorediate with round to elliptical soralia containing dark brown to creamy white granular soredia 3

2. Thallus lacking soredia, often within the wood and not evident 4

3. Soralia usually speckled brown when mature, uniformly dark brown when young, becoming irregular and sometimes coalescing; ascomata pale brown, broad, irregular in shape, infrequent; spores (8-)12.0-16.5 x (4-)6.0-7.5 µm; with stictic acid, sometimes with norstictic acid in smaller amounts; widespread, boreal-montane, rarely temperate .... *Xylographa vitiligo* (Ach.) J. R. Laundon

3. Soralia entirely dark brown, scattered and punctiform (very small); ascomata narrowly fusiform to linear, up to 2 mm long, very dark brown to black; spores (11-)12-16(-22) x (5.0-)5.9-7.8(-9.5) µm; with norstictic acid, sometimes with stictic acid; western and western Great Lakes region .... *Xylographa septentrionalis* T. Sprib.

4. Spores narrow, small, 10-13(-14) x 3.0-5.0 µm 5

4. Spores broader than 5.0 µm 6

5. Thallus relatively thick, becoming areolate; ascomata yellow-brown, short, often branched once (Y-shaped), somewhat sunken into thallus areole; containing norstictic or, rarely, stictic acid; mostly on northern Atlantic and Pacific coasts, on maritime logs and lignum ....*Xylographa opegraphella* Nyl.

5. Thallus mostly within the wood, rarely forming areoles; ascomata dark brown to almost black, linear to elongate-fusiform, not branched; containing stictic acid or no substances; mostly western montane .... *Xylographa stenospora* T. Sprib. & Resl

6. Ascomata mostly long, narrow, length to width ratio 3-15:1 7

6. Ascomata short and broad, length to width ratio 1.5-3.5:1., pale to dark brown, sometimes branched or forming star-like clusters; Pacific Northwest 9

7. Containing norstictic acid (major) ± stictic acid; spores 11-12.5 x 5.5-6.7 µm; ascomata tend to be broad, fusiform, breaking up into "dotted lines" consisting of short segments of interrupted disks, dark brown to black with pale margins; northwestern montane ....*Xylographa rubescens* Räsänen

7. Contains no substances or stictic acid with norstictic acid only in traces, ascomata narrow, linear, black to brown, not breaking into short segments; spores 11-16 x 5.8-7.6 µm; thallus within the wood 8

8. Ascomata blunt at one end and pointed at the other, which is the growing tip, developing in parallel lines, always linear, never forming star-like clusters; containing stictic acid or no substances; widespread temperate to boreal ....***Xylographa parallela* (Ach:Fr.) Fr.**

8. Ascomata pointed and growing at both ends, sometimes branched or forming star-like clusters; containing stictic acid, sometimes with a trace of norstictic acid; widespread especially in dry montane regions but perhaps not common ....*Xylographa pallens* (Nyl.) Harm.

9. (6) Ascomata elliptical, unbranched; containing confriesiic acid and, rarely, stictic acid; spores 9.5-11.3(-12.8) x 5-6.3 µm ....*Xylographa truncigena* (Th. Fr.) Minks *ex* Redinger.

9. Ascomata elliptical to Y-shaped or curved; containing stictic and sometimes norstictic acid, lacking confriesiic acid; spores (9.5-)11.0-13.5(-15) x 6.2-7.4 µm .... *Xylographa hians* Tuck.

Glossary

adnate. Tightly attached to the surface, like species of *Dirinaria* or *Bulbothrix*.

alga (*-gae*). Green photosynthetic organism containing chloroplasts and nuclei and belonging to the Kingdom Protoctista.

algal layer. Layer of algal cells in a stratified lichen thallus (fig. 4). Sometimes used synonymously with "photobiont layer" (a layer of either green algae or cyanobacteria).

amphithecium. The portion of a lecanorine apothecium external to the exciple (fig. 13c), usually containing algae, constituting the thalline margin.

amyloid. Containing carbohydrates that turn blue in an iodine solution (IKI).

anastomosing. Forming a net-like, interconnected growth (fig. 16c).

anisotomic. Dividing in unequal dichotomies to produce a distinguishable main branch with side branches, as in *Cladonia rangiferina*.

apical. At the tip or summit (i.e., the apex).

apothecium (*-cia*). A disk- or cup-shaped ascoma, usually with an exposed hymenium (figs. 12, 13).

appressed. Flattened and closely adnate; as in *Xanthoparmelia cumberlandia*.

areolate. Broken up into areoles, often appearing tile-like (fig. 2b; *Lecidea tesselata*).

areole. a) A small, round to irregular or angular patch of thallus; b) a patch of vegetative tissue (containing algae and cortex) on the podetial surface of some *Cladonia* species.

ascending. Lifting from the surface and becoming free from it, at least in part, like the lobes of species of *Tuckermanopsis*.

ascocarp. See *ascoma*.

ascohymenial. Pertaining to a type of ascoma having a particular development resulting in a hymenium with true paraphyses or remnant paraphysoid tissue, but not pseudoparaphyses (figs. 13a-e).

ascolocular. Pertaining to a type of ascoma in which the asci arise within a uniform mass of fungal tissue and are separated in maturity, not by true paraphyses, but by paraphysoides or pseudoparaphyses (fig. 13f).

ascoma (*-mata*). The fruiting body of an Ascomycete; the structure that bears the asci, which in turn contain the ascospores (figs. 12, 17). Apothecia and perithecia are types of ascomata.

ascomycete. A fungus that produces its sexual spores within an ascus.

ascospore. A spore produced in an ascus (figs. 14, 15).

ascus (*asci*). The sac-like structure in Ascomycetes in which the ascospores are formed (fig. 14). The sexual fusion of nuclei and reduction division occur within the ascus.

axial body. Conical, vertically oriented, nonamyloid structure in the tholus of certain types of asci such as the *Bacidia*- and *Biatora*-types (figs. 14a, b); also called an apical cushion or masse axiale.

bacilliform. Stick- or rod-shaped; cylindrical, usually straight.

basidiomycete. A fungus that produces its sexual spores as external buds on a club-like basal cell (the basidium). Mushrooms, bracket fungi, and coral fungi, among others, belong to the Basidiomycetes, one of the main classes in the Kingdom Fungi.

basidium. A club-shaped cell in the fruiting body of a Basidiomycete in which sexual fusion and reduction division occurs, with the subsequent budding off of 2 to 4 spores (basidiospores).

biatorine. Referring to a type of apothecium having a relatively soft, clear or lightly pigmented (not carbonized) margin containing no photobiont cells (fig. 13b).

blastidium (*-dia*). A granule-sized fragment of a lichen thallus that is formed by a budding off of the thallus margin (fig. 19e).

blue-green algae. Cyanobacteria.

boreal. In reference to a northern region dominated by conifer forests; in North America, the region forms a belt from the Maritime Provinces to Alaska.

bryophytes. Mosses, liverworts (hepatics), and hornworts.

C. Bleaching solution (sodium hypochlorite) or undiluted commercial bleach (e.g., Clorox© or Javex©), used as a reagent in spot tests for revealing certain lichen substances (see table 2, pages 106-107 in *Lichens of North America*).

calcareous. Containing lime or chalk (calcium carbonate), producing vigorous bubbling (CO_2) in the presence of a strong acid. Calcareous rocks include limestone, dolomite, and marble; some sandstones and soils can also be calcareous.

campylidium (*-dia*). Helmet-shaped conidia-bearing structure found in many tropical, foliicolous lichens.

capitate. Referring to a type of rounded, almost hemispherical structure (usually a soralium); see *Melanelia sorediata*.

capitulum. The tiny spherical or cup-shaped apothecium formed at the summit of a slender stalk; found in stubble lichens, *Calicium, Chaeonotheca* and related genera (fig. 12f).

carbonaceous. Opaque black, usually brittle, used in reference to tissue such as the exciple. It is hard to distinguish individual cells in carbonaceous tissue.

cartilaginous. Tough, pliable cartilage- or sinew-like tissue. The term is usually used in reference to supporting tissue (see fig. 8).

cephalodium (*-dia*). A small gall-like growth that contains cyanobacteria and occurs within the tissues or on the surface of some lichens with green algal photobionts (plate 16; *Placopsis lambii*).

Chlorococcales. An order of green algae having taxa with individual spherical cells, lacking a central pyrenoid.

chlorococcoid. A green alga resembling a member of the Chlorococcales.

chloroplast. The structure in a green cell that contains chlorophyll, the substance responsible for photosynthesis.

Chroococcidiopsis-group. Cyanobacteria with small cells that are solitary or in packets of 2 to 5 and appear to be lacking gelatinous envelopes; sheaths, if present, are very thin and hard to see, unchanged with K; cells (5.0-)6.5-10 µm in diameter; in *Peltula.*

Chroococcus. A genus of cyanobacteria with large cells, 15-40 µm in diameter, in colonies of 2 to 4 in a gelatinous, layered or unlayered, relatively thin colorless envelope (fig. 1b). Individual cells do not have a clearly defined gelatinous sheath as in *Gloeocapsa.*

cilia. Hair-like appendages on the margins of the thallus or apothecia of many foliose and fruticose lichens (fig. 10; plate 13).

CK. A spot test for revealing diffractaic acid, performed by wetting the tested area with C followed by the application of K. A positive reaction is orange.

Coccomyxa. A genus of green algae with small, ellipsoid cells that contain a single chloroplast lying against one wall.

columella. A vertical projection or axis of sterile tissue found in some ascomata.

conidium (-dia). An asexual spore usually formed in large numbers within special structures such as pycnidia and campylidia. Conidia sometimes serve as male sexual cells (spermatia) (fig.18).

consoredium. A round cluster of tiny individual soredia that resembles a large soredium or granule.

continuous. Unbroken, or broken very little by cracks (fig. 2c; *Lecanora caesiorubella*).

coralloid. Composed of, or having, minutely branched cylindrical outgrowths, as in *Caloplaca coralloides* or *Sphaerophorus tuckermanii.*

cortex. The outer protective layers of a lichen thallus or apothecium, completely fungal in composition, often composed of hyphae with thick, gelatinized walls (figs. 4, 13).

corticate (-ical). Having a cortex; pertaining to a cortex.

corticolous. Growing on bark.

crenulate. Having a scalloped margin with rounded teeth or lobes (fig. 3).

crisped. Having a ruffled, wavy, or twisted margin, as in many species of *Cetrelia* and *Parmotrema.*

crustose. A thallus type which is generally in contact with the substratum at all points and lacks a lower cortex; cannot be removed intact from its substrate without removing a portion of the substrate as well (figs. 2a-d; plate 10).

cryptolecanorine. a) A kind of apothecium that is mostly sunken into the thallus and thereby lacks a prominent margin; b) the partial, alga-containing margin in such apothecia (e.g., *Aspicilia* or *Ionaspis*).

cuff-shaped. In reference to laminal soralia that burst through the upper cortex of a hollow lobe, leaving a hole in the center of the soralium (fig. 20f), seen in *Menegazzia subsimilis.*

cyanobacteria. Photosynthetic, chlorophyll-containing organisms related to bacteria (in the Kingdom Monera), without organized nuclei or chloroplasts; sometimes called blue-green algae.

cyanolichen. A lichen with cyanobacteria as the photobiont.

cyphella (*-ae*). A specialized, depressed pore in a lichen thallus, lined with small, loosely packed, spherical cells (fig. 11c; plate 15); characteristic of the genus *Sticta*.

decorticate. Having had a cortex which has now fallen away or decomposed.

diahyphae. Conidia of a special type formed at the tips of tiny stalks especially in the Gomphillaceae (e.g., *Gyalideopsis*).

dichotomous. Branching into two equal parts, as in the letter "Y" (see, e.g., *Speerschneidera*).

disjunct (*-tion*). A population that is geographically remote from other occurrences of the same species (as in the distribution pattern of *Lasallia caroliniana*).

dorsiventral. With distinguishable upper and lower surfaces.

effigurate. Having a definite, usually somewhat lobed margin (fig. 2d), as in *Dimelaena oreina*.

ellipsoid. Oval in outline; more or less football-shaped (figs. 15a-d).

endemic. Found only in a certain, usually limited, region; in this book, used in reference to lichens found only in North America or smaller regions, e.g., the Appalachian Mountains (see map of *Hypotrachyna croceopustulata* in LNA).

endolithic. Growing within the upper layers of a rock, i.e., under and around the rock crystals, often with little or no thallus visible on the outer rock surface (plate 32).

endophloeodal. Growing largely within the upper layers of bark tissue (fig. 9b), as in the thallus of *Graphis scripta*.

epihymenium. The uppermost portion of the hymenium formed by the expanded and (or) branched tips of paraphyses; often pigmented and sometimes containing tiny granules (fig. 13); considered here synonymous with *epithecium*.

epispore. See *halo*.

epithecium. See *epihymenium*.

excavate. Hollowed out, concave or depressed; often in reference to a type of soralium (fig. 20h).

exciple. An area in an apothecium external to and below the hypothecium, forming the apothecial margin in lecideine or biatorine apothecia (figs. 13a-b); much reduced in lecanorine apothecia (fig. 13c).

excipulum. In this book, refers to the perithecial wall (fig. 17); often considered synonymous with *exciple*.

f. Form; a formal subdivision of a species, subspecies, or variety usually applied to a relatively minor morphological, chemical, or ecological variant.

family. A taxonomic category consisting of closely related genera.

farinose soredia. Very fine, powdery soredia, as on the podetial surface of *Cladonia deformis*.

fibril. A short branch, usually perpendicular to the main filament, in *Usnea* (fig. 37a; see *U. diplotypus* or *U. longissima*).

filamentous. Hair-like.

fissural. Resembling a gaping slit, often oval or fusiform in shape; usually in reference to a type of soralium (fig. 20d).

flexuose. Wavy.

flocculent. Having a loose, cottony, fibrous texture, like the nubby surface of an old sweater; as seen, e.g., on the surface of *Cladonia portentosa*.

foliicolous. Growing on leaves of vascular plants (especially in the tropics; e.g., *Calopadia fusca*).

foliose. Pertaining to a more or less "leafy" lichen thallus, distinctly dorsiventral, and varying in its attachment to the substrate from completely adnate to umbilicate (figs. 2f-g; plate 9).

fruiting body. The sexual reproductive structure of a lichen fungus (apothecium, perithecium, mushroom, etc.); in most lichens, the ascoma.

fruticose. Pertaining to a lichen thallus that is stalked, pendent, or shrubby, and normally with no clearly distinguishable upper and lower surfaces (figs. 2h-j; plate 9).

fusiform. Narrow, tapering at both ends, usually with pointed tips; spindle- or cigar-shaped (fig. 15f).

genus. A group of closely related species, presumably with the same ancestor; comprises the first word of the two-word name of every organism.

Gloeocapsa. Cyanobacteria consisting of groups of 2 to 8 spherical cells enclosed within a thick gelatinous matrix. Individual cells have their own, often colored, gelatinous sheaths that usually are K+ purple (fig. 1c).

granular (*-ose*). (a) Having granules or granule-like particles (as in the thallus of *Bacidia rubella*); (b) pertaining to soredia that are large enough to be easily distinguished under a dissecting microscope (as in *Cladonia chlorophaea*); see *farinose soredia*.

granule. A spherical or nearly spherical corticate particle.

graphid. A lichen with a lirellum-type apothecium, resembling *Graphis*.

halo. A transparent gelatinous covering, often irregular in thickness, surrounding an ascospore (fig. 15k); technically called a *perispore*, and referred to as an *epispore* by some lichenologists; revealed by an India ink preparation (see "Microscopic Study," in Chapter 13 in *LNA*).

halonate. Having a gelatinous perispore or "halo" (fig. 15k).

haustorium (*-ria*). A special branch of a mycobiont hypha that penetrates or otherwise attaches itself to the photobiont cell for the purpose of food absorption.

heath. A vegetation type consisting of extensive areas of low shrubs and few trees.

hemiamyloid. Turning dark blue with iodine (IKI) when pretreated with K, but negative or red-orange with IKI in tissues not pretreated with K.

herbarium. A collection of dried plants (or lichens).

Heterococcus. A genus of yellow-green algae (Xanthophyceae) found in some species of *Verrucaria* as single-celled or short filamentous units, but which in culture becomes filamentous and branched (as in fig. 1i).

heterocyst. A specialized, thick-walled, colorless cell in certain cyanobacteria such as *Nostoc*, and the site of most nitrogen fixation (fig. 1d).

holdfast. The relatively thick and, in many cases, only attachment point of some lichens, especially *Usnea* and umbilicate lichens such as *Umbilicaria*.

hymenium. The spore-bearing layer of an ascoma, consisting of asci and paraphyses, paraphysoides, or pseudoparaphyses (fig. 13).

hypha (*-ae*). The filamentous elements of a fungus, often modified and resembling round or angular cells (fig. 5).

hyphophore. A stalked, often umbrella-shaped structure bearing a mass of conidia in certain lichens, especially those growing on leaves. See *Gomphillus*.

hypothallus. A specialized tissue developing as a basal layer on certain foliose, squamulose, and crustose lichens (e.g., *Anzia, Coccocarpia, Lepraria*; fig. 6g); sometimes considered synonymous with *prothallus*, but not in this book.

hypothecium. The tissue just below the hymenium (and subhymenium) but above the exciple (fig. 13); often with a distinctive color or texture as in *Buellia*, but sometimes merging with the exciple as in *Biatora vernalis*.

IKI. A 0.5-1.5 per cent solution of iodine in 10 per cent potassium iodide; Lugol's solution (Chapter 13, *LNA*). Lower percentages of IKI are sometimes used for special types of staining procedures.

imbricate. Overlapping in a shingle-like fashion; usually pertaining to scales or squamules, as in *Psora himalayana*.

inflated. Swollen and hollow, like the lobes of *Hypogymnia enteromorpha*.

involucrellum. The covering or cap external to the excipulum and present on many perithecia; usually black and carbonaceous (fig. 17).

isidium (*-dia*). A minute thalline outgrowth that is corticate and contains photobiont cells. Isidia are easily detached from the thallus and serve as vegetative reproductive units (fig. 19d; plate 18).

isotomic. Dividing in regular dichotomies so that a main stem is not easily distinguishable, as in *Cladonia stellaris*.

jugae. Carbonized chunks of sterile hyphal tissue either superficial or originating in the black medullary tissue of several species of *Verrucaria*, and appearing on the thallus surface as black dots, lines, or ridges.

K. A 10 per cent solution of potassium hydroxide (KOH) used in various microscopic preparations, or in spot tests for revealing certain lichen substances (see table 2, pages 106-107 in *LNA*); can be substituted with household lye (sodium hydroxide, NaOH), 10 pellets per 20-30 ml (1-1.5 ounces) of water.

KC. A spot test for revealing certain lichen substances, performed by wetting the tested area with K followed by the application of C (see table 2, pages 106-107 in *LNA*).

K/I. An analytical staining procedure for the hymenium or asci performed by pretreating the tissue with K, removing the K by replacing it with water, and then replacing the water with 0.5-1.5 per cent IKI.

labriform. (a) Lip-shaped; (b) pertaining to soralia, generally formed by an upturned thallus margin or a bursting hollow thallus lobe, sorediate on the lower or inside (i.e., exposed) surface, as in *Hypogymnia physodes* (fig. 20e).

laminal. On the upper surface of a thallus (laminal soralium: fig. 20a).

lax. Loose, not compact; usually referring to the medulla.

lecanorine. Pertaining to an apothecium having a distinct thallus-like margin containing a photobiont, as in the genus *Lecanora* (figs. 12b, 13c).

lecideine. Pertaining to an apothecium with no photobiont cells in the margin, and in which the exciple is at least partially carbonized forming a black apothecial margin, as in the genus *Lecidea* (figs. 12a, 13a); see also *biatorine*.

leprose. Composed entirely of soredia, in reference to a thallus surface or the thallus itself (fig. 2a); as in *Lepraria*.

lichen. An association of a fungus and a photosynthetic symbiont (photobiont) resulting in a stable vegetative body with a specific structure, in which the fungus encloses the photobiont.

lichenicolous. Growing on or in a lichen, pertaining to fungi and lichens at various levels of parasitism with respect to the host lichen.

lichenized. Pertaining to a fungus, alga, or cyanobacterium living within a lichen association.

lignicolous. Growing on bare wood (lignum), as on a log or a wooden fence.

lirella (*-ae*). An elongated, sometimes branched apothecium, as in *Graphis* (fig. 12e).

lobe. A rounded or somewhat elongated division or projection of a thallus margin; measured at its widest point (fig. 3).

lobule. A small, often scale-like lobe growing from a foliose thallus either along its margin or from the surface, sometimes also appearing along apothecial margins, generally of the same color and character as the parent thallus (fig. 19f; plate 19). Lobules that are constricted at the base and function as propagules are often called *phyllidia*.

lobulate. Having many lobules.

locule. The cell cavity in an ascospore; the locule sometimes developing a distinctive shape due to unevenly thickened spore walls (figs. 15c, d, i, j, l; plate 925).

macrolichen. A foliose or fruticose lichen.

maculate. Spotted or blotched by *maculae*, which are pale round or reticulate areas caused by gaps in the photobiont layer below the cortex (fig. 4; plate 13).

marginal. Along the thallus margins (fig. 20b: marginal soralia).

mazaedium (*-dia*). A dry, powdery mass of ascospores and paraphyses formed by the disintegration of the asci in the ascoma of lichens in the order Caliciales, such as *Chaenotheca* and *Cyphelium* (fig. 12f).

mealy. Coarsely granular, like corn meal.

medulla. The internal layer in a thallus or lecanorine apothecium, generally composed of loosely packed fungal hyphae (figs. 4, 13).

microlichen. A crustose or squamulose lichen. Dwarf fruticose lichens such as *Polychidium* and *Ephebe* are often considered to be microlichens.

morphology. Physical characteristics, including external shape and internal anatomy.

morphotype. A variant of a species differing in some minor features of the morphology.

muriform. Spores divided into several cells by both longitudinal walls and crosswalls; the cells therefore look like stones or bricks in a wall (figs. 15g,h).

muscicolous. Growing over bryophytes, especially moss (e.g., like *Biatora vernalis*).

mutualism. A type of symbiosis in which both components benefit from the association.

mycobiont. The fungal symbiont or partner in a lichen.

Myrmecia. A genus of green algae with round to slightly pear-shaped single cells containing a flat or (in lichen thalli) strongly folded and twisted chloroplast. Unlike *Trebouxia*, which it closely resembles, the chloroplast has no central pyrenoid.

nonamyloid. Not turning blue with iodine (IKI), whether or not the tissue has been pretreated with K. (See also *hemiamyloid*.)

Nostoc. A genus of blue-green algae found in many lichens; producing bead-like chains or filaments of cells including thick-walled, colorless heterocysts (fig. 1d); when lichenized, the filaments may be few-celled.

nucleus (*-clei*). A spherical structure within the cells of fungi, protoctists, and plants, that contains genetic material, i.e., the chromosomes and genes.

oceanic. Pertaining to a climate characterized by mild wet winters, cool moist summers, and frequent fogs.

ocular chamber. A dome-shaped indentation of the lower edge of the tholus in certain types of ascus tips such as those of *Lecanora*-type asci (fig. 14e).

ostiole. The small, usually round, apical pore in various types of ascomata, especially perithecia, through which the ascospores escape (fig. 17).

papillae. Small rounded or cylindrical bumps like "goose pimples," on the cortex of certain lichens, especially *Usnea* (fig. 37a). Papillae do not contain medullary tissue.

paraphysis (*paraphyses*). A sterile fungal filament, sometimes branched, attached at the base and free at the summit, associated with asci in the hymenium (figs. 13, 16).

paraphysoids. Sterile hymenial tissue between the asci, usually abundantly branched with frequent anastomoses (figs. 13f, 16c).

PD. Para-phenylenediamine, a reagent used in spot tests for revealing certain lichen substances (see table 2, pages 106-107 in *LNA*).

peltate. Attached at the center of the lower surface; umbrella-like.

pendent. Hanging straight down and usually soft and pliable (fig. 2j; *Usnea longissima*); grades into *almost pendent* forms that are bushy and relatively stiff at the base but pendent over most of their length, as in *Usnea diplotypus*.

periphyses. Short, hair-like hyphae that sometimes line the inner walls of a perithecium (or an apothecium opening by a pore) near the ostiole (fig. 17a).

perispore. See *halo.*

perithecium (*-cia*). A flask-shaped ascoma opening by a pore at the summit (fig. 17); may be prominent, but is more often partially or completely embedded in the thallus tissue.

photobiont. The photosynthetic component (symbiont) in a lichen thallus, either algae in the strict sense (e.g., green algae) or cyanobacteria (blue-green algae) (fig. 1).

phyllidium (*-dia*). See *lobule.*

phyllocladium (*-dia*). A minute cylindrical, lobed, or scale-like outgrowth on the stalks and branches of species of *Stereocaulon* (fig. 8g).

podetium (*-tia*). A stalk formed by a vertical extension of lower apothecial tissues (usually the hypothecium and stipe). The stalk can become secondarily invested with an algal layer and cortex (as in *Cladonia*) or can remain almost free of vegetative tissue (as in *Dibaeis*); the fertile tissue or apothecia can be present (as in the red-fruited *Cladonia bellidiflora*) or absent (as in *C. coniocraea*); podetia can be either short and unbranched (fig. 2h) or quite tall and highly branched (fig. 2i).

polarilocular. Pertaining to spores having two cell cavities separated by a relatively thick septum through which a narrow canal passes (fig. 15d); characteristic of *Caloplaca* and related genera.

polyphyllous. In species of *Umbilicaria* or *Dermatocarpon*: composed of numerous, crowded lobes rather than a single, peltate or umbilicate thallus (see *U. polyphylla*).

primary squamules. Small, scale-like lobes forming the basal or primary thallus of *Cladonia* species.

primary thallus. A squamulose or crustose thallus from which fruticose stalks or podetia arise as secondary components. Examples are found in *Cladonia, Pilophorus*, and *Baeomyces*.

propagule. A reproductive unit, either sexual (such as an ascospore) or vegetative (like a soredium or isidium).

prosoplechtenchyma. Fungal tissue consisting of coalesced, rather elongate hyphal cells often with thick, gelatinized walls (figs. 5e, g). (Compare with *pseudoparenchyma.*)

prothallus. The purely fungal, white or darkly pigmented border of many crustose thalli, often visible as a fungal mat between the areoles or granules (fig. 2b; also see *Lecanora thysanophora* and *Rhizocarpon geographicum*); NOT synonymous with *hypothallus*, which see.

pruina. Powdery, frost-like deposit (usually white or gray, rarely yellow or reddish), typically composed of calcium oxalate or pigment crystals, dead cortical tissue, or some mixture of them; often occurs on a thallus (plate 11) or apothecial surface (see *Lecanora rupicola*).

pruinose. Having a frosted appearance due to a deposit of pruina.

pseudocyphella (*-ae*). A tiny round to elongate dot, pore, or slit caused by a break in the cortex and the extension of medullary hyphae to the surface (figs. 4 and 11a,b; plate 14); it usually takes the color of the medulla (e.g., white or yellow), or can be dark at the surface.

pseudoparaphyses. Sterile hyphal threads that originate in the ceiling of a ascolocular perithecium-like ascoma, then grown downward and fuse with tissue at the base. They can be highly branched and anastomosing and difficult to distinguish from paraphysoides (fig. 16c).

pseudoparenchyma. Fungal tissue that appears cellular in section due to short, rounded to almost square cells of highly branched, irregularly oriented fungal filaments (figs. 5d, 5f); sometimes called *paraplectenchyma*.

pseudostroma (-ata). A wart-like mass of vegetative fungal tissue containing some material from the substrate (usually bark), and supporting fruiting bodies of various kinds, as in *Trypethelium* species and *Glyphis cicatricosa*.

pustule. A more or less hollow wart or verruca, small and knobby as in *Variolaria pustulata*, or broad and blister-like as in *Lasallia*.

pustulate. Having many pustules.

pycnidium (*-dia*). A small globular or flask-shaped body in which conidia are formed, embedded in a thallus or entirely superficial, often closely resembling a perithecium (fig. 18).

pyrenocarp (*-ous*). A lichen in which the fruiting bodies are perithecia, e.g., *Verrucaria* or *Pyrenula*.

pyrenoid. A round, conspicuous body associated with the production and storage of starch and found within or attached to the chloroplast of some green algae. In *Trebouxia*, it is in the center of the chloroplast and is very conspicuous (often mistaken for the nucleus, which is almost invisible without staining and is in the colorless cytoplasm).

reticulate. Net-like and interconnected, like the branches of *Ramalina menziesii*.

rhizine. A purely hyphal extension of the lower cortex that generally serves to attach a foliose thallus to its substrate; of various lengths, thicknesses, colors, and degrees of branching (figs. 4 and 6).

rhizohypha(*-ae*). Loose hyphae that attach a crustose thallus with a cottony hypothallus to the substrate (e.g., in species of *Lepraria* and *Chrysothrix*).

rimose. Having a minutely cracked appearance (fig. 2c).

saprophyte. An organism that lives on decaying organic matter.

saxicolous. Growing on rock, stone, pebbles, concrete, or brick.

scabrose. Having a minutely roughened, almost crusty surface, generally caused by an accumulation of dead cortical material (see *Peltigera scabrosa*); often intergrades with pruina, which is more powdery than crusty.

schizidium (*-dia*). A lichen fragment consisting of the upper layers of a thallus (with the cortex and photobiont) and serving as a vegetative reproductive unit; formed by a scaling off of the thallus surface (as in *Fulgensia*) or a breakdown of pustules (as in some species of *Hypotrachyna*) (fig. 19g).

Scytonema. A genus of cyanobacteria with a filamentous thallus (fig. 1e) that branches by breaking through its gelatinous sheath (false branching).

septate. Divided by a crosswall or septum. A spore can be *transversely septate* (figs. 15b-f, i-j) dividing a long spore into shorter cells, or both *transversely* and *longitudinally septate* (figs. 15g-h), creating a muriform spore.

septum (*-ta*). A crosswall in a fungal hypha or spore (figs. 5, 15b-j).

sessile. Sitting on the surface, without a stalk of any kind, as with the apothecia of *Lecanora pacifica*.

siliceous. Pertaining to a rock or soil rich in silica and lacking calcium; examples are granite and gneiss, and quartz sand.

s. lat., "*sensu lato,*" i.e., in the broad sense, using the name to possibly include populations, morphotypes, or chemotypes that are sometimes not included under this name; usually used in reference to species or genera that are somewhat heterogeneous and may include other taxa.

soralium (*-lia*). An area of a thallus in which the cortex has broken down or cracked and soredia are produced; can be in many forms (fig. 20; plate 17); sometimes contains isidia as well as soredia, as in *Lobaria pulmonaria* and *Usnea diplotypus*.

soredium (*-dia*). A vegetative propagule of a lichen consisting of a few algal cells entwined and surrounded by fungal filaments, and without a cortex; generally produced in localized masses called soralia, or covering large diffuse areas on a thallus (figs. 19h-j).

sp. Abbreviation of "species," generally used where the species is unknown or unspecified.

species. The basic evolutionary unit of an organism; named with two words, the first being the genus to which the species belongs, and the second the species' own name or "epithet." With respect to lichens, the name of a species refers to its fungal component. Lichen species are defined by discontinuities in various morphological, chemical, ecological, and geographic characteristics, and now, almost routinely, by analysis of the actual genetic material (e.g., DNA) of the fungal component.

spore. A single- or multicelled reproductive body capable of giving rise to a new organism; as used here, refers specifically to an ascospore (fig. 15).

spp. Several unspecified species.

squamule. A small, scale-like lobe or areole, lifting from the surface at least at the edges (as in *Acarospora fuscata* or *Peltula* species), and sometimes strongly ascending and almost foliose (as in some species of *Cladonia*).

squamulose. Composed of or characterized by having squamules (fig. 2e).

squarrose. With short, stiff, perpendicular branches; having the general appearance of a bottle-brush, as in certain types of rhizines (figs. 6d-e).

subsp. Subspecies, a formal subdivision of a species, used for important morphologically, chemically, or ecologically distinct segregates, usually somewhat geographically isolated.

s. str. "*sensu stricto,*" i.e., in the strict sense, applying the name in the narrowest sense, including the type specimen.

stereome. A tough, cartilaginous cylinder forming the supporting tissue for species of *Cladonia* and *Cladina* (fig. 8f).

Stichococcus. A genus of small, unicellular green algae having short, cylindrical (rod-shaped) cells (figure 1g).

Stigonema. A genus of filamentous cyanobacteria having "true branching" resulting from perpendicular division of cells within the filament (fig. 1a). (Compare with *Scytonema*, fig. 1e). Lichens with *Stigonema* (e.g., *Ephebe*) do not look very different from the free-living cyanobacteria.

stratified. Layered; in reference to lichen thalli having distinguishable layers of tissue including a cortex, photobiont layer, medulla, and often a lower cortex (fig. 4).

striate. With fine longitudinal lines, ridges, or grooves.

stroma (*-ata*). See *pseudostroma*.

sub- (a) Partially; (b) incompletely; (c) approaching; (d) under: as in *subfoliose* (not quite foliose), *submarginal* (close to the margin), *subpendent* (almost pendent), or *subhymenium* (the layer often distinguishable just below the hymenium and above the hypothecium).

substrate. The surface upon which a lichen grows; a nutritional relationship is not implied and rarely occurs in lichens.

superficial. Used with reference to apothecia that sit on the surface of the thallus (i.e., sessile, not immersed or stalked).

taxon (*-a*). A unit in a classification scheme; most commonly used with reference to a genus, species, or subdivision of a species (subspecies, variety, or form).

taxonomy. The study of taxa, especially with respect to their identification and classification, and the correct application of their names.

terete. Round in cross section (cylindrical) like the branches of *Bryoria* species.

terricolous. Growing on soil, sand, or peat.

thalline. Pertaining to the lichen thallus; similar to the thallus in appearance or structure.

thalline margin. See *amphithecium*.

thallospore. A granular vegetative propagule derived from thallus tissue (not conidia) and consisting of agglomerations of 2-several cells, often darkly pigmented; seen at the edge of some crustose lichens (as part of the prothallus) or on the lower surface of some umbilicate lichens.

thallus. In lichens, the vegetative body consisting of both algal and fungal components (fig. 2).

thin-layer chromatography. A method used for the identification of chemical compounds, specifically lichen substances (plate 89, *LNA*).

tier. A platform-like expansion or flat cup on the podetia of some species of *Cladonia* (e.g., *Cladonia verticillata*), often proliferating from the center or margins with one or more new branches.

tholus. The thickened tip of an ascus, frequently staining in iodine (IKI or K/I) in various ways (fig. 14).

TLC. Thin-layer chromatography, a technique for separating and identifying lichen substances (see p. 108 in LNA).

tomentose. Having tomentum; with a downy or woolly appearance.

tomentum. A covering of fine hair or fuzz usually caused by a superficial growth of colorless hyphae (plate 12).

Trebouxia. A genus of single-celled green algae with one distinctive, disk-shaped chloroplast almost filling the cell. The chloroplast has a lobed or scalloped margin and a single round to oval pyrenoid at the center. *Trebouxia* is the most common green photobiont in lichens. When followed by a query (?) in the text, it is used in the broad sense, including *Asterochloris* (fig. 1).

Trebouxioid. Resembling *Trebouxia* or *Asterochloris*.

Trentepohlia. A genus of filamentous green algae found in many crustose lichens (fig 1j); when lichenized, the alga often produces very short filaments or is single-celled. The cells are often not perfectly round, and the walls often shine (are birefringent) in polarized light under the microscope. The orange-red pigmented globules, common in the cells of unlichenized individuals, are infrequent or absent in lichenized individuals.

tubercle. A wart-like protuberance that contains some medullary tissue; characteristic of *Usnea ceratina* and *Ramalina paludosa*.

tuberculate. a) Having the general form of a tubercle, usually in reference to small, round soralia (as in *Bryoria fuscescens*); b) having many tubercles, as in some species of *Usnea*.

umbilicate. Attached by a single, central holdfast (an umbilicus) on the lower surface of the thallus (fig. 2g).

umbilicus. A short, thick, purely fungal, central attachment organ present on certain foliose lichens such as *Umbilicaria* and *Dermatocarpon*.

umbo. The central bump (or depression) on the upper surface of an umbilicate lichen, corresponding to the position of the umbilicus.

UV. Ultraviolet light. A number of lichen substances fluoresce and therefore can be detected in long wave UV light (365 nm) (plate 90); short wave UV (254 nm) is used in thin layer chromatography for analyzing plates made with gels containing certain fluorescent dyes.

vagrant. Growing unattached to rock or soil and therefore able to roll freely over the ground (plate 34).

var. Variety; a formal subdivision of a species or subspecies, usually used for recurring, genetically based variants in morphology, chemistry, or habitat.

vein. In lichens, broad or narrow ridges or thickenings, often pigmented, on the lower surface of some lichens such as *Peltigera* (e.g., *P. kristinsonii*), but not functioning as conducting tissue.

verruca (*-cae*). A conspicuous, wart-like, thalline protuberance (e.g., *Pertusaria plittiana*).

verrucose. With a rough, warty surface (e.g., *Ochrolechia oregonensis*).

verruculose. Minutely verrucose, like the thallus of *Lecanora circumborealis*.

wood. Lignum; trunks, logs, and stumps having no bark.

Xanthocapsa. A genus of cyanbacteria consisting of groups of 2-8 spherical cells enclosed within a thick brownish to yellow-brown gelatinous matrix. Gelatinous sheaths of individual cells usually are K–.

Literature Cited

Ahti, T. 2007. Further studies on the *Cladonia verticillata* group (Lecanorales) in East Asia and western North America. *Bibliotheca Lichenologica* 96: 5-19.

Ahti, T., and S. Stenroos. 2013. Cladoniaceae. *Nordic Lichen Flora* 5: 1-116. Museum of Evolution, Uppsala University for Nordic Lichen Society, Uppsala

Amtoft, A., F. Lutzoni, and J. Miadlikowska. 2008. *Dermatocarpon* (Verrucariaceae) in the Ozark Highlands, North America. *Bryologist* 111: 1-40.

Aptroot, A. 2012. A world key to the species of *Anthracothecium* and *Pyrenula*. *Lichenologist* 44: 5-53.

Arcadia, L. I. 2012. (2071) Proposal to conserve the name *Lichen leucomelos* (*Heterodermia leucomelos*) with that spelling (lichenised Ascomycota). *Taxon* 61: 681.

Arup, U. 1992. *Caloplaca marina* and *C. rosei*, two difficult species in North America. *Bryologist* 95: 148-160.

Arup, U., and E. S. Berlin. 2011. A taxonomic study of *Melanelixia fuliginosa* in Europe. *Lichenologist* 43: 89-97.

Arup, U., U. Søchting, and P. Frödén. 2013. A new taxonomy of the family Teloschistaceae. *Nordic Journal of Botany* 31: 16-83.

Barton, J., and J. Lendemer. 2014. *Micarea micrococca* and *M. prasina*, the first assessment of two very similar species in eastern North America. *Bryologist* 117: 223-231.

Bendiksby, M., and E. Timdal. 2013. Molecular phylogenetics and taxonomy of *Hypocenomyce* sensu lato (Ascomycota: Lecanoromycetes): Extreme polyphyly and morphological/ecological convergence. *Taxon* 62: 940-956.

Benneti, M. N., and J. A. Elix. 2012. The true identity of *Bulbothrix goebelii* (Zenker) Hale and the re-establishment of some of its synonyms as accepted species. *Lichenologist* 44: 813-826.

Breuss, O. 2002a. *Endocarpon, in* Nash et al., *Lichen Flora of the Greater Sonoran Desert Region*, Vol. 1, pp. 181-187.

Breuss, O. 2002b. *Placidium, in* Nash et al., *Lichen Flora of the Greater Sonoran Desert Region*, Vol. 1, pp. 384-393.

Breuss, O., and B. McCune. 1994. Additions to the pyrenolichen flora of North America. *Bryologist* 97: 365-370.

Brodo, I. M. 1981. *Lichens of the Ottawa Region*. Syllogeus no. 29. National Museum of Natural Sciences, Ottawa. 137 pp.

Brodo, I. M. 1988. *Lichens of the Ottawa Region,* 2nd ed. Ottawa Field-Naturalists' Club Special Publication No. 3. 115 pp.

Brodo, I. M. 1995. Notes on the lichen genus *Placopsis* (Ascomycotina, Trapeliaceae) in North America. *Bibliotheca Lichenologica* 57: 59-70.

Brodo, I. M. 2010. The lichens and lichenicolous fungi of Haida Gwaii (Queen Charlotte Islands), British Columbia, Canada. 5. A new species of *Lecanora* from shoreline rocks. *Botany* 88: 352-358.

Brodo, I. M., and A. Aptroot. 2005. Corticolous species of *Protoparmelia* (lichenized Ascomycotina) in North America. *Canadian Journal of Botany* 83: 1075-1081.

Brodo, I. M., and J. C. Lendemer. 2012. On the perplexing variability of reproductive modes in the genus *Ochrolechia*: Notes on *O. africana* and *O. arborea* in eastern North America. *Opuscula Philolichenum* 11: 120-134.

Brodo, I. M., and R. Santesson. 1997. Lichens of the Queen Charlotte Islands, British Columbia, Canada. 3. Marine species of *Verrucaria* (Verrucariaceae, Ascomycotina). *Journal of the Hattori Botanical Laboratory* 82: 27-37.

Brodo, I. M., S. D. Sharnoff, and S. Sharnoff. 2001. *Lichens of North America*, Yale University Press, New Haven and London. 795 pp.

Büdel, B., and T. H. Nash III. 2002. *Peltula, in* Nash et al., *Lichen Flora of the Greater Sonoran Desert Region*, Vol. 1, pp. 331-340.

Bungartz, F., A. Nordin, and U. Grube. 2007. *Buellia, in* Nash et al., *Lichen Flora of the Greater Sonoran Desert Region*, Vol. 3, pp. 113-179.

Clauzade, G., and C. Roux. 1985. *Likenoj de Okcidenta Europo. Ilustrita Determinlibro*. Bulletin de la Societé Botanique du Centre-Ouest, Nouvelle Série, Numero Special 7. Royan, France. 893 pp.

Clerc, P. 2011. Notes on the genus *Usnea* Adanson (lichenized Ascomycota). III. *Bibliotheca Lichenologica* 106: 41-51.

Coppins, B. J. 2009. *Micarea, in* C. W. Smith et al., *The Lichens of Great Britain and Ireland*, pp. 583-606.

Czarnota, P. 2007. *The lichen genus* Micarea *(Lecanorales, Ascomycota) in Poland.* Polish Botanical Studies 23. 199 pp.

Degelius, G. 1974. The lichen genus *Collema* with special reference to the extra-European species. *Symbolae Botanicae Upsalienses* 20(2): 1-215.

Dibben, M. J. 1980. *The Chemosystematics of the Lichen Genus Pertusaria in North America North of Mexico.* Publications in Biology and Geology No. 5, Milwaukee Public Museum Press, Milwaukee. 162 pp.

Divakar, P. K., A. Crespo, J. Núñez-Zapata, A. Flakus, H. J. M. Sipman, J. A. Elix, and H. T. Lumbsch. 2013. A molecular perspective on generic concepts in the *Hypotrachyna clade* (Parmeliaceae, Ascomycota). *Phytotaxa* 132: 21-38.

Egea, J. M., and P. Torrente. 1993. The lichen genus *Bactrospora. Lichenologist* 25: 211-255.

Egea, J. M., P. Torrente, and B. D. Ryan. 2004. *Bactrospora, in* Nash et al., *Lichen Flora of the Greater Sonoran Desert Region,* Vol. 1, pp. 32-37.

Elix, J. A., and G. Kantvilas. 2007. The genus *Chrysothrix* in Australia. *Lichenologist* 39: 361-369.

Ertz, D., and J. M. Egea. 2007. *Opegrapha, in* Nash et al., *Lichen Flora of the Greater Sonoran Desert Region*, Vol. 3, pp. 255-266.

Ertz, D., and A. Tehler. 2011. The phylogeny of Arthoniales (Pezizomycotina) inferred from nucLSU and RPB2 sequences. *Fungal Diversity* 49: 47-71.

Esslinger, T. L. 1977. A chemosystematic revision of the brown *Parmeliae. Journal of the Hattori Botanical Laboratory* 42: 1-211.

Esslinger, T. L. 2002. *Neofuscelia, in* Nash et al., *Lichen Flora of the Greater Sonoran Desert Region*. Vol. 1, pp. 289-295.

Esslinger, T. L. 2007. A synopsis of the North America species of *Anaptychia* (Physciaceae). *Bryologist* 110: 788-797.

Esslinger, T. L. 2014. A cumulative checklist for the lichen-forming, lichenicolous and allied fungi of the continental United States and Canada. North Dakota State University: http://www.ndsu.edu/pubweb/~esslinge/chcklst/chcklst7.htm (Posted 23 March 2014), Fargo, North Dakota.

Fink, B. 1935. *The Lichen Flora of the United States*. Completed for Publication by Joyce Hedrick. University of Michigan Press, Ann Arbor. 426 pp.

Fletcher, A., B. J. Coppins, and O. W. Purvis. 2009a. *Hymenelia, in* Smith et al., *The Lichens of Great Britain and Ireland*, pp. 432-434.

Fletcher, A., O. L. Gilbert, S. Clayden, and A. M. Fryday. 2009b. *Rhizocarpon, in* Smith et al., *The Lichens of Great Britain and Ireland*, pp. 792-808.

Frödén, P., B. D. Ryan, and I. Kärnefelt. 2004. *Teloschistes, in* Nash et al., *Lichen Flora of the Greater Sonoran Desert Region,* Vol. 2, pp. 524-529.

Fryday, A. M. 2000. On *Rhizocarpon obscuratum* (Ach.) Massal., with notes on some related species in the British Isles. *Lichenologist* 32: 207-224.

Fryday, A. M. 2002. A revision of the species of the *Rhizocarpon hochstetteri* group occurring in the British Isles. *Lichenologist* 34: 451-477.

Fryday, A. M. 2005. The genus *Porpidia* in northern and western Europe, with special emphasis on collections from the British Isles. *Lichenologist* 37: 1-35.

Fryday, A. M. 2008. The genus *Fuscidea* (Fuscideaceae, lichenized Ascomycota) in North America. *Lichenologist* 40: 295-328.

Fryday, A. M., and B. J. Coppins. 1996. A new crustose *Stereocaulon* from the mountains of Scotland and Wales. *Lichenologist* 28: 513-519.

Galloway, D. J. 2005. *Placopsis fusciduloides* (Ascomycota: Agyriaceae), a new lichen from Aotearoa New Zealand, British Columbia, and Bolivia. *Australasian Lichenology* 57: 16-20.

Galloway, D. J., and M. A. Thomas. 2004. *Sticta, in* Nash et al., *Lichen Flora of the Greater Sonoran Desert Region*, Vol. 2, pp. 513-524.

Gilbert, O. P. 2009. *Frutidella, in* Smith et al., *The Lichens of Great Britain and Ireland*, p. 405.

Gilbert, O. P., B. J. Coppins, and P. M. Jørgensen. 2009. *Lempholemma, in* Smith et al., *The Lichens of Great Britain and Ireland,* pp. 527-530.

Giralt, M., and P. Clerc. 2011. *Tetramelas thiopolizus* comb. nov. with a key to all known species of *Tetramelas*. *Lichenologist* 43: 417-425.

Gowan, S. P. 1989. The lichen genus *Porpidia* (Porpidiaceae) in North America. *Bryologist* 92: 25-59.

Gowan, S. P. and T. Ahti. 1993. Status of the lichen genus *Porpidia* in eastern Fennoscandia. *Annales Botanici Fennici* 30: 53-75.

Goward, T., B. McCune, and D. Meidinger. 1994. *The Lichens of British Columbia. Illustrated Keys. Part 1 - Foliose and Squamulose Species*. Special Report Series, 8, Research Program, B.C. Ministry of Forests, Victoria. 181 pp.

Goward, T., T. Spribille, T. Ahti, and C. J. Hampton-Miller. 2012. Four new sorediate species in the *Hypogymnia austerodes* group (lichens) from northwestern North America, with notes on thallus morphology. *Bryologist* 115: 84-100.

Heiðmarsson, S., and O. Breuss. 2004. *Dermatocarpon in* Nash et al., *Lichen Flora of the Greater Sororan Desert Region*, Vol. 2, pp. 88-93.

Halonen, P., P. Clerc, T. Goward, I. M. Brodo, and K. Wulff. 1998. Synopsis of the genus *Usnea* (lichenized Ascomycetes) in British Columbia, Canada. *Bryologist* 101: 36-60.

Harris, R. C. 1973. The corticolous pyrenolichens of the Great Lakes Region. *Michigan Botanist* 12: 3-68.

Harris, R. C. 1979. Four species of *Thelopsis* Nyl. (lichenized Ascomycetes) new to North America. *Bryologist* 82: 77-78.

Harris, R. C. 1989. A sketch of the family Pyrenulaceae (Melanommatales) in eastern North America. *Memoirs of the New York Botanical Garden* 49: 74-107.

Harris, R. C. 1990. *Some Florida Lichens*, published by the author, Bronx, N.Y. 109 pp.

Harris, R. C. 1995. *More Florida Lichens*, published by the author, Bronx, New York. 192 pp.

Harris, R. C., and D. Ladd. 2005. *Preliminary draft: Ozark lichens*. Prepared for the 14th Tuckerman Lichen Workshop, Eureka Springs, Arkansas. Distributed by authors, litho.

Harris, R. C., and D. Ladd. 2008. The lichen genus *Chrysothrix in* the Ozark ecoregion, including a preliminary treatment for eastern and central North America. *Opuscula Philolichenum* 5: 29-42.

Henssen, A., and H. M. Jahns. 1974. *Lichenes, Eine Einführung in die Flechtenkunde*. Georg Thieme Verlag, Stuttgart. 467 pp.

Hines, J. W., and P. L. Hines. 2007. *The Macrolichens of New England.* Memoirs of the New York Botanical Garden 96: 584 pp.

Jahns, H. M. 1973. Anatomy, morphology and development. *in* V. Ahmadjian and M. E. Hale (eds.), *The Lichens.* Academic Press, New York and London, pp. 3-58.

James, P. W., and M. F Watson. 2009. *Mycoblastus, in* Smith et al., *The lichens of Great Britain and Ireland*, pp. 615-618.

James, P. W., T. Duke, and D. J. Coppins. 2009. *Strangospora, in* Smith et al., *The Lichens of Great Britain and Ireland*, pp. 867-869.

Jørgensen, P. M. 2000 [2001]. Survey of the lichen family Pannariaceae on the American continent, north of Mexico. *Bryologist* 103: 670-704.

Jørgensen, P. M., and T. H. Nash III. 2004. *Leptogium, in* Nash et al., *Lichen Flora of the Greater Sonoran Desert Region*, Vol. 2, pp. 330-350.

Jørgensen, P. M., M. Otálora, and M. Wedin. 2013. (2235) Proposal to conserve the name *Leptogium* (lichenized *Ascomycota*) with a conserved type. *Taxon* 62: 1333-1334.

Joshi, S., D. K. Upreti, and S. Nayaka. 2012. Two new species in the lichen genus *Phlyctis* (Phlyctidaceae) from India. *Lichenologist* 44: 363-369.

Kantvilas, G., and J. A. Elix. 1994. *Ramboldia* a new genus in the lichen family Lecanoraceae. *Bryologist* 97: 296-304.

Kantvilas, G., and J. A. Elix. 2007. The genus *Ramboldia* (Lecanoraceae): a new species, key and notes. *Lichenologist* 38: 135-141.

Kalb, K., B. Staiger, J. A. Elix, U. Lange and H. T. Lumbsch. 2008: A new circumscription of the genus *Ramboldia* (Lecanoraceae, Ascomycota) based on morphological and molecular evidence. *Nova Hedwigia* 86: 23-42.

Knopf, J.-G., and C. Leuckert. 2004. *Lecidella, in* Nash et al., *Lichen Flora of the Greater Sonoran Desert Region*, Vol. 2, pp. 309-320.

Knudsen, K. 2007a. *Acarospora, in* Nash et al., *Lichen Flora of the Greater Sonoran Desert Region*, Vol. 3, pp. 1-38.

Knudsen, K. 2007b. *Polysporina, in* Nash et al., *Lichen Flora of the Greater Sonoran Desert Region*, Vol. 3, pp. 276-278.

Knudsen, K., and J. C. Lendemer. 2007. Studies in lichens and lichenicolous fungi: notes on some North American taxa. *Mycotaxon* 101: 81-87.

Knudsen, K., J. C. Lendemer, and R. C. Harris. 2011. Studies in lichens and lichenicolous fungi – no 15: miscellaneous notes on species from eastern North America. *Opuscula Philolichenum* 9: 45-75.

Knudsen, K., and B. D. Ryan. 2007. *Strangospora, in* Nash et al., *Lichen Flora of the Greater Sonoran Desert Region*, Vol. 3, pp. 299-301.

Knudsen, K., and S. M. Standley. 2007. *Sarcogyne, in* Nash et al., *Lichen Flora of the Greater Sonoran Desert Region*, Vol. 3, pp. 289-296.

Krzewicka, B. 2012. A revision of *Verrucaria* s.l. (Verrucariaceae) in Poland. *Polish Botanical Studies* 27: 3-143.

Kukwa, M. 2011. *The lichen genus* Ochrolechia *in Europe*. Fundacja Rozwoju Uniwersytetu Gdańskiego, Gdańsk. 309 pp.

Leavitt , S. D., F. Fernández-Mendoza, S. Pérez-Ortega, M. Sohrabi, P. K. Divakar, H. T. Lumbsch, and L. L. St. Clair. 2013. DNA barcode identification of lichen-forming fungal species in the *Rhizoplaca melanophthalma* species-complex (Lecanorales, Lecanoraceae), including five new species. *Mycokeys* 7: 1-22.

Lendemer, J. C. 2007. The occurrence of *Endocarpon pallidulum* and *Endocarpon petrolepidium* in Eastern North America. *Evansia* 24: 103-107.

Lendemer, J. C. 2009. A synopsis of the lichen genus *Heterodermia* (Physciaceae, lichenized Ascomycota) in eastern North America. *Opuscula Philolichenum* 6: 1-36.

Lendemer, J. C. 2010. Preliminary Keys to the Typically Sterile Crustose Lichens in North America 34 pp., +32 color plates. On-line: http://sweetgum.nybg.org/southeastlichens/biblio_detail.php?irn=250001.

Lendemer, J. C. 2013. A monograph of the crustose members of the genus *Lepraria* Ach. s. str. (Stereocaulaceae, Lichenized Ascomycetes) in North America north of Mexico. *Opuscula Philolichenum* 12: 27-141.

Lendemer, J. C., and R. C. Harris. 2008. Keys to Lime-Loving Lichens. New York Botanical Garden, litho.

Lendemer, J. C., and B. P. Hodkinson. 2010. A new perspective on *Punctelia subrudecta* (Parmeliaceae) in North America: previously rejected morphological characters corroborate molecular phylogenetic evidence and provide insight into an old problem. *Lichenologist* 42: 405-421.

Lendemer, J. C., and B. P. Hodkinson. 2013. A radical shift in the taxonomy of *Lepraria* s.l.: molecular and morphological studies shed new light on the evolution of asexuality and lichen growth form diversification. *Mycologia* 105: 994-1018.

Lendemer, J. C., R. C. Harris, and E. A. Tripp. 2013. *The lichens and allied fungi of Great Smoky Mountains National Park.* Memoirs of the New York Botanical Garden, 104: 156 pp.

Linda in Arcadia. 2012 (see Arcadia).

Lücking, R. 2008. *Foliicolous Lichenized Fungi*. Flora Neotropica Monograph 103. Organization for Flora Neotropica and The New York Botanical Garden Press, Bronx, New York. 866 pp.

Lücking, R., W. R. Buck, and E. Rivas Plata. 2007. The lichen family Gomphillaceae (Ostropales) in eastern North America, with notes on hyphophore development in *Gomphillus* and *Gyalideopsis*. *Bryologist* 110: 622-672.

Lutzoni, F. M., and I. M. Brodo. 1995. A generic redelimitation of the *Ionaspis-Hymenelia* complex (lichenized Ascomycotina). *Systematic Botany* 20: 224-258.

Malíček, J. 2014. A revision of the epiphytic species of the *Lecanora subfusca* group (*Lecanoraceae*, Ascomycota) in the Czech Republic. *Lichenologist* 46: 489-513.

Marbach, B. 2000. Corticole und lignicole Arten der Flechtengattung *Buellia* sensu lato in den Subtropen und Tropen. *Bibliotheca Lichenologica* 74: 1-384.

McCune, B., and L. Geiser. 1997. *Macrolichens of the Pacific Northwest*. Oregon State University Press/USDA–Forest Service, Corvallis. 386 pp.

McCune, B., and L. Geiser. 2009. *Macrolichens of the Pacific Northwest. Second Edition*. Oregon State University Press, Corvallis. 464 pp.

McCune, B., and S. Altermann. 2009. *Letharia gracilis* (Parmeliaceae), a new species from California and Oregon. *Bryologist* 112: 375-378.

McCune, B. and L. Geiser. 1997. *Macrolichens of the Pacific Northwest*. Oregon State University Press/U.S.D.A. Forest Service, Corvallis. 386 pp.

McCune, B., and T. Goward. 1995. *Macrolichens of the Northern Rocky Mountains*. Mad River Press, Eureka, California. 208 pp.

McCune, B., and R. Rosentreter. 2007. *Biotic Soil Crust Lichens of the Columbia Basin*. Monographs in North American Lichenology, Vol. 1. Northwest Lichenologists, Ltd., Corvallis. 105 pp.

McDonald, T., J. Miadlikowska, and F. Lutzoni. 2003. The lichen genus *Sticta* in the Great Smoky Mountains: a phylogenetic study of morphologial, chemical, and molecular data. *Bryologist* 106: 61-79.

Moberg, R. 1997. The lichen genus *Physcia* in the Sonoran Desert and adjacent regions, *in* L. Tibell, and I. Hedberg (eds.), *Lichen Studies Dedicated to Rolf Santesson*. Symbolae Botanicae Upsalienses, Acta Universitatis Upsaliensis, Uppsala, pp. 163-186.

Moberg, R., and T. A. Nash III. 2002. *Heterodermia, in* Nash et al., *Lichen Flora of the Greater Sonoran Desert Region*, Vol. 1, pp. 207-219.

Moncada, B, R. Lücking, and L. Betancourt-Macuase. 2013. Phylogeny of the Lobariaceae (lichenized Ascomycota: Peltigerales), with a reappraisal of the genus *Lobariella*. *Lichenologist* 45: 203-263.

Moncada, B., B. Reidy, and R. Lücking. 2014. A phylogenetic revision of Hawaiian *Pseudocyphellaria* sensu lato (lichenized Ascomycota: Lobariaceae) reveals eight new species and a high degree of inferred endemism. *Bryologist* 117: 119-160.

Myllys, L., S. Velmala, H. Lindgren, D. Glavich, T. Carlberg, L-S. Wang, and T. Goward. 2014. Taxonomic delimitation of the genera *Bryoria* and *Sulcaria*, with a new combination *Sulcaria spiralifera* introduced. *Lichenologist* 46: 737-752.

Næsborg (see Reese Næsborg).

Nash, T. H., III. 2002. Previously reported but currently excluded species from the Greater Sonoran Desert Region, *in* Nash et al., *Lichen Flora of the Greater Sonoran Desert Region*, Vol. 1, pp. 509-510.

Nash, T. H., III, B. D. Ryan, C. Gries, and F. Bungartz (eds.). 2002. *Lichen Flora of the Greater Sonoran Desert Region,* Vol. 1. Lichens Unlimited, Arizona State University, Tempe, Arizona. 532 pp.

Nash, T. H., III, B. D. Ryan, P. Diederich, C. Gries, and F. Bungartz (eds.). 2004. *Lichen Flora of the Greater Sonoran Desert Region,* Vol. 2. Lichens Unlimited, Arizona State University, Tempe, Arizona. 742 pp.

Nash, T. H. III, C. Gries, and F. Bungartz (eds.). 2007. *Lichen Flora of the Greater Sonoran Desert Region,* Vol. 3. Lichens Unlimited, Arizona State University, Tempe, Arizona. 567 pp.

Orange, A., O. W. Purvis, and P. W. James. 2009a. *Polyblastia, in* Smith et al., *The Lichens of Great Britain and Ireland,* pp. 722-728

Orange, A., O. W. Purvis, and P. W. James. 2009b. *Porina, in* Smith et al., *The Lichens of Great Britain and Ireland*, pp. 729-737.

Orange, A., O. W. Purvis, and P. W. James. 2009c. *Protothelenella, in* Smith et al., *The Lichens of Great Britain and Ireland*, pp. 755-757.

Orange, A., O. W. Purvis, and P. W. James. 2009d. *Thelenella, in* Smith et al., *The Lichens of Great Britain and Ireland*, pp. 877-879.

Orange, A., M. F. Watson, P. W. James, and D. M. Moore. 2009d. *Thelocarpon, in* Smith et al., *The Lichens of Great Britain and Ireland*, pp. 884-888.

Otálora, M. A. G., P. M. Jørgensen, and M. Wedin. 2013. A revised generic classification of the jelly lichens, *Collemataceae. Fungal Diversity* (on-line: DOI 10.1007/s13225-013-0266-1)

Owe-Larson, B., A. Norden, and L. Tibell. 2007. *Aspicilia, in* Nash et al, *Lichen Flora of the Greater Sonoran Desert Region*, vol. 3, pp. 61-108.

Ozenda, P. 1963. *Lichens*. Handbuch de Pflanzenanatomie. Band VI, Teil 9, Abteilung: Spezieller Teil. 199 pp.

Pérez-Ortega, S., T. Spribille, Z. Palice, J. A. Elix and C. Printzen. 2010. A molecular phylogeny of the *Lecanora varia* group, including a new species from western North America. *Mycological Progress* 9: 523-535.

Poelt, J. 1969. *Bestimmungsschlüssel Europäischer Flechten*. J. Cramer, Lehre. 757 pp.

Ponzetti, J., and B. McCune. 2006. A new species of *Bactrospora* from northwestern North America. *Bryologist* 109: 85-88.

Printzen, C. 2001. Corticolous and lignicolous species of *Lecanora* (Lecanoraceae, Lecanorales) with usnic or isousnic acid in the Sonoran Desert Region. *Bryologist* 104: 382-409.

Printzen, C. 2005. *Biatora, in* Nash et al., *Lichen Flora of the Greater Sonoran Desert Region*, Vol. 2, pp. 37-39.

Printzen, C., T. Spribille, and T. Tønsberg. 2008. *Myochroidea*, a new genus of corticolous, crustose lichens to accommodate the *Lecidea leprosula* group. *Lichenologist* 40: 195-207.

Printzen, C. and T. Tønsberg. 1999. The lichen genus *Biatora* in northwestern North America. *Bryologist* 102: 692-713.

Printzen, C., and T. Tønsberg. 2004. New and interesting *Biatora* species. *Symbolae Botanicae Uppsaliense* 34:1.

Purvis, O. W., B. J. Coppins, D. L. Hawksworth, P. W. James, and D. M. Moore (eds.) 1992. *The Lichen Flora of Great Britain and Ireland.* British Lichen Society, Natural History Museum Publications, London. 710 pp.

Reese Næsborg, R. 2008. Taxonomic revision of the *Lecania cyrtella* group based on molecular and morphological evidence. *Mycologia* 100: 397–416.

Ryan, B. D. 2002. *Rhizoplaca, in* Nash et al., *Lichen Flora of the Greater Sonoran Desert Region*, Vol. 1, pp. 442-448.

Ryan, B. D. 2004a. *Lignoscripta, in* Nash et al., *Lichen Flora of the Greater Sonoran Desert Region*, Vol. 2, pp. 350-351.

Ryan, B. D. 2004b. *Xylographa, in* Nash et al., *Lichen Flora of the Greater Sonoran Desert Region*, Vol. 2, pp. 612-616.

Ryan, B. D. 2004c. *Lobothallia, in* Nash et al., *Lichen Flora of the Greater Sonoran Desert Region*, Vol. 2, pp. 352-357.

Ryan, B. D., H. T. Lumbsch, M. I. Messuti, C. Printzen, L. Śliwa, and T. H. Nash III. 2004. *Lecanora, in* Nash et al., *Lichen Flora of the Greater Sonoran Desert Region*, Vol. 2, pp. 176-286.

Ryan, B. D., T. H. Nash III, and J. Hafellner. 2004. *Protoparmelia, in* Nash et al., *Lichen Flora of the Greater Sonoran Desert Region*, Vol. 2, pp. 425-430.

Schultz, M. 2006. *Cryptothele rhodosticta* new to Scandinavia from Norway. *Graphis Scripta* 18: 49-51.

Sheard, J. 2010. *The lichen genus* Rinodina *(Ach.) Gray (Lecanoromycetidae, Physciaceae) in North America, north of Mexico.* NRC Research Press, Ottawa, Ontario, Canada. 246 pp.

Śliwa, L. 2007. A revision of the *Lecanora dispera* complex in North America. *Polish Botanical Journal* 52: 1-70.

Smith, C. W., A. Aptroot, B. J. Coppins, A. Fletcher, O. L. Gilbert, P. W. James, and P. A. Wolseley (eds.). 2009. *The Lichens of Great Britain and Ireland.* British Lichen Society and Natural History Museum, London. 1046 pp.

Spribille, T., B. Klug, and H. Mayrhofer. 2011. A phylogenetic analysis of the boreal lichen *Mycoblastus sanguinarius* (Mycoblastaceae, lichenized Ascomycota) reveals cryptic clades correlated with fatty acid profiles. *Molecular Phylogenetics and Evolution* 59: 603-614.

Spribille, T., P. Resl, T. Ahti, S. Pérez-Ortega, T. Tønsberg, H. Mayrhofer, and H.T. Lumbsch. 2014. Molecular systematics of the wood-inhabiting, lichen-forming genus *Xylographa* (Baeomycetales, Ostropomycetidae) with eight new species. *Acta Universitatis Upsaliensis, Symbolae Botanicae Upsalienses* 37(1): 1-87. Uppsala.

Tehler, A., D. Ertz, and M. Irestedt. 2013. The genus *Dirina* (*Roccellaceae, Arthoniales*) revisited. *Lichenologist* 45: 427-476.

Thomson, J. W. 1997. *American Arctic Lichens. 2. The Microlichens*. The University of Wisconsin Press, Madison. 675 pp.

Thomson, J. W. 2002. *Staurothele, in* Nash et al., *Lichens of the Greater Sonoran Desert Region*, Vol. 1, pp. 468-472.

Thüs, H., and M. Schultz. 2009. *Fungi: Lichens, Pt. 1 (Süßwasserflora Von Mitteleuropa / Freshwater Flora of Central Europe)*. Spektrum Akademischer Verlag, Heidelberg, Germany. 223 pp.

Tibell, L. 1999. Calicioid lichens and fungi. *Nordic Lichen Flora* 1: 20-94.

Timdal, E. 1984. The genus *Hypocenomyce* (Lecanorales, Lecideaceae) with special emphasis on the Norwegian and Swedish species. *Nordic Journal of Botany* 4: 83-108.

Timdal, E., and J. Holtan-Hartwig. 1988. A preliminary key to *Rhizocarpon* in Scandinavia. *Graphis Scripta* 2: 41-54.

Tønsberg, T., and I. M. Brodo. 1992. *Enterographa zonata* and *Opegrapha gyrocarpa* new to North America. *Bryologist* 95: 225-226.

Velmala, S., L. Myllys, T. Goward, H. Holein, and P. Halonen. 2014. Taxonomy of *Bryoria* section *Implexae* (Parmeliaceae, Lecanoromycetes) in North America and Europe, based on chemical, morphological and molecular data. *Annales Botanici Fennici* 51: 345-371.

Vobis, G. 1980. *Bau und Entwicklung der Flechten-Pycnidien und ihrer Conidien. Bibliotheca Lichenologica* 14. J. Cramer, Vaduz. 141 pp.

Watling, R., and R. G. Woods. 2009. *Lichenomphalia, in* Smith et al., *The Lichens of Great Britain and Ireland*, pp. 553-556.

Wedin, M., F. Högnabba, and T. Goward. 2009. A new species of *Sphaerophorus,* and a key to the family Sphaerophoraceae in western North America. *Bryologist* 112: 368-374.

Westberg, M. 2007a. *Candelariella* (Candelariaceae) in western United States and northern Mexico; the species with biatorine apothecia. *Bryologist* 110: 365-374.

Westberg, M. 2007b. *Candelariella* (Candelariaceae) in western United States and northern Mexico: the polysporous species. *Bryologist* 110: 375-390

Westberg, M. 2007c. *Candelariella* (Candelariaceae) in western United States and northern Mexico: the 8-spored, lecanorine species. *Bryologist* 110: 391-419.

Westberg, M., C. A. Morse, and M. Wedin. 2011. Two new species of *Candelariella* and a key to the Candelariales (lichenized Ascomycetes) in North America. *Bryologist* 114: 325-334.

Wirth, V. 1995. *Die Flechten Baden-Württembergs, Teil 1 & 2*. Eugen Ulmer GmbH., Stuttgart. 1006 pp.

Index to Names of Genera and Species

Accepted names are in *italics*; synonyms are in ordinary Roman font. Page numbers in **boldface** indicate where the major generic treatments begin.

A

C

N

O

P

R

S

Z

Notes

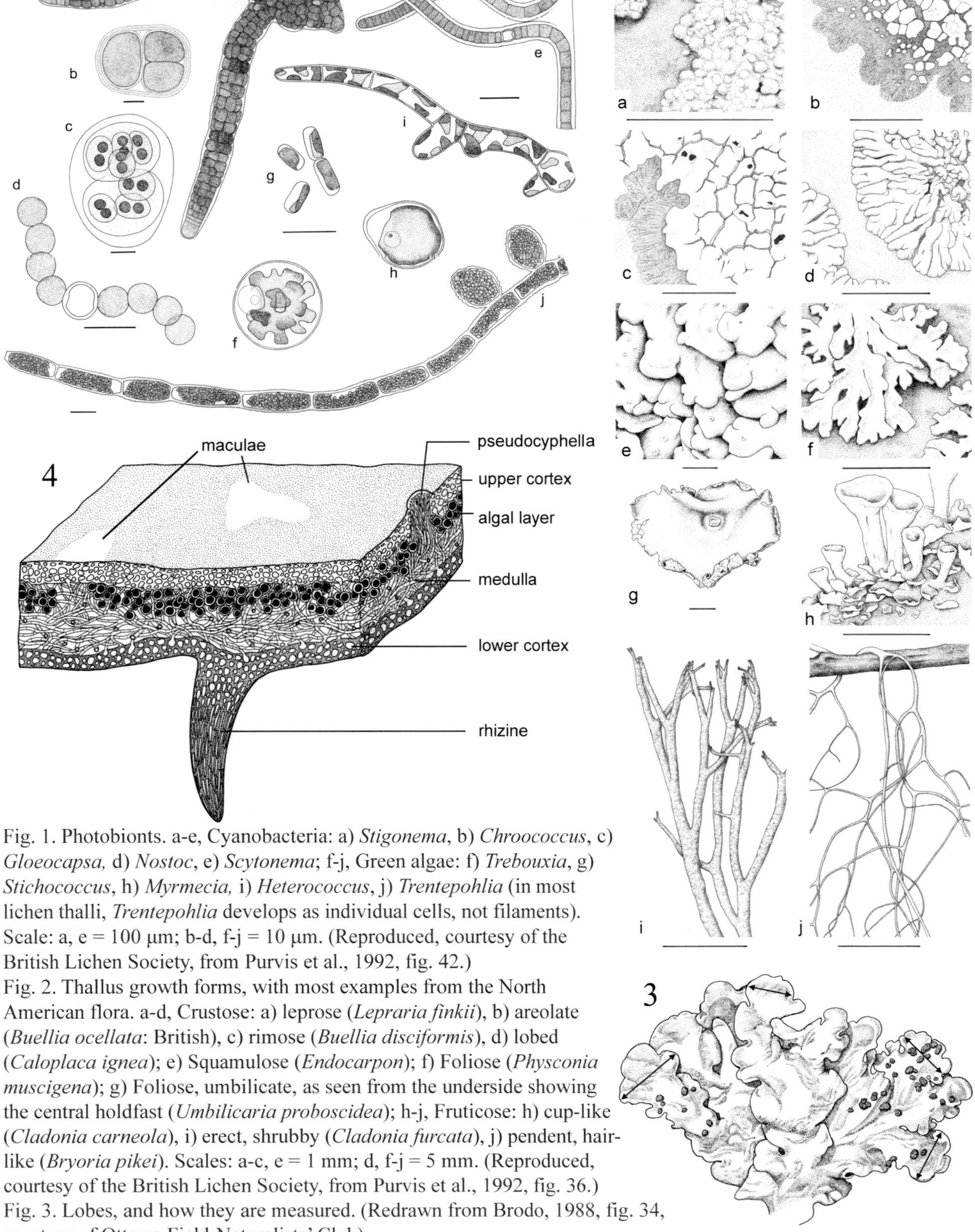

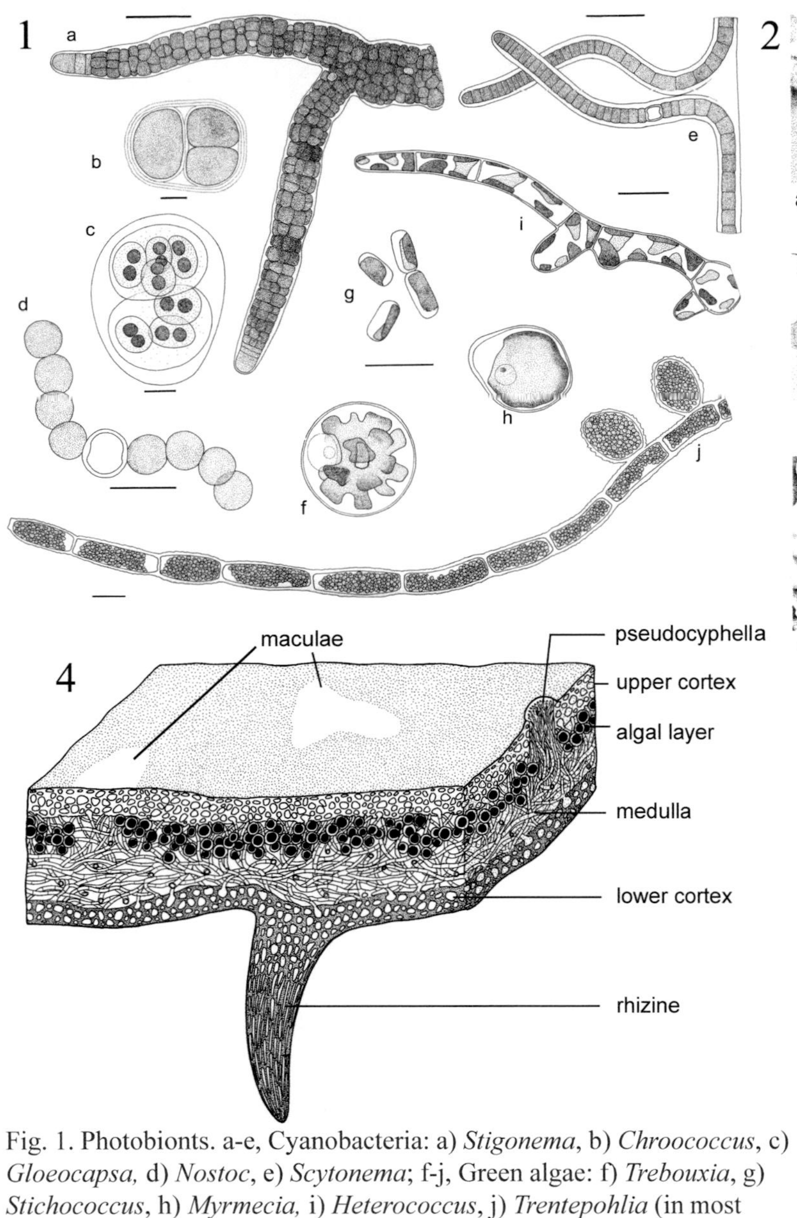

Fig. 1. Photobionts. a-e, Cyanobacteria: a) *Stigonema*, b) *Chroococcus*, c) *Gloeocapsa,* d) *Nostoc*, e) *Scytonema*; f-j, Green algae: f) *Trebouxia*, g) *Stichococcus*, h) *Myrmecia,* i) *Heterococcus*, j) *Trentepohlia* (in most lichen thalli, *Trentepohlia* develops as individual cells, not filaments). Scale: a, e = 100 µm; b-d, f-j = 10 µm. (Reproduced, courtesy of the British Lichen Society, from Purvis et al., 1992, fig. 42.)

Fig. 2. Thallus growth forms, with most examples from the North American flora. a-d, Crustose: a) leprose (*Lepraria finkii*), b) areolate (*Buellia ocellata*: British), c) rimose (*Buellia disciformis*), d) lobed (*Caloplaca ignea*); e) Squamulose (*Endocarpon*); f) Foliose (*Physconia muscigena*); g) Foliose, umbilicate, as seen from the underside showing the central holdfast (*Umbilicaria proboscidea*); h-j, Fruticose: h) cup-like (*Cladonia carneola*), i) erect, shrubby (*Cladonia furcata*), j) pendent, hair-like (*Bryoria pikei*). Scales: a-c, e = 1 mm; d, f-j = 5 mm. (Reproduced, courtesy of the British Lichen Society, from Purvis et al., 1992, fig. 36.)

Fig. 3. Lobes, and how they are measured. (Redrawn from Brodo, 1988, fig. 34, courtesy of Ottawa Field-Naturalists' Club)

Fig. 4. Stratified foliose lichen thallus, a 3-dimensional view. In the pseudocyphella, the hyphae of the medulla reach the surface through a small hole in the cortex; maculae show as pale spots on the thallus surface because of an interruption in the algal layer just below it.

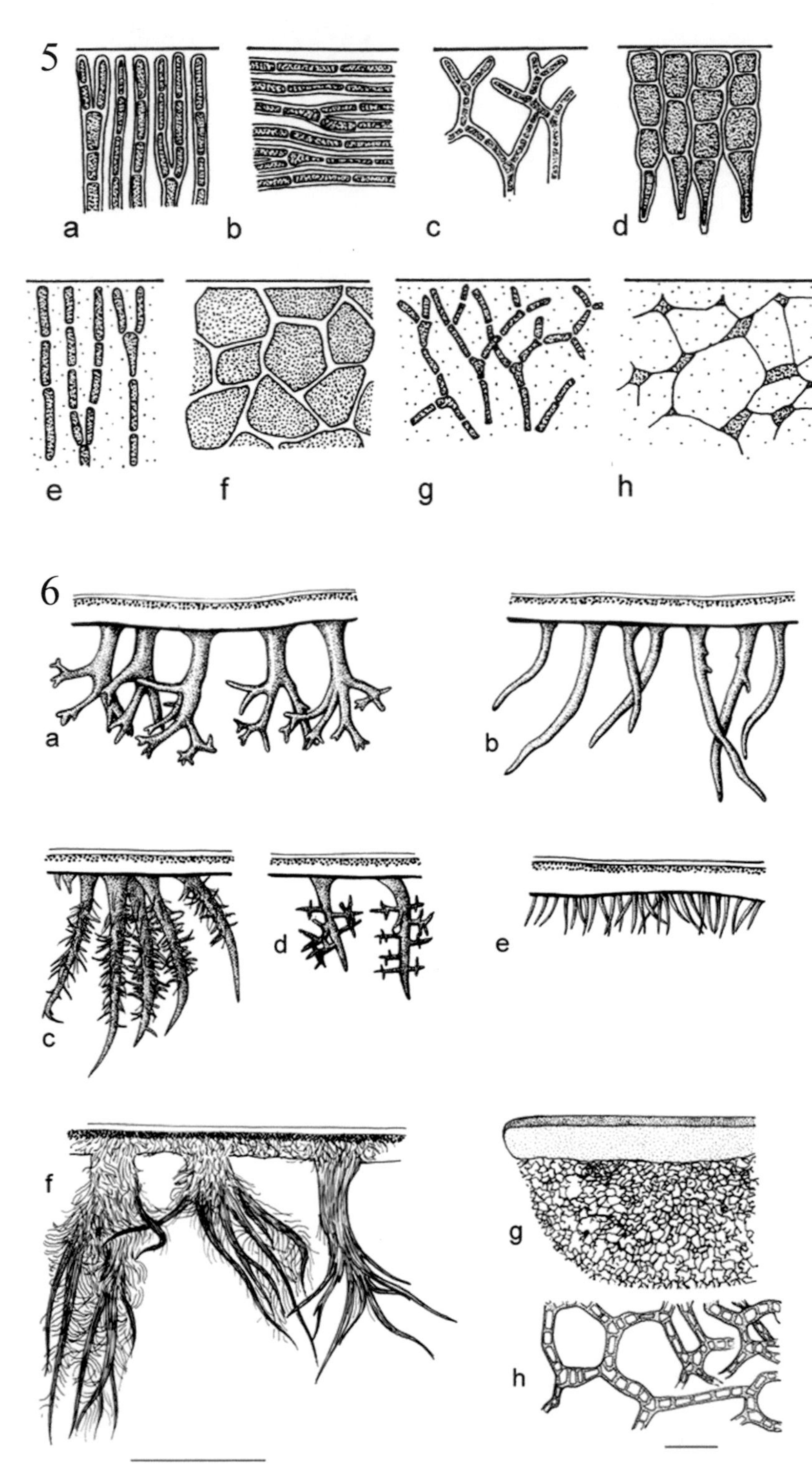

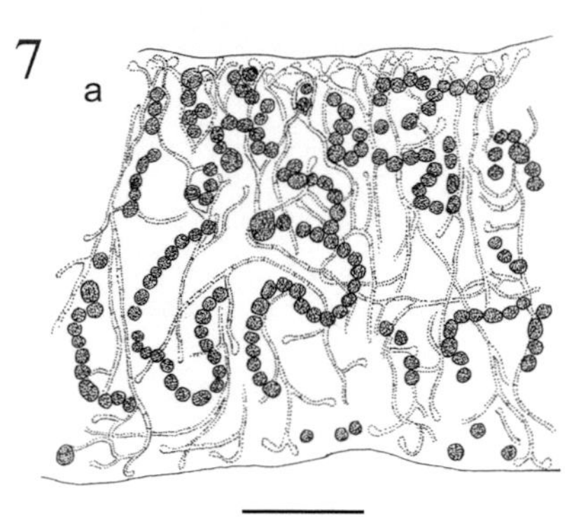

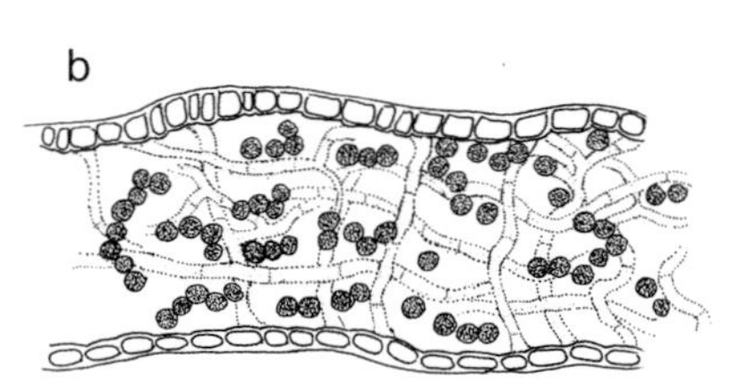

Fig. 5. Tissue types produced by different kinds of hyphal growth. a) hyphae perpendicular to the surface forming a palisade-like cortex (as in *Roccella*); b) distinct hyphae lying parallel to the surface (cortex of *Anaptychia*); c) net-like, branched hyphae (close to surface of *Collema*); d) pseudoparenchyma formed by palisade-like hyphae (cortex of *Allocetraria*); e) prosoplectenchyma, palisade-like here, but can also be longitudinally arranged as in the cortex of *Bryoria*; f) pseudoparenchyma of equal-sized cells (upper cortex of *Peltigera*); g) net-like hyphae in a type of prosoplectenchyma (part of the cortex of *Ramalina*); h) small cells in an amorphous, swollen matrix (as in the cortex of some species of *Lecanora*). (Reprinted, by permission from Georg Thieme Verlag, from Henssen and Jahns, 1974, fig. 3.5.)

Fig. 6. Rhizines and tomentum. a) forked or dichotomously branched rhizines (*Hypotrachyna*); b) unbranched rhizines (*Parmelia saxatilis*); c) squarrose rhizines (*Physconia detersa*); d) squarrose rhizines (*Parmelia squarrosa*); e) tomentum (*Lobaria pulmonaria*); f) fibrous, tufted rhizines (*Peltigera canina*); g) hypothallus of *Anzia colpodes*; h) same, at higher magnification, showing the black anastomosing hyphae. Scale: a-g = 0.5 mm; h = 300 µm. (b-c, redrawn, courtesy of the British Lichen Society, from Purvis et al., 1992, figs. 39b, c.)

Fig. 7. Thallus structure in the jelly lichens: a) *Collema* (lacking a cortex); b) *Leptogium* (upper and lower cortices present). Scale = 30 µm. (Reprinted, courtesy of the Canadian Museum of Nature, from Brodo, 1981, figs. 33, 35.)

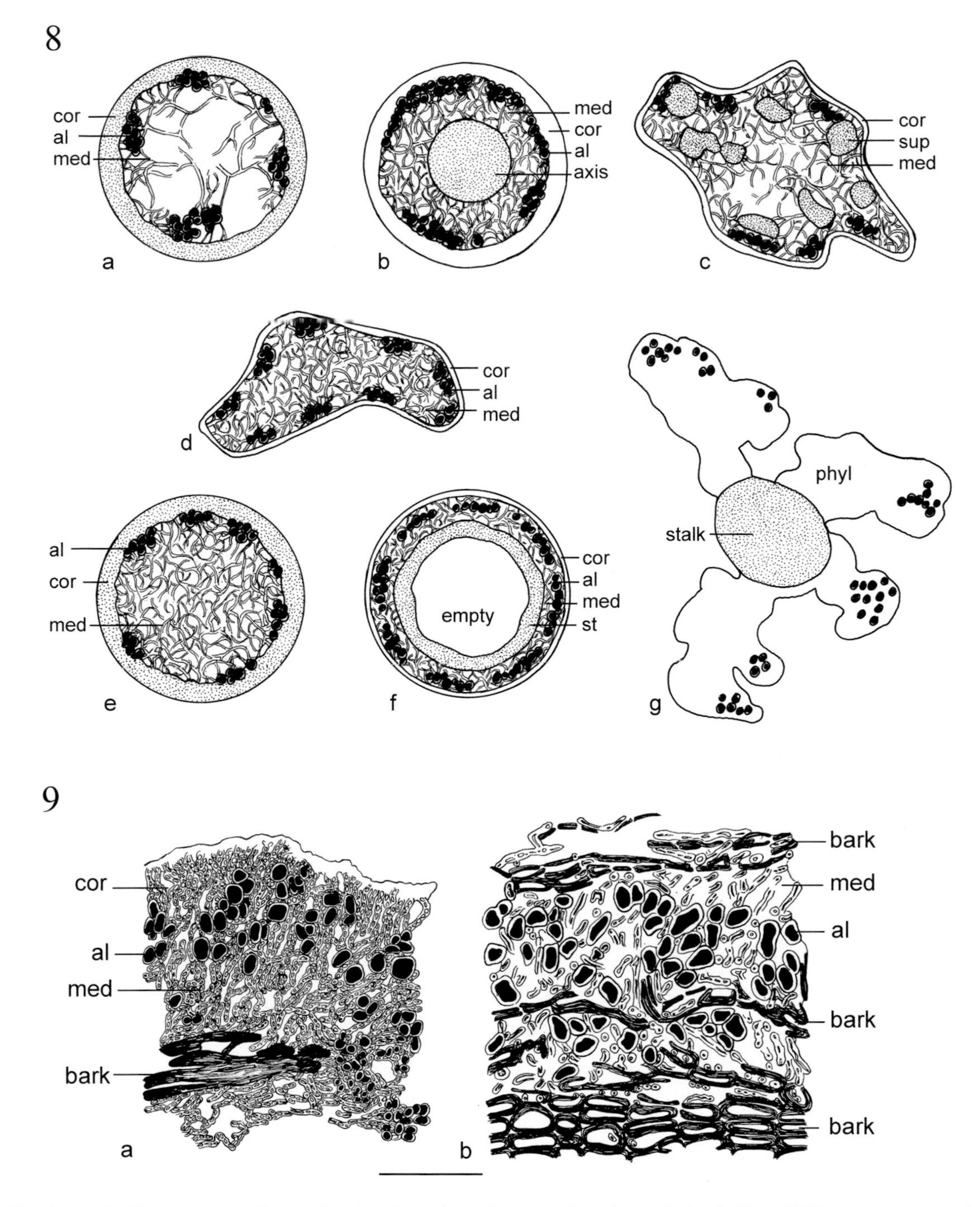

Fig. 8. Fruticose thallus cross sections, showing location of supporting tissue (stippled) in different genera. a-d, pendent lichens: a) in cortex, as in *Bryoria* and *Alectoria*; b) in central cord, as in *Usnea*; c) in separate strands, as in *Letharia*; d) with no special supporting tissue, as in *Evernia*. e-g, erect lichens: e) in cortex, as in *Sphaerophorus*; f) in cylindrical stereome, as in *Cladonia*; g) constituting entire stalk to which vegetative phyllocladia are attached, as in *Stereocaulon*. al, algae; axis, central axis; cor, cortex; med, medulla; phyl, phyllocladium; st, stereome; sup, supporting strands. Schematic, at various magnifications.

Fig. 9. Crustose lichen structure: a) on bark surface (epiphloeodal), *Lecanora*; b) within bark layers (endophloeodal), *Graphis*. Scale = ca. 50 µm. (Reprinted, by permission from Borntraeger-Cramer Verlag [www.schweizerbart.de], from Ozenda, 1963, figs. 28D, 29A.) *al*, algae; *bark*, cells of bark; *cor*, cortex; *med*, medulla.

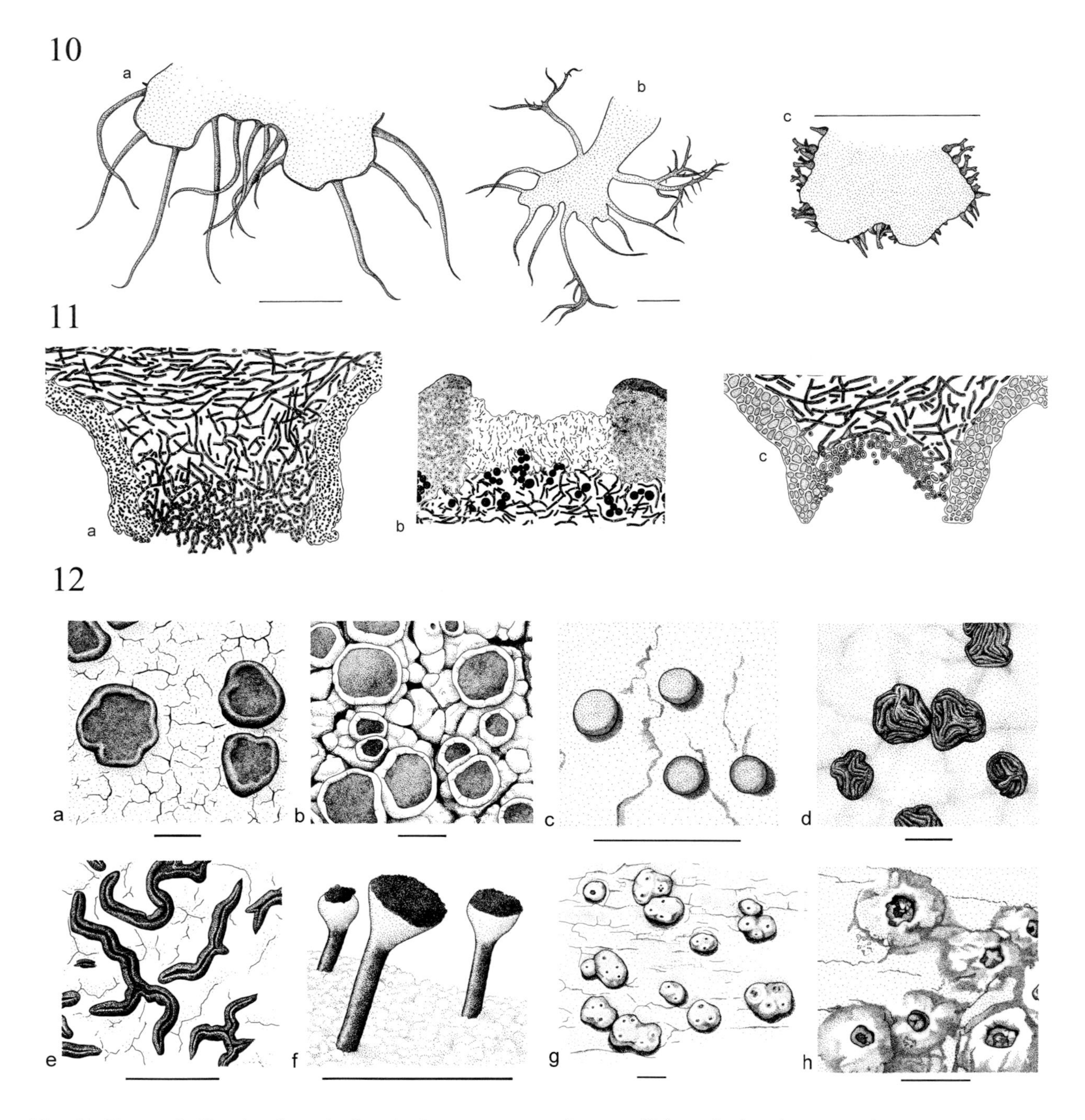

Fig. 10. Types of cilia. a) unbranched, as in *Parmotrema perforatum*; b) branched, as in *Heterodermia speciosa*; c) bulbous, as in *Bulbothrix*. Scale = 1 mm.

Fig. 11. Cyphellae and pseudocyphellae. a-b, pseudocyphellae: a) on lower surface of thallus, slightly raised, *Pseudocyphellaria*; b) on upper surface, depressed, *Bryocaulon divergens*. c) young cyphellum on lower thallus surface, *Sticta*. Magnification unknown. (Reprinted, by permission from Elsevier Publishers, from Jahns, 1973, figs. 89, 90, 94.)

Fig. 12. Types of apothecia. a) lecideine, lacking a thalline margin (*Porpidia macrocarpa*); b) lecanorine, with a thalline margin (as in *Lecanora cenisia*); c) arthonioid (as in *Arthonia patellulata*); d) gyrose, with concentric sterile ridges (*Umbilicaria hyperborea*); e) lirellae (as in *Opegrapha* and *Graphis*); f) stalked mazaedia (*Calicium viride*); g) pertusarioid (*Pertusaria macounii*); h) double-walled (*Thelotrema lepadinum*). Scales = 1 mm. (a-b and d-f, reproduced, courtesy of the British Lichen Society, from Purvis et al., 1992, fig. 28; g, reprinted, courtesy of the Canadian Museum of Nature, from Brodo, 1981, fig. 24.)

Fig. 13. Apothecia seen in section. a) lecideine, exciple with a dark outer layer and pale inner layer, lacking algae; b) biatorine, exciple pale and often radiating, lacking algae; c) lecanorine, exciple thin, surrounded by a thalline margin containing algae; d) double, both the exciple and thalline margins are well developed and distinct; e) lirella, like an elongate lecideine apothecium, usually lacking algae; f) arthonioid, asci arising in a rather uniform tissue consisting of paraphysoids. *amph,* amphithecium; *cor,* cortex; *epihym*, epihymenium; *exc*, exciple; *hym*, hymenium (containing asci and paraphyses); *hyp*, hypothecium; *med*, medulla; *par*, paraphysoid tissue; *subhym*, subhymenium; *thall*, thalline margin. Schematic, at various magnifications.

Fig. 14. Ascus types. a) *Bacidia*-type; b) *Biatora*-type; c) *Catillaria*-type; d) *Fuscidea intercincta*-type; e) *Lecanora*-type; f) *Lecidea*-type; g) *Porpidia*-type; h) *Rhizocarpon*-type; i) *Schaereria*-type; j) *Tremolecia*-type; k) *Xanthoria parietina* (*Teloschistes*-type); l) *Trapelia coarctata*; m) *Candelariella vitellina*; n) *Arthonia radiata*; o) *Thelocarpon epibolum*; p) *Gyalecta truncigena*. *aw*, ascus wall; *ax*, axial body; *oc*, ocular chamber; *tho*, tholus. (a-c and e-p, reproduced, courtesy of the British Lichen Society, from Purvis et al., 1992, fig. 33.)

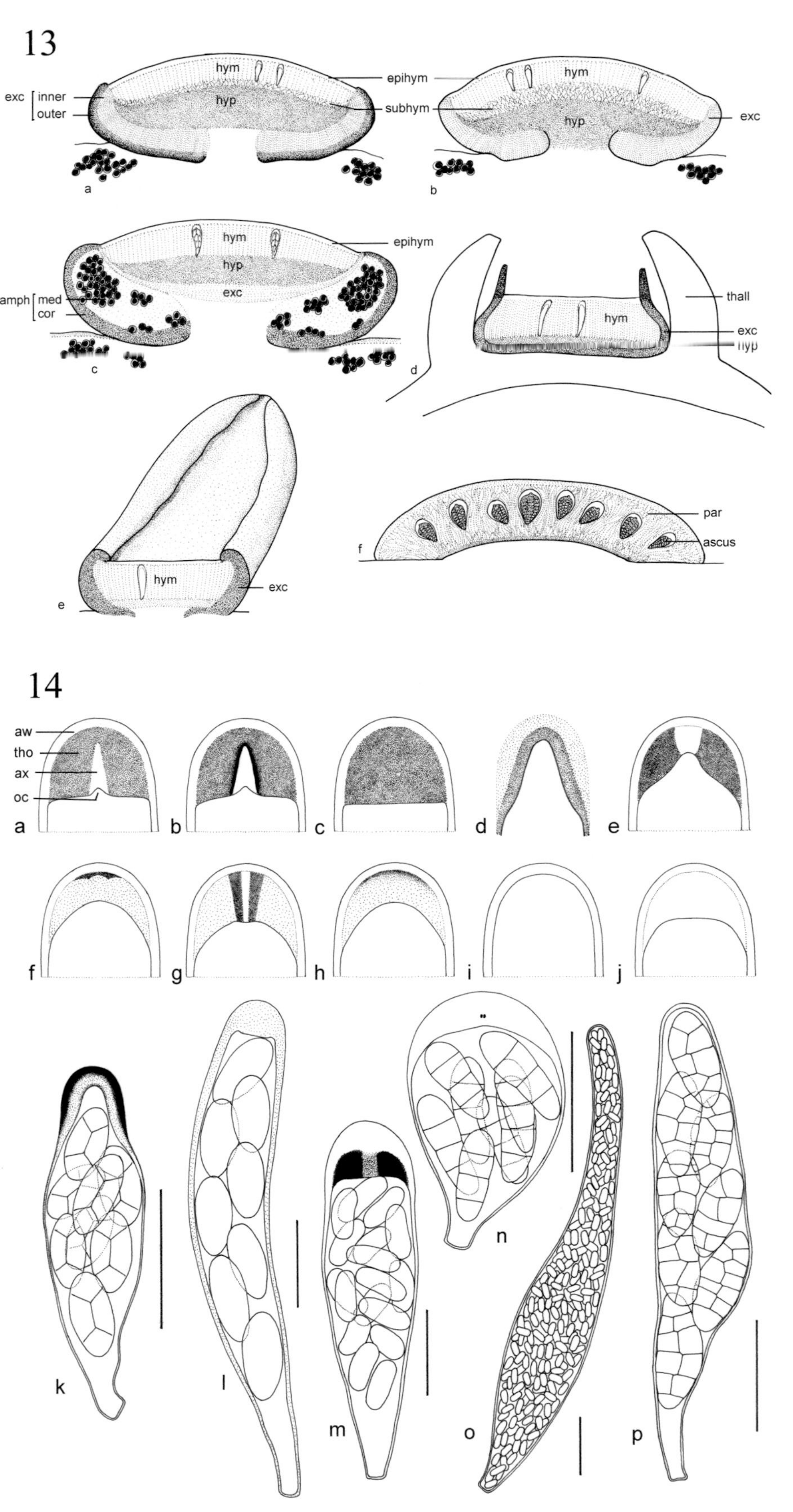

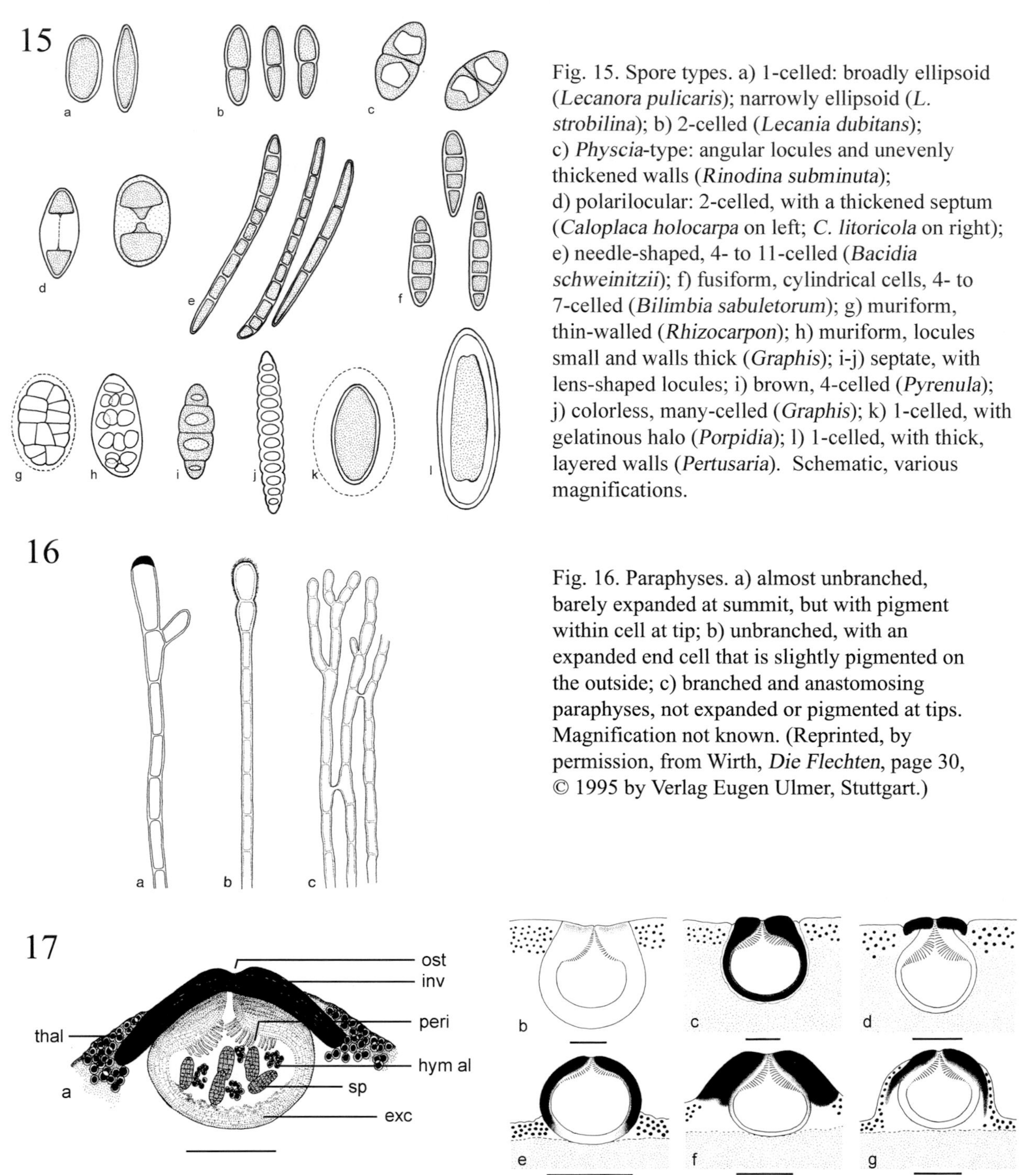

Fig. 15. Spore types. a) 1-celled: broadly ellipsoid (*Lecanora pulicaris*); narrowly ellipsoid (*L. strobilina*); b) 2-celled (*Lecania dubitans*); c) *Physcia*-type: angular locules and unevenly thickened walls (*Rinodina subminuta*); d) polarilocular: 2-celled, with a thickened septum (*Caloplaca holocarpa* on left; *C. litoricola* on right); e) needle-shaped, 4- to 11-celled (*Bacidia schweinitzii*); f) fusiform, cylindrical cells, 4- to 7-celled (*Bilimbia sabuletorum*); g) muriform, thin-walled (*Rhizocarpon*); h) muriform, locules small and walls thick (*Graphis*); i-j) septate, with lens-shaped locules; i) brown, 4-celled (*Pyrenula*); j) colorless, many-celled (*Graphis*); k) 1-celled, with gelatinous halo (*Porpidia*); l) 1-celled, with thick, layered walls (*Pertusaria*). Schematic, various magnifications.

Fig. 16. Paraphyses. a) almost unbranched, barely expanded at summit, but with pigment within cell at tip; b) unbranched, with an expanded end cell that is slightly pigmented on the outside; c) branched and anastomosing paraphyses, not expanded or pigmented at tips. Magnification not known. (Reprinted, by permission, from Wirth, *Die Flechten*, page 30, © 1995 by Verlag Eugen Ulmer, Stuttgart.)

Fig. 17. Perithecia in section. a) perithecium of *Staurothele fissa*. *exc*, exciple; *hym al*, hymenial algae (only found in *Staurothele* and *Endocarpon*); *inv*, involucrellum; *ost*, ostiole; *peri*, periphyses; *sp*, spore; *thal*, thallus tissue. b-g, types of perithecia. b) immersed in foliose thallus, involucrellum absent, exciple pale (*Dermatocarpon miniatum*); c) immersed in limestone, involucrellum absent, exciple black and friable (as in *Bagliettoa calciseda*); d) immersed in limestone, involucrellum lid-like, exciple mostly pale (*Verrucaria* sp.); e) prominent, involucrellum absent, exciple black and friable above, pale at base (*Thelidium* sp.); f) involucrellum well developed, exciple pale below (as in *Pyrenula pseudobufonia*); g) immersed in thalline wart, involucrellum present, exciple pale below (as in *Staurothele fissa*). Scale = 200 µm. (a, reproduced, by permission from the Canadian Museum of Nature, from Brodo, 1981, fig. 26; b-g, reproduced, courtesy of the British Lichen Society, from Purvis et al., 1992, fig. 41.)

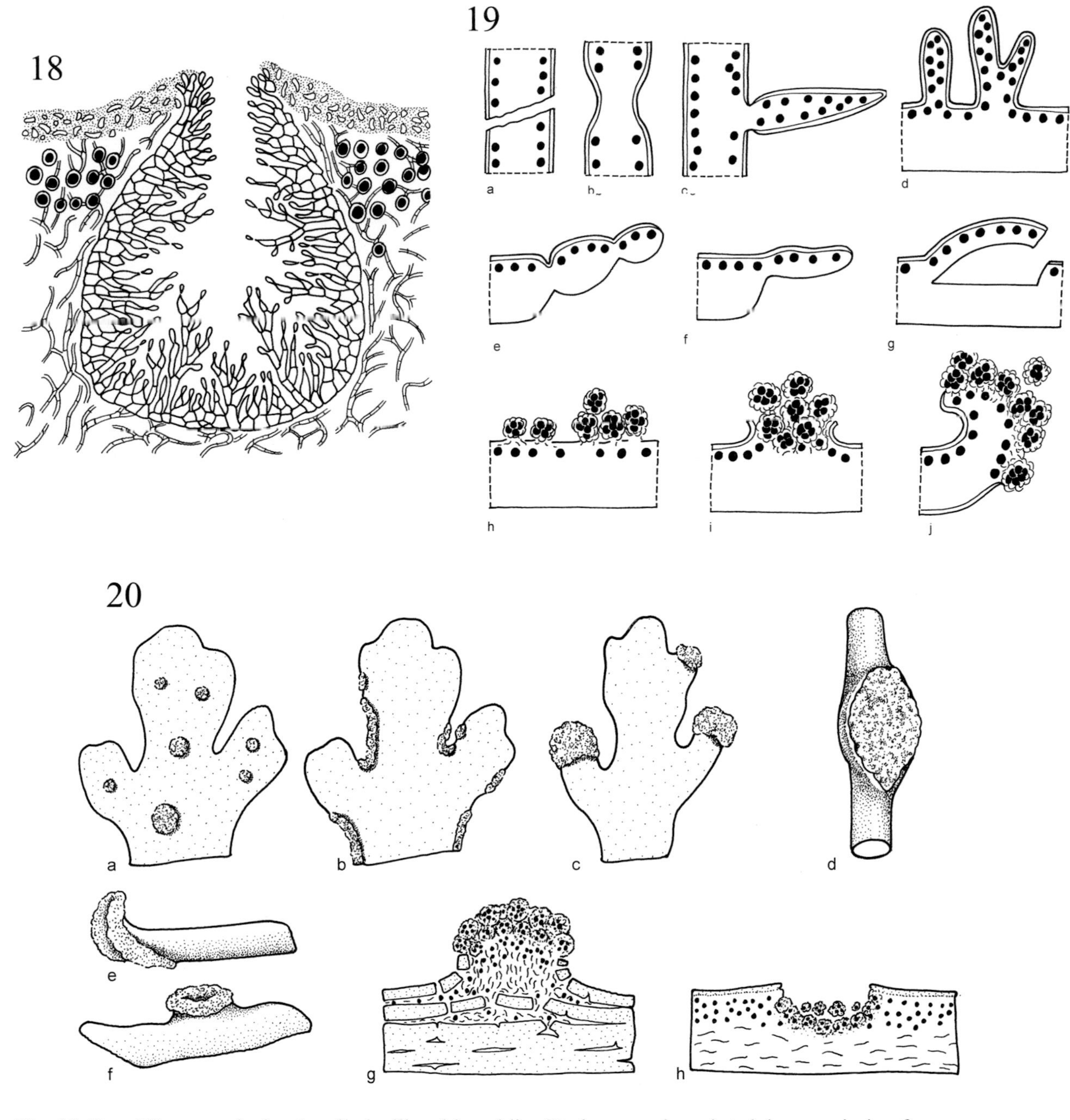

Fig. 18. Pycnidium, producing 1-celled, ellipsoid conidia. (Redrawn and reprinted, by permission from Borntraeger-Cramer Verlag [www.schweizerbart.de], from Vobis [1980], fig. 5d.)

Fig. 19. Vegetative (nonsexual) propagules. a) thallus fragmentation; b) weakened segment for fragmentation; c) lateral spinule, narrowed at base; d) isidia, with cortex intact; e) blastidia, budding off granules from thallus margin; f) lobule; g) schizidium, a fragment of the upper cortex and algal layer; h) soredia formed from eroded thallus surface; i) soredia breaking through cortex and forming a soralium; j) soredia formed from lower cortex of recurved lobe (labriform soralium). (Redrawn, courtesy of the British Lichen Society, from Purvis et al., 1992, fig. 29.)

Fig. 20. Soralia. a) laminal; b) marginal; c) terminal; d) fissural; e) labriform; f) cuff-shaped; g) hemispherical or erumpent; h) excavate. (a-c, e-h, redrawn, courtesy of the British Lichen Society, from Purvis et al., 1992, fig. 43.)

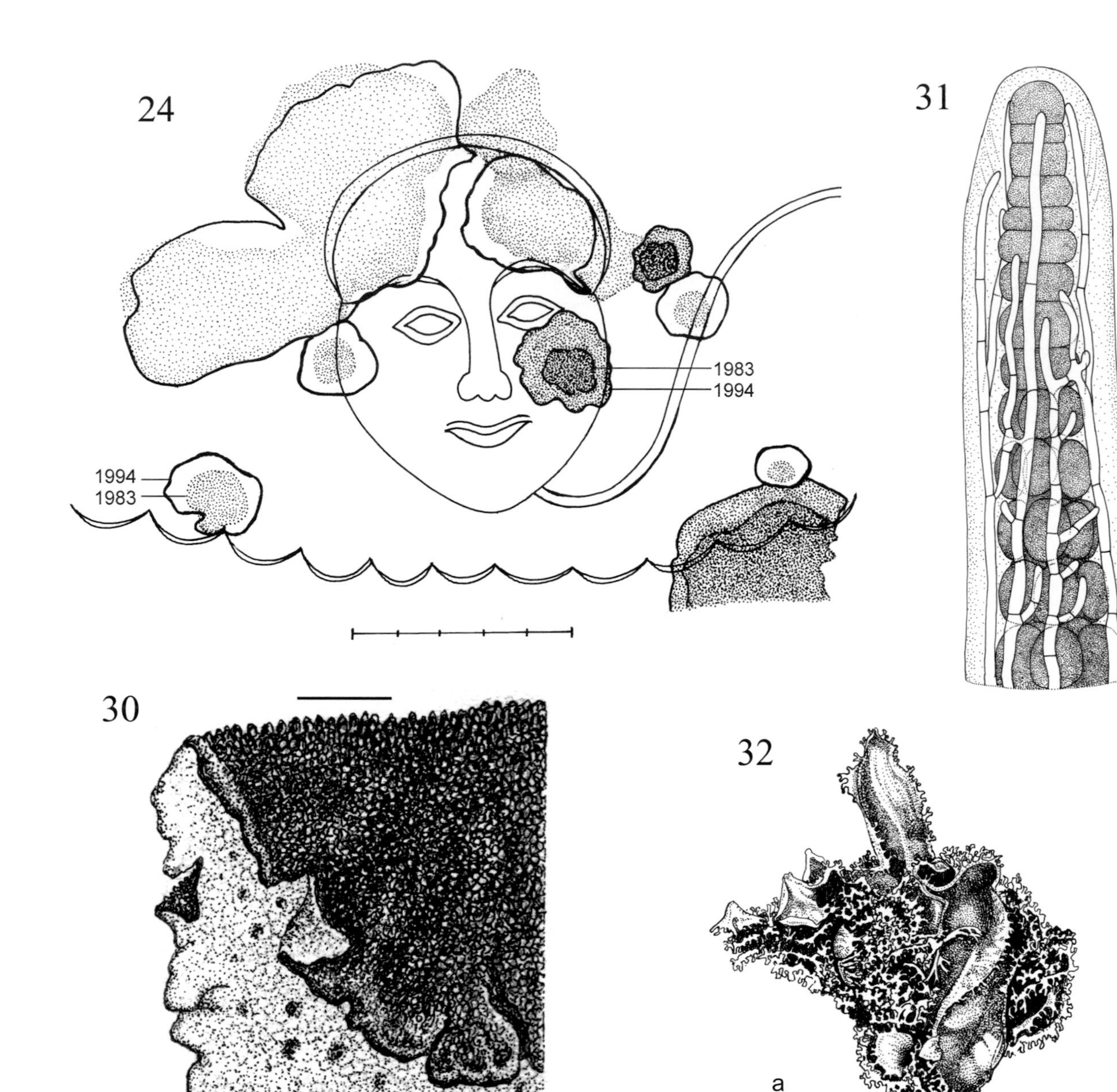

Fig. 24. Tracings from gravestones showing growth increments (compare with plates 26A and 26B in LNA). The heavily stippled, dark lichen is *Aspicilia* sp.; the paler one is *Dimelaena oreina*. The inner, more darkly stippled thalli were traced in 1983; the outer margins represent the lichen in 1994.

Fig. 30. The lower surface of *Dermatocarpon reticulatum* showing the rough, papillate texture. Scale = 1 mm. (Drawing by Alexander Mikulin. Reprinted, by permission from Mad River Press, from McCune and Goward, 1995, page 95.)

Fig. 31. A microscopic view of *Ephebe* showing the hyphal sheath over the cyanobacterial filament. Scale = 20 μm. (Reproduced, courtesy of the British Lichen Society, from Purvis et al., 1992, fig. 44a.)

Fig. 32. *Leptogium lichenoides*. a) dry thallus; b) moistened and flattened lobes showing the finely divided, isidia-like margins. Scale = 1 mm. (Reproduced, courtesy of the Canadian Museum of Nature, from Brodo, 1981, fig. 37.)

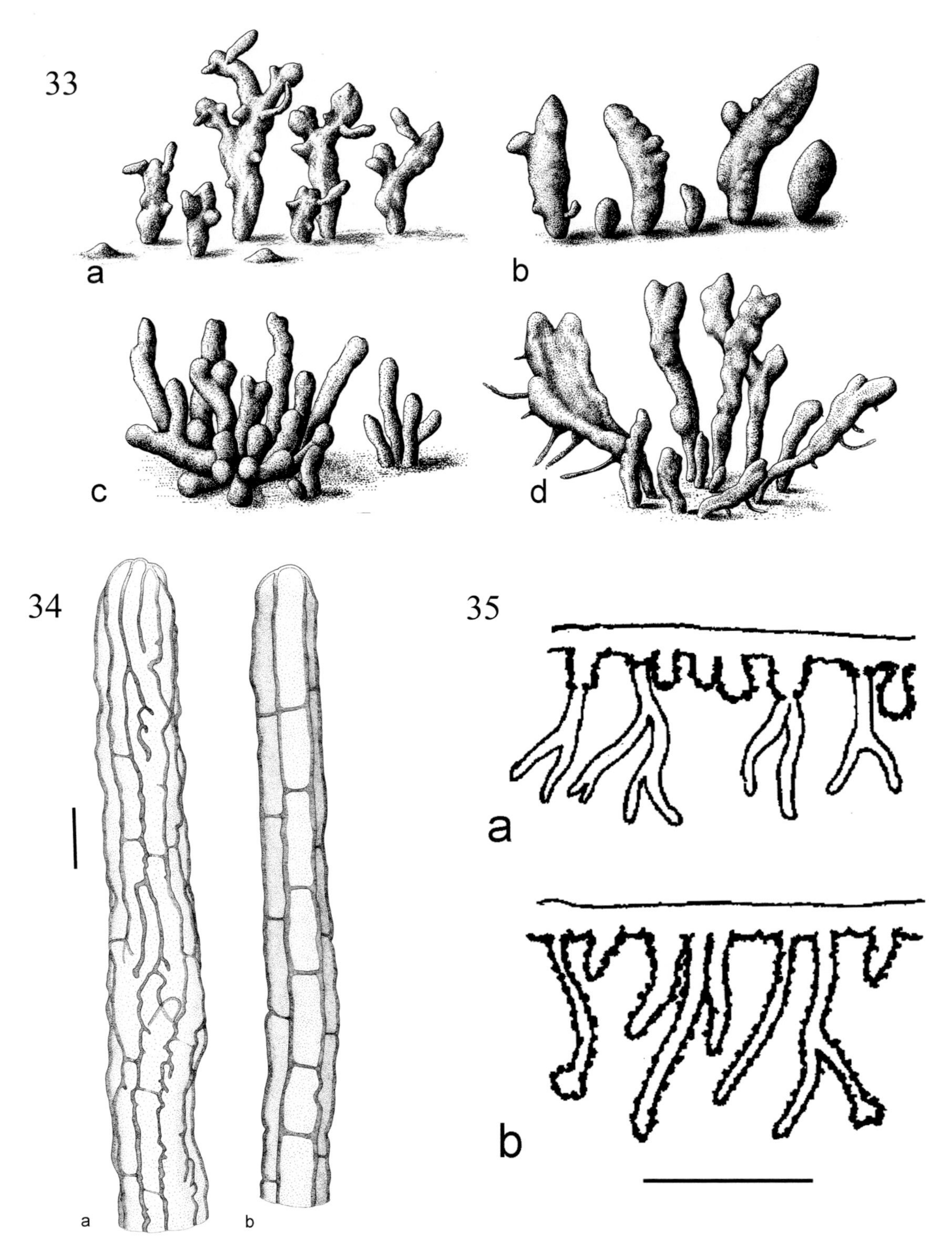

Fig. 33. The isidia of *Melanelia s. lat* species. a) *Melanelixia elegantula*; b) *Melanohalea exasperatula*; c) *Melanelixia fuliginosa*; d) *Melanohalea subelegantula.* Scale = 1 mm. (Drawing by Alexander Mikulin. Reprinted, by permission of the USDA–Forest Service, from McCune and Geiser, 1997, pages 159-161, 164.

Fig. 34. *Cystocoleus* compared with *Racodium* under the microscope. a) *C. ebeneus*; b) *R. rupestre*. Scale = 10μm. (Reproduced, courtesy of the British Lichen Society, from Purvis et al., 1992, fig. 13.)

Fig. 35. The rhizines of *U. vellea* (a) compared with those of *Umbilicaria americana* (b). Scale = 1 mm. (Drawing by Alexander Mikulin. Reprinted, by permission, from McCune and Goward, 1995, page 170.)

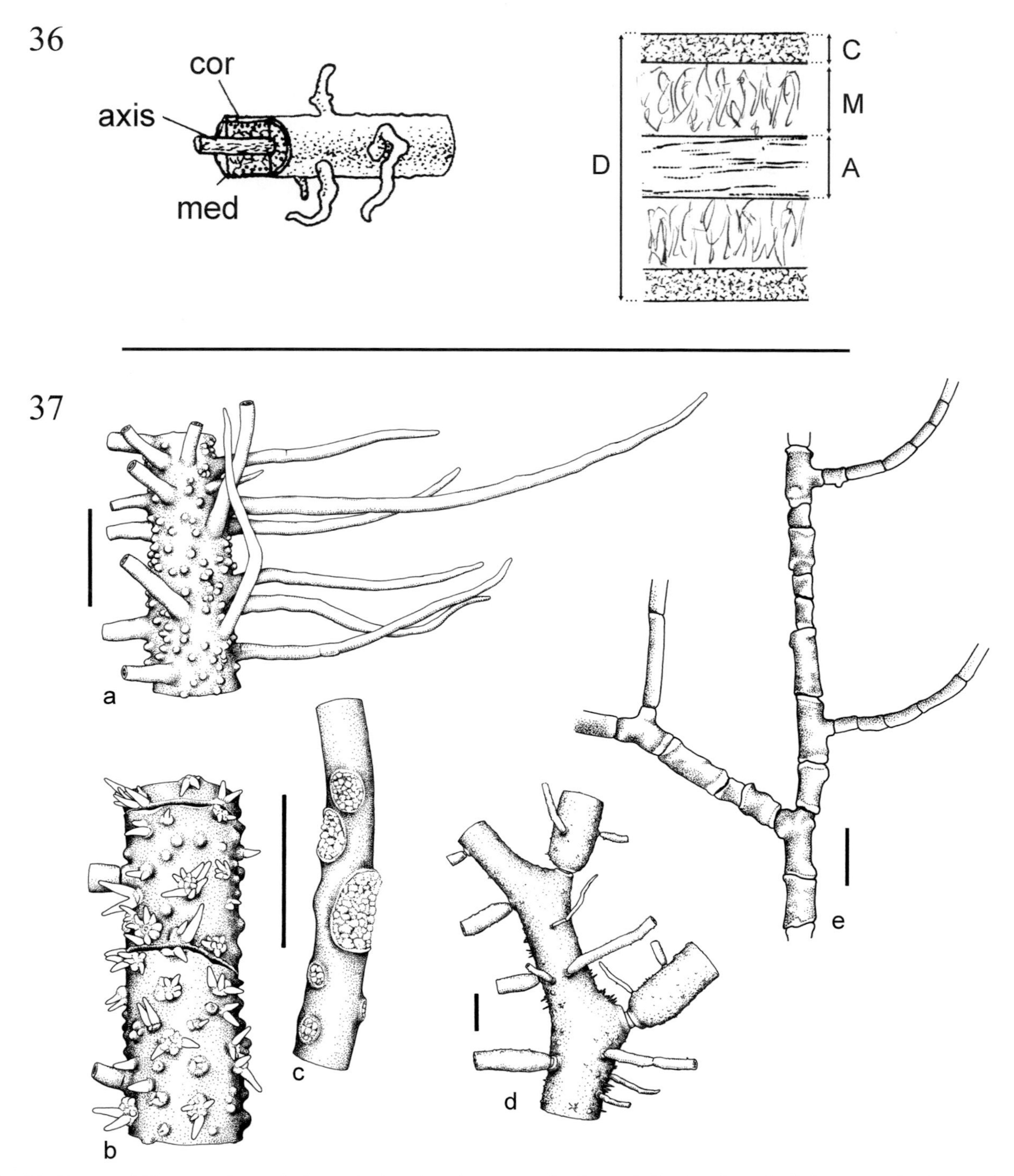

Fig. 36. *Usnea.* a) The structure of a branch showing the cortex (*cor*), medulla (*med*), and slightly stretched central axis (*ax*). (Reproduced, courtesy of the Canadian Museum of Nature, from Brodo, 1981, fig. 67b.); b) an *Usnea* filament in side view indicating how the cortex, medulla, and axis are measured. *A*, axis; *M*, medulla; *C*, cortex; *D*, diameter of entire filament. A/D X 100=axis percent, M/D X 100=medulla percent; C/D X 100=cortex percent. (Drawing by I. Brodo.)

Fig. 37. Thallus features in *Usnea.* a) Papillae and fibrils (as in *U. diplotypus*); b) clusters of isidia arising from small soralia (*U. rubicunda*); c) soralia (*U. glabrescens*); d) inflated branches constricted at the base (as in *U. fragilescens*); e) bone-like articulations (*U. trichodea*). (a-d, reproduced, courtesy of the British Lichen Society, from Purvis et al., 1992, fig. 26.)

Plate 9. The gray foliose lichen, *Parmotrema hypotropum*, shares a twig in Texas with two fruticose species, *Teloschistes flavicans* (orange) and *Usnea parvula.*

Plate 10. Crustose lichens, like these in central California, come in many colors and shapes.

Plate 11. White pruina on the lobes of *Physconia detersa.*

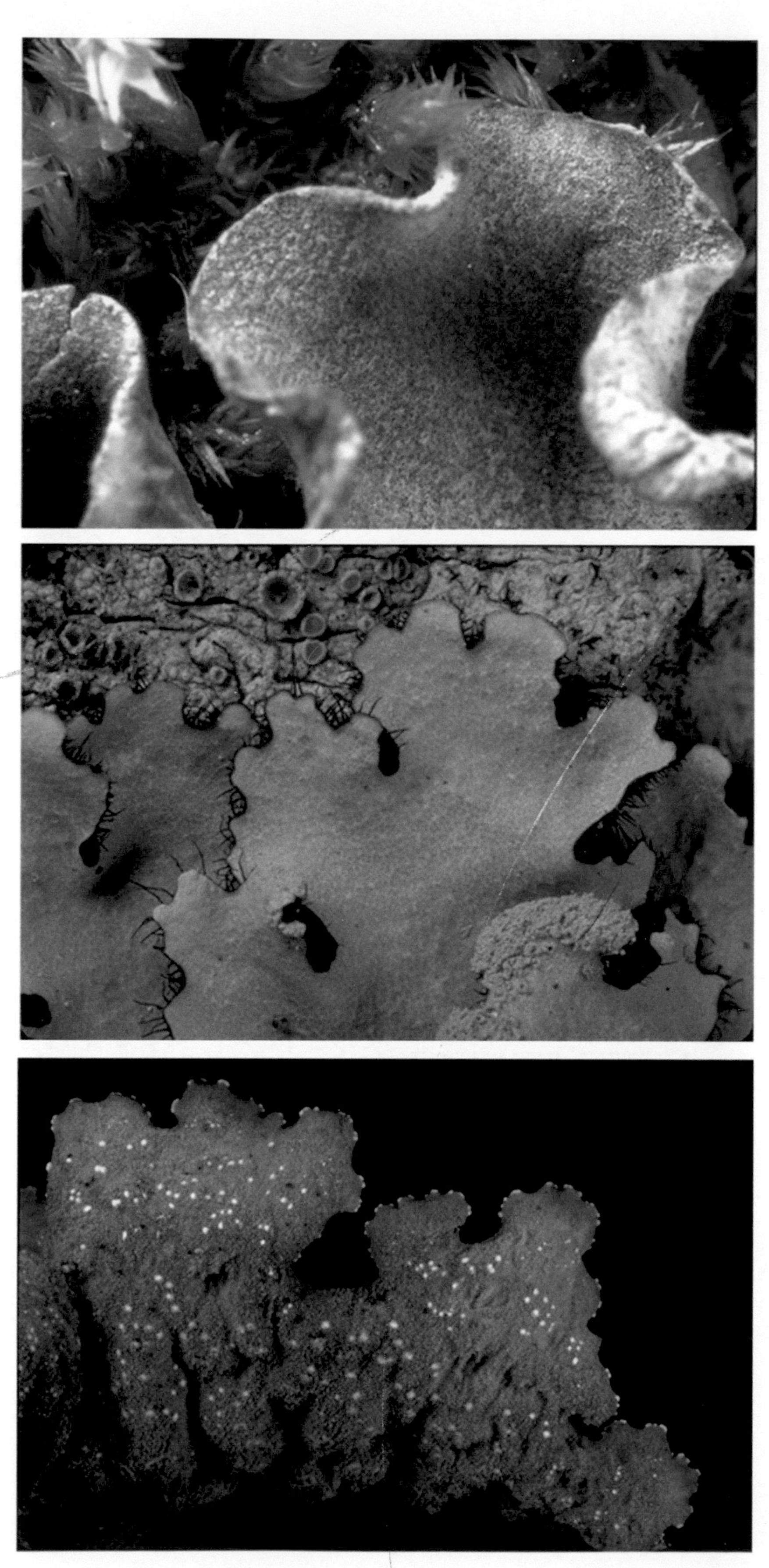

Plate 12. The fuzzy tomentum on the upper surface of *Peltigera rufescens* is particularly thick.

Plate 13. A network of white maculae on *Parmotrema reticulatum.* Unbranched black cilia are on the lobe margins.

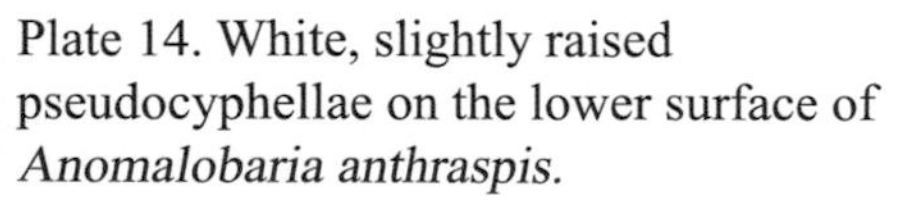

Plate 14. White, slightly raised pseudocyphellae on the lower surface of *Anomalobaria anthraspis.*

Plate 15. Cyphellae on the lower surface of the thallus characterize the genus *Sticta*.

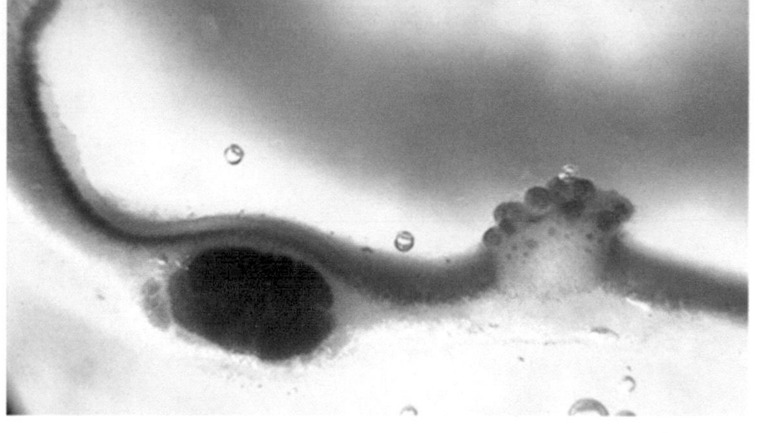

Plate 16. In this section of the thallus of *Lobaria pulmonaria*, a cephalodium forms a large dark lump (on the left) below the green algal layer. Soredia burst through the upper cortex in a soralium on the right.

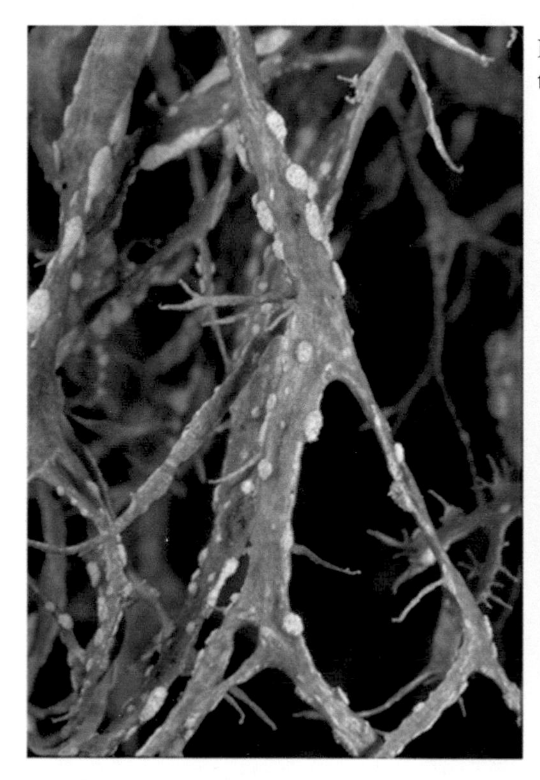

Plate 17. Oval soralia filled with powdery soredia occur on the margins and surface of branches of *Ramalina farinacea*.

Plate 18. Branched isidia on *Platismatia herrei*.

Plate 19. Scalelike, overlapping lobules are abundant on the upper surface of *Lobaria tenuis*.

Plate 32. White fungal hyphae and the green photobiont of an endolithic crustose lichen can be seen within the upper layers of this section of sandstone. On the rock surface, only the large black apothecia are visible.

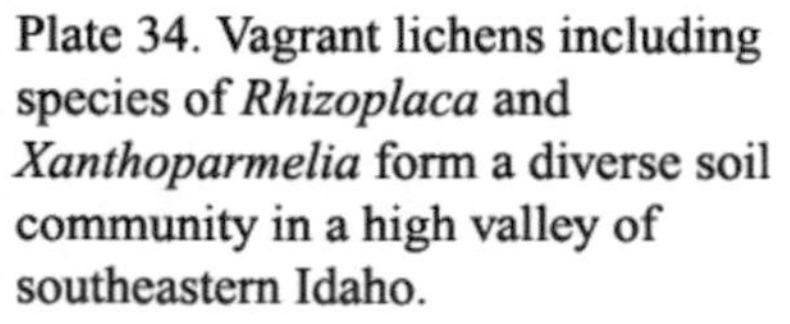

Plate 34. Vagrant lichens including species of *Rhizoplaca* and *Xanthoparmelia* form a diverse soil community in a high valley of southeastern Idaho.

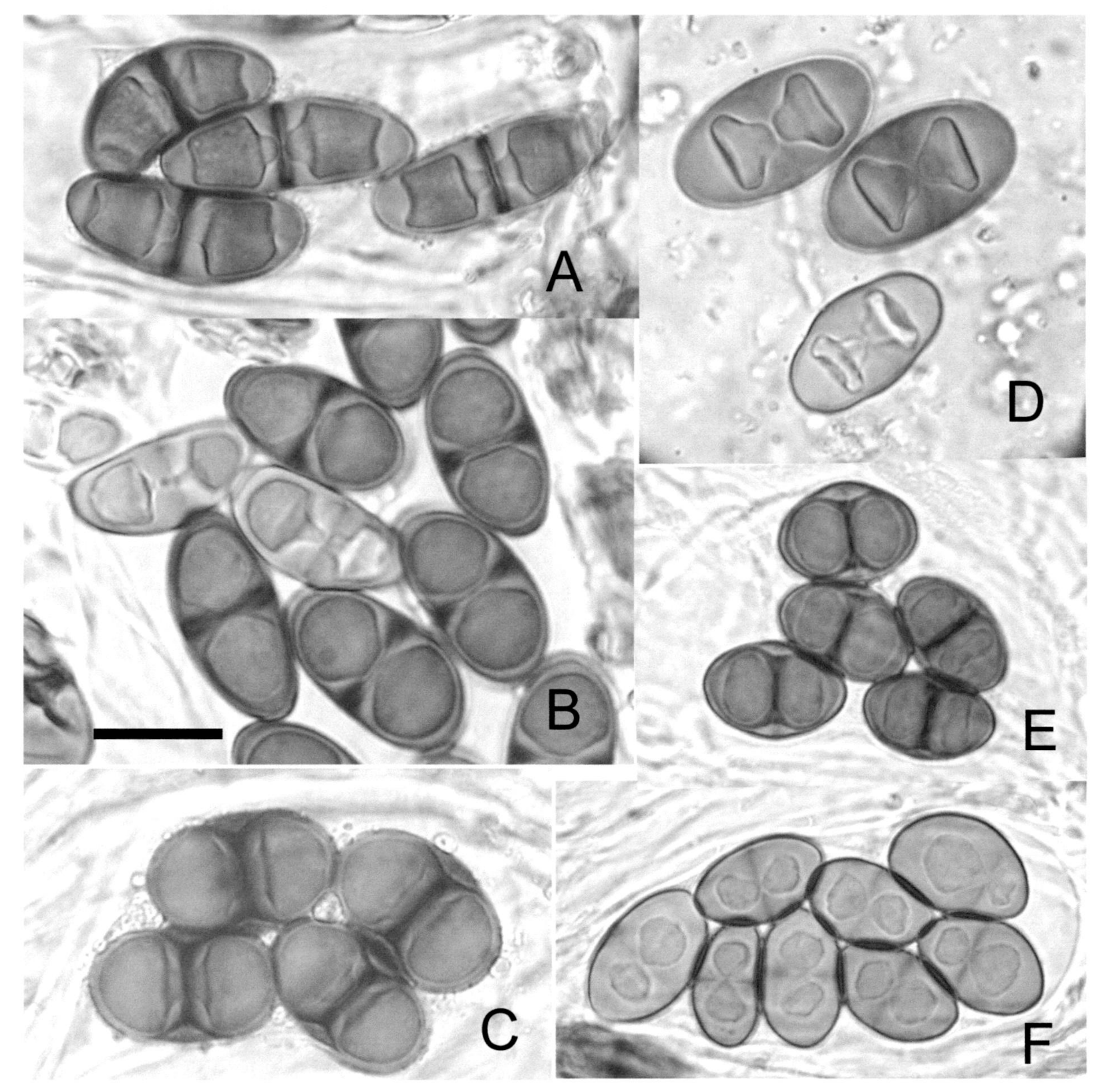

Plate 925. Spore types in the genus *Rinodina*. A. *Physcia*-type, *R. turfacea*; B. *Physconia*-type, *R. archaea*; C. *Bischoffii*-type, *R. bischoffii*; D. *Mischoblastia*-type, *R. oxydata*; E. *Dirinaria*-type, *R. marysvillensis*; F. *Pachysporaria*-type, *R. papillata*. Photographs from Sheard (2010), provided by the author. (Reprinted by permission from Canadian Science Publishing [NRC Research Press].)

Based on the acclaimed reference *Lichens of North America*, this resource for the classroom, field, and laboratory presents updated and expanded keys for the identification of over 2,000 species of lichens indigenous to the continent, twice the number covered by the previous keys. The book includes a glossary illustrated with photographs by Sylvia Duran Sharnoff and Stephen Sharnoff and drawings by Susan Laurie-Bourque, all from the original book. The revised keys are an indispensable identification tool for botanists, students, scientists, and enthusiasts alike.

"This updated and much-expanded identification guide accomplishes a feat no one else has dared attempt. It is extremely useful for nonprofessionals and professionals alike."
RICHARD HARRIS, New York Botanical Garden

"Brodo is responsible for the renewal of North American lichenology. He updates the keys to North American lichens to accommodate rapid changes in lichen systematics and taxonomy in the beginning of the twenty-first century. Using these new keys with *Lichens of North America*, individuals, study groups, and classes can identify many species in any region of North America."
KERRY KNUDSEN, University of California, Riverside

Considered a world authority on lichens and their biology, **Irwin M. Brodo** is emeritus research scientist at the Canadian Museum of Nature, Ottawa, Ontario.

PRAISE FOR *Lichens of North America:*

"This is a book to love. It is a reminder that many of the most environmentally important organisms are also at first glance the most humble in appearance. *Lichens of North America* will open a whole new world for naturalists, professional and amateur, to explore and enjoy."
EDWARD O. WILSON, author of *Consilience: The Unity of Knowledge* and coauthor of *The Ants*

"*Lichens of North America* is one of the great natural history books of this or any age—a work in which art and science are perfectly united in tribute to one of the most beautiful, important, and least described parts of nature."
DAVID EHRENFELD, Rutgers University

"A gold mine of information. The book has superb photos and excellent keys and range maps, and is packed with information. It will be the 'lichen reference bible' for a long time."
NOBLE S. PROCTOR, Southern Connecticut State University

Cover illustration: *Physcia aipolia*. Photo courtesy of Irwin M. Brodo.

Yale UNIVERSITY PRESS
New Haven and London
yalebooks.com
yalebooks.co.uk
ISBN 978-0-300-19573-6